International Conference e-Health (EH2020)

Held at MCCSIS 2020 (Held online due to COVID-19)

Lisbon, Portugal
21-25 July 2020

Editors:

Mário Macedo
Piet Kommers
Pedro Isaías
Luis Rodrigues

ISBN: 978-1-7138-2274-5

Printed from e-media with permission by:

Curran Associates, Inc.
57 Morehouse Lane
Red Hook, NY 12571

International Association for Development of the Information Society (IADIS)
Rua Sao Sebastiao Da Pedreira
No. 100, 30, 1050-209
Lisbon, Portugal

Phone: 351 21 3151373
Fax: 351 21 3151244

secretariat@iadis.org

Additional copies of this publication are available from:

Curran Associates, Inc.
57 Morehouse Lane
Red Hook, NY 12571 USA
Phone: 845-758-0400
Fax: 845-758-2633
Email: curran@proceedings.com
Web: www.proceedings.com

INTERNATIONAL CONFERENCE

E-HEALTH 2020

part of the

MULTI CONFERENCE ON COMPUTER SCIENCE

AND INFORMATION SYSTEMS 2020

MULTI CONFERENCE ON COMPUTER SCIENCE AND INFORMATION SYSTEMS

MCCSIS 2020

21-25 July

Proceedings of the International Conference

e-Health 2020

Edited by
Mário Macedo

iadis
international association for development of the information society

PROCEEDINGS OF THE INTERNATIONAL CONFERENCE

E-HEALTH 2020

JULY 21 - 23, 2020

Organised by

iadis

international association for development of the information society

As a member of Crossref (a non-profit membership organization for scholarly publishing working with the purpose to make content easy to find, link, cite and assess) each published paper in this book of proceedings will be allocated a DOI (Digital Object Identifier) number for its fast and easy citation and indexation.

Volume Editor:
Mário Macedo

Computer Science and Information Systems Series Editors:
Piet Kommers and Pedro Isaías

Associate Editor: Luís Rodrigues

TABLE OF CONTENTS

FULL PAPERS

SHORT PAPERS

REFLECTION PAPER

POSTER

FOREWORD

These proceedings contain the papers and poster of the 12th International Conference on e-Health 2020, which was organised by the International Association for Development of the Information Society, from 21- 23 July, 2020. This conference is part of the Multi Conference on Computer Science and Information Systems 2020, 21 - 25 July, which had a total of 625 submissions. Due to an unprecedented situation caused by the COVID-19 pandemic, this year the conference was hosted virtually.

The use of ICTs (Information and Communication Technologies) in Healthcare Services is the main mechanism to improve efficiency and effectiveness. Nowadays ICTs are being developed to achieve the following objectives:
– To integrate heterogeneous systems;
– To develop frameworks to make all data meaningful, accessible and available everywhere and permanently;
– To develop AIDC (Automatic Identification and Data Collectors) systems;
– To develop intelligent systems to support clinical and management decisions;
The use of these technologies also improves the quality of patient care and reduces clinical risk. At the same time, the patient will be part of the healthcare process, having more information about diseases and access to his/her electronic health record.

The e-Health (EH) 2020 conference aims to draw together information systems, practitioners and management experts from all quadrants involved in developing computer technology to improve healthcare quality.

Submissions were accepted under the following 3 main areas in the field of e-health within specific topics:

-Research Issues
-Management Issues
-Applications

The EH 2020 received 134 submissions from more than 19 countries. Each submission has been anonymously reviewed by an average of four independent reviewers, to ensure that accepted submissions were of a high standard. Consequently, only 23 full papers were approved which means an acceptance rate of 17%. A few more papers were accepted as short papers, reflection paper and poster. An extended version of the best papers may be published in the IADIS International Journal on WWW/Internet (ISSN: 1645-7641) and also in other selected journals, including journals from Inderscience.

Besides the presentation of full, short and reflection papers and poster, the conference also included a keynote presentation and one tutorial from internationally distinguished researchers. We would therefore like to express our gratitude to Professor Gordon McVie, Chair of the European Alliance for Personalised Medicine, UK, for accepting our invitation to be the keynote speaker and Prof. Mário Macedo, Universidade Atlântica, Portugal for the Tutorial presentation.

This volume has taken shape as a result of the contributions from a number of individuals. We are grateful to all authors who have submitted their papers to enrich the conference proceedings. We wish to thank all members of the organizing committee, delegates, invitees and guests whose contribution and involvement are crucial for the success of the conference.

Last but not the least, we hope that everybody enjoyed the presentations, and we invite all participants for the next years' edition of this conference.

Mário Macedo, Universidade Atlântica, Portugal
e-Health 2020 Program Chair

Piet Kommers, University of Twente, The Netherlands
Pedro Isaias, The University of New South Wales (UNSW – Sydney), Australia
MCCSIS 2020 General Conference Co-Chairs

July 2020

PROGRAM COMMITTEE

E-HEALTH PROGRAM CHAIR

Mário Macedo, Universidade Atlântica, Portugal

MCCSIS GENERAL CONFERENCE CO-CHAIRS

Piet Kommers, University of Twente, The Netherlands
Pedro Isaias, The University of New South Wales (UNSW – Sydney), Australia

COMMITTEE MEMBERS

Abdel-Badeeh M.Salem, Ain Shams University, Egypt
Andreas Schrader, University of Lübeck, Germany
Antonio Lanatà, University of Pisa, Italy
Asa Smedberg, Stockholm University, Sweden
Athina Lazakidou, University of Peloponnese, Greece
Bert-Jan van Beijnum, University of Twente, Netherlands
Bridget Kane, Karlstad University, Sweden
Craig Kuziemsky, MacEwan University, Canada
Cristian Moral, Universidad Politécnica de Madrid, Spain
David Isern, University Rovira I Virgili, Spain
Edward de Quincey, Keele University, United Kingdom
Edward Kacki, EFMI, Poland
Elena Villalba, Universidad Politécnica de Madrid, Spain
Elizabeth Krupinski, Emory University, United States
Eric Campo, LAAS-CNRS, France
Frederic Bousefsaf, Université de Lorraine, France
Gayo Diallo, University of Bordeaux, France
Georg Duftschmid, Medical University of Vienna, Austria
George Okeyo, Jomo Kenyatta University of Agriculture and Techno, Kenya
Jan Aidemark, Linnéuniversitet, Sweden
Jan Vejvalka, Charles University, Czech Republic
Jesuk Ko, Universidad Mayor de San Andres (UMSA), Bolivia
Jose Luis Sanchez, Laboratorio de Computadoras y Control, Spain
Karla Navarro, Faculty of Engineering and Information Technology, Australia
Laurent Billonnet, Ensil - Université de Limoges, France
Lucy Hederman, Trinity College Dublin, Ireland
Malina Jordanova, Bulgarian Academy of Sciences, Bulgaria
Manolis Tsiknakis, Computational Medicine Laboratory of ICS-FORTH, Greece
Mario Cannataro, University, Italy
Maurice Mars, University of KwaZulu-Natal, South Africa
Mohy Uddin, King Abdullah International Medical Research Cente, Saudi Arabia
Ovidiu Stan, Technical University of Cluj Napoca, Romania
Panos Liatsis, Khalifa University, United Arab Emirates
Paula De Toledo, Universidad Carlos III de Madrid, Spain
Pedro Pablo Escobar, Intelymec Group, UNCPBA, Argentina
Roberto Hornero, University of Valladolid, Spain
Sabah Mohammed, Lakehead University, Canada
Sandro Scattareggia Marchese, Signo Motus Srl, Italy
Silvia Alayon, Universidad de La Laguna, Spain
Vivian Vimarlund, Linköpings University and Jönköping International, Sweden
Walter Gall, Medical University of Vienna, Austria
Wendy Maccaull, St. Francis Xavier University, Canada

KEYNOTE LECTURE

TOWARDS INTELLIGENT PERSONALISED CANCER TREATMENT VIA ARTIFICIAL INTELLIGENCE

Professor Gordon McVie,
Chair of the European Alliance for Personalised Medicine, UK

ABSTRACT

If any disease were susceptible to mathematical unravelling, it should be cancer. After all, there is a cancer genome, and nearly a cancer epigenome, and maybe a cancer proteome. Research in this area has been funded more generously than any area of medicine since Nixon's War against Cancer. So what's stopping us? There are simply too few tangible markers which accurately predict propensity to get cancer, or prognosis, or best treatment strategies, or resistance to those therapies. There is no device on any clever Apple or Google wrist watch which can shine a light on any of these issues. But biomarkers are emerging slowly, and will make a positive impact in the next decade.

Meanwhile most European cancer doctors can't access the existing molecular information buried in the common cancers due to cost. The explosion of links of gene mutations to specially designed medicines or vaccines has been unaffordable outside the big cancer centres.

But while prices come down, AI research can feast on simpler challenges. Each cancer patient will have a large file full of simple cheap information- H&E stained pathology slides, CT scans, maybe MRI too, and numerous haematological, biochemical and immunological data, obtained in under a minute for miniscule monies. Data on treatment by surgery, radiotherapy and systemic therapeutics, with accompanying side effects and accurate dates of death, are highly likely to unravel algorithms which will help make cancer treatments more intelligent, and more personalised.

The main block in this simple science is the ignorance of doctors who are frequently bereft of up to date mathematics, buried as they too frequently are in a silo which contains information about one cancerous organ and one modality of therapy. So while the AI specialists get busy dissecting subcellular characteristics of all cancers, simple cross talk between clinicians, imprisoned in their comfortable silos, and those AI experts, will yield low cost applicable results in the short term. My talk will illustrate successes and failures in AI in cancer.

TUTORIAL

INTEGRATED ARCHITECTURE FRAMEWORK OF STANDARDS FOR HEALTHCARE SERVICES

Prof. Mário Macedo,
Universidade Atlântica, Portugal

ABSTRACT

The new challenges of healthcare services regarding the requirements for a cross-border network of information and collaboration of care providers, politicians and researchers are stressing to a new foundation for health services.

A critical review of the available models and standards, some of them originally developed to the industry and others specific to health care services should be the foundations for the new paradigm called Health 4.0.

In addition, the medicinal products data traceability and the data architecture proposed by International Organization of Standardization will digitalize the data flow of the medicines supply chain. This upcoming will enhance the quality, accessibility and ubiquity of the Health 4.0 as a whole.Some technology paradigms are changing, the monolithic applications are being replaced by microservices. Will the microservices be the future landscape of Health 4.0?

If so, how will the return of investments and risk be evaluated? Some models to measure the systems reliability are already on the field. However, the microservices technologies have some new challenges due to their cross dependencies and potential cycled operations.

FULL PAPERS

DESIGNING THE VALUE PROPOSITION OF SYSTEMIC EHEALTH INNOVATION ISEREADOM CASE STUDY

Angela Martin[1], Véronique Chirie[2] and Emmanuelle Chifflot[3]
[1]*ALTRAN frog Design, 96 Avenue Charles de Gaulle, 92200 Neuilly-sur-Seine, France*
[2]*TASDA, 163 Cours Berriat 38000 Grenoble, France*
[3]*Département de l'Isère, 15 av du Doyen Louis Weil, 38000 Grenoble, France*

ABSTRACT

The goal of this paper is to provide an assessment tool of the acceptability and feasibility of an eHealth ecosystem innovation and how to deploy it on the field. The Value Proposition Canvas is applied to IsereADOM case study as a "score-board" to track if/ and how the innovation project actually eases pains and creates gains when testing it with customers. Beyond the interest of making readable and understandable the impact of such a complex project, the focus of this work is on the integration of the project value proposition at the national, regional, local level to overcome barriers but also pull out the best levers of the ecosystem.

KEYWORDS

ICT, Evaluation, Innovation, Value Proposition Analysis, Territorial Cooperation, eHealth, Well Aging

1. INTRODUCTION

While eHealth innovation projects have a great potential to address healthcare disparities, the great majority are not deployed beyond the pilot stage. Moreover, the organization of the health industry is unique. Stakeholders are particularly dependent on each other and their multiplicity enhances the complexity of business models. We have already proved in our previous work that, by taking a complex system modelling approach, we can provide a tool, for both entrepreneurs and investors, that offers a comprehensive view of value streams by the comparison of different business models based on the **economic** shared value creation during the go to market process (Martin *et al.*, 2018).

In the present paper, we extend to non-economic value from Organisational Complex System Modelling framework (OCSM): quality of care, social, medical (Jean *et al.*, 2016; Martin *et al.*, 2019). In fact, for Porter (2011), generating **social** and **medical value** is a competitive advantage.

The goal of this paper is to provide a design thinking framework of the central component of a business model: the Value Proposition (Osterwalder *et al.*, 2010; 2015). The focus here is to deep dive into the way organizations function and create value by using IsereADOM case study. The paper is organized as follows: in the next section, a literature review examines business model design and challenges and the methodology. The case study data and methodology (Value Proposition Canvas and the impact of the case study integration into its ecosystem) are described and discussed in the third section. Finally, a conclusion ends this paper.

2. THEORETICAL BACKGROUND

The field of business models has developed dramatically over the last ten years both in academia and in business and therefore we intend to shed some light on its central element i.e. the Value Proposition. According to the literature review, business models are classified into four categories: value proposition, value network, value finance and value architecture (Zott *et al.*, 2011).

Among the frameworks provided, the Value Proposition Canvas introduced by Osterwalder and Pigneur (2014) is one of the first to create a tool that specifically helps entrepreneurs to design the pivotal element of their business models. It gives a simple and accessible way of examining whether the value propositions of a company's business model correlates with the actual needs of the customers it wishes to serve (Osterwalder *et al.*, 2015). Whereas the Business Model Canvas (Osterwalder and Pigneur, 2010) focused on how to create value for businesses, the Value Proposition Canvas focuses on how to create value to **customers**. By combining qualitative research and the impact measurement of the innovation on the ecosystem stakeholders, we design and test value propositions in an iterative search for what stakeholders want.

3. METHODOLOGY APPLIED TO THE CASE STUDY

By using a Value Proposition Canvas we focus on whether the innovation answers real customer pain points to ensure its adoption before the scaling up. In addition, the circumscription of the project value proposition at the national, regional, local level is necessary to overcome barriers but also pull out the best levers of the ecosystem.

3.1 Project IsereADOM

IsereADOM is a French eHealth demonstrator of efficiency of a set of services relating to information, support and care of frail or dependent people living at home. The project goal is to create more fluidity and decompartmentalising for home support. It is conducted by the Isère County council in consortium with financial partners (CARSAT, ARS AURA, AG2R LA MONDIALE, APICILE, CDC, EUROPE, REGION ARA), the assistant project owner (TASDA), and the industrial consortium gathered Altran, IMA, ORANGE (and their subcontractors). The project has been conducted with the support from different stakeholders involved in home healthcare, either with national and remote offers (such as medical monitoring or remote assistance), or with local offers (such as care givers, nurses).

Carried out over 5 years (2.5 years of development, 2 years of field trials, 0.5 year of evaluation), the project involved nearly 150 actors, from more than 20 different companies. All IsereADOM components have been evaluated. 13 thematic reports were produced, about public services, personalized supports for frail people, project management, economic balance, IT solutions' diagnosis, etc. The assessment results are based on online surveys, focus groups, individual and telephone interviews, statistics analyses on the use of digital tools, medico-economic analyses, feedback sharing from the actors involved. The web portal reached more than 15,000 visitors, mobilizing more than 1,000 service providers for the directory of local services. Personalized support was carried out with 200 seniors, with the mobilization of around 100 professionals.

Finally, IsereADOM consists in:

• A development toolbox, which have been applied to further advantage in the form of computer modules (such as a web directory of local services, a digital link notebook, etc.), formalized business processes (such as "processing requests arriving at the call centre", "carry out sentinel monitoring", etc.), communication and training tools.

• Acknowledged expertise in setting up a complex organisational innovation, change management support, integration into existing organizations and tools, management of arbitrations with locally elected bodies. The "engineering" component is an essential point of the project, essential to make available to other territories in case of future deployment.

IsereADOM assessment data gathered is used for designing the Value Proposition Canvas (here after) and for measuring the impact of the project at the local, regional and National level (see 3.3).

3.2 Value Proposition Canvas

The Value Proposition Canvas (VPC) describes the features of a specific value proposition in the business model in a more structured and detailed approach (Osterwalder *et al.*, 2014). The VPC (figure 1) breaks the value proposition down into *products & services*, *pain relievers* and *gain creators* (on the left side). On the right side, the customer (segment) profile where the customer understanding is clarified: the *jobs* the customers are trying to get done in their lives and their work, the *pains* describing bad outcomes, risks and obstacles related to customer jobs and finally the *gains*, the outcomes they want to achieve or the concrete benefits they are seeking. The goal is to achieve a *fit* between the two sides of the model. Three type of *fit* are possible depending on the maturity of the project:

1) *Problem-solution fit*: evidence that the customers care about the jobs, pains and gains IsereADOM intends to address with the value proposition.

2) *Product & services-market fit* evidence that customers want the value proposition of IsereADOM.

3) *Business model-fit* evidence that the business model for IsereADOM value proposition is scalable and profitable.

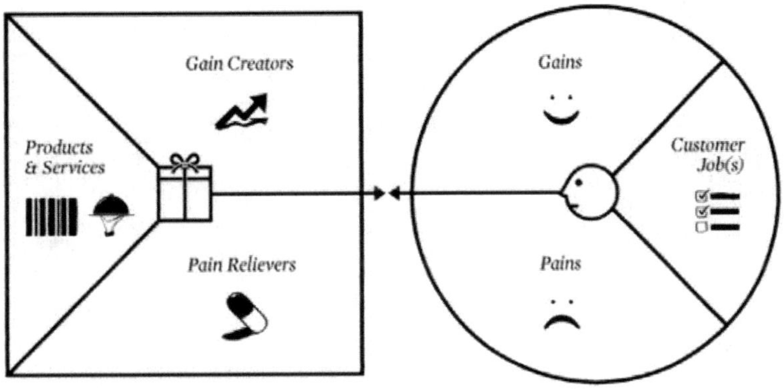

Figure 1. Value Proposition Canvas

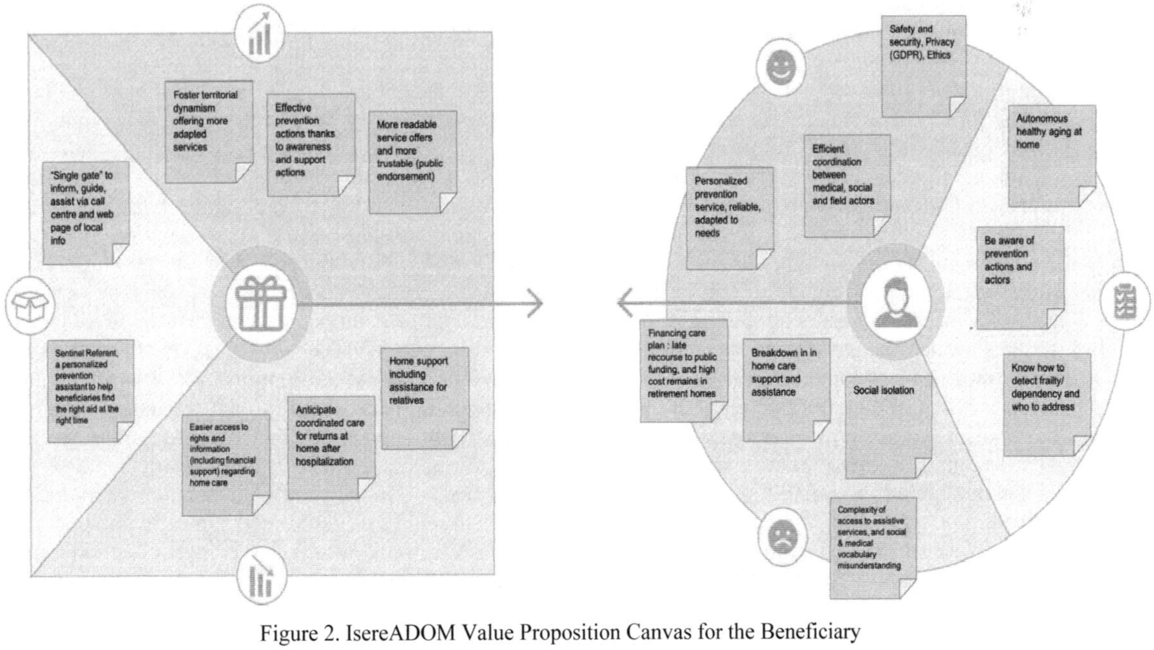

Figure 2. IsereADOM Value Proposition Canvas for the Beneficiary

Based on IsereADOM assessment data gathered, we bring evidence that IsereADOM creates value for the customers by investigating the *product & services market fit*. The *fit* is achieved when IsereADOM services produce pain relievers and gain creators that match one or more of the jobs, pains and gains that are important for the two segments identified as IsereADOM customers are: the *beneficiary* of the services i.e. the elderly living at home (figure 2) and their designated caregiver called the *Sentinel Referent* (see figure 3 below).

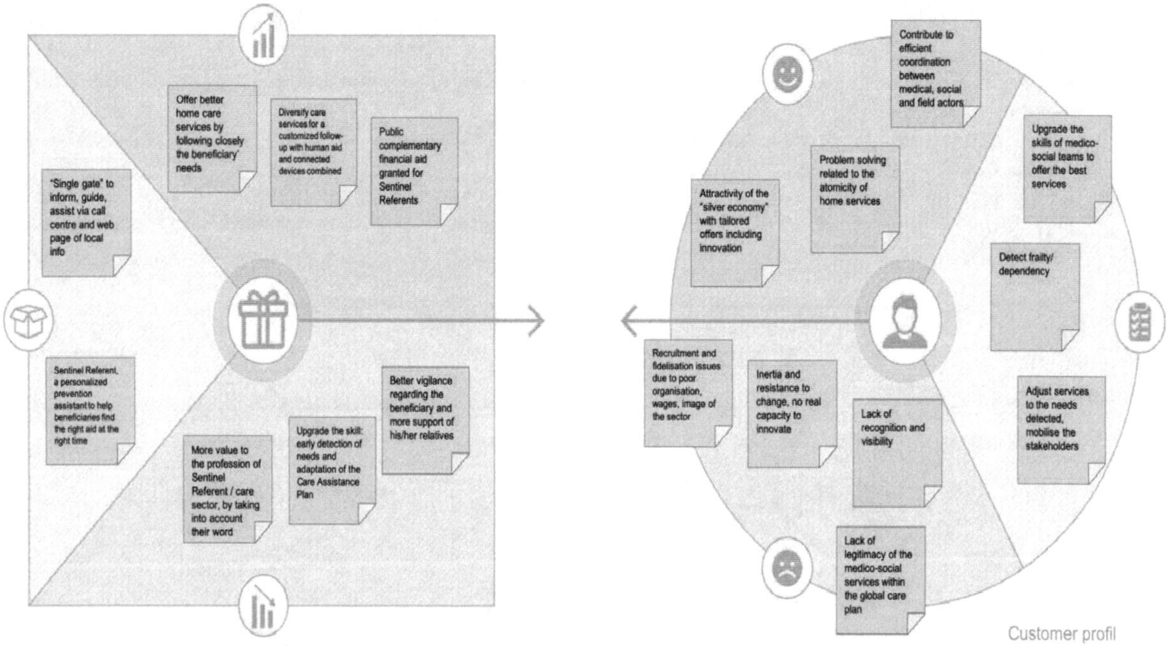

Figure 3. IsereADOM Value Proposition Canvas for Sentinel Referent

The *beneficiary customer profile* is described on the *right side* of figure 2. Their jobs to be done are: autonomous healthy aging at home and so they need to increase their awareness regarding *prevention* actions by detecting frailty/ dependency and the local stakeholders (actors) to address. The pains are: social isolation, the complexity of access to assistive services, the misunderstanding of social and medical vocabulary that may lead to a breakdown in home care support and assistance and the risk of premature exhaustion of the family caregiver. A particular highlight is on the financing of the dependency by a late demand of public funding, and a very high "out of pocket" for people living in retirement homes. The gains that the beneficiary would like to obtain are: personalized prevention services that are not only reliable but also adapted to their needs offering more comfort and social bonds; services compliant to safety, security, privacy and ethics. An efficient coordination between medical, social and field actors may avoid breakdowns and social isolation.

The *sentinel referent customer profile* is described on the *right side* of figure 3. The jobs they are trying to perform and complete are: detect the frailty/ dependency early signs of the beneficiary by adjusting services to the needs detected, mobilising the "right" stakeholders, upgrade the quality of their services by improving the skills of medico-social teams on the field. What annoys the sentinel referents or simply prevent them to perform their job is: the lack of recognition and visibility accompanied by the lack of legitimacy of the medico-social services within the global care plan, the inertia and resistance to change with no real capacity to innovate, recruitment and fidelisation issues due to the poor organisation, low wages and bad image of the caregiver job. The sentinel referents expect to contribute to problem solving related to the atomicity of home services stakeholders and to a more efficient coordination between medical, social and field actors. Furthermore, they desire a better attractivity of the "silver economy" (destined to seniors) with tailored innovation offers.

The use of the customer profiles helped IsereADOM visualize what matters to the beneficiary and sentinel referent. As a "score-board" it helps to give a global vision of the needs and responses produced, which are multiple, complementary, different.

On the *left side* of the VPC (figure 2 and 3), the services offered by IsereADOM are described:

1) A "single gate" access to general information and orientation on local services (personal services, housing, transportation, leisure, etc.) and local prevention activities (health and well-being); a call center and a website (with a database) gathering information on local resources and facilitating everyone's access to this information; a personalized support component for people living at home with decreasing independence.

2) Thanks to the designation of a Sentinel Referent (SR) using a digital link notebook, the care giver identifies the senior status change and calls on a doctor or the Department referent to adapt the care/aid plan. The "sentinel referent process" involves services like: teleconsultation, tele-assistance, telephone coaching combined with human intervention. The goal is to detect frailty/ dependency in order to search for the right assistance at the right time.

The value proposition of IsereADOM services is what figure 2 and 3 have in common. What differs is the *benefits* customers can expect from IsereADOM because the pains they struggle with and the gains they try to get depend on the jobs to be done in their lives (beneficiaries) or their work (sentinel referents). So two main benefits or *gain creators* arise from using IsereADOM services. From the beneficiary's point of view, more prevention (less avoidable hospitalization, complexity, ...) to support autonomy, thanks to a good shared knowledge of the beneficiary and his/her situation at home. This benefit must result in both a better quality of life but also more efficient means of care. The VPC (figure 2) shows how IsereADOM brings a global approach of the "well aging" with home support, thanks to information, coordination, access to the adapted local services or local prevention action, within the GDPR framework. From the SR point of view, a modernization of the medico-social sector, with real recognition of the SR added value (the detailed knowledge of the beneficiary), including in coordination and the promotion of his/her profession. This leads to better working conditions, more ease of recruitment and better quality service and so one. The VPC in figure 3 shows how IsereADOM brings the opportunity to innovation, sustainability, change for more efficient, thanks to a systemic approach with tools (ICT solutions), process and a foundation of public funding.

Hence, the answer provided by IsereADOM to support autonomous healthy aging at home creates some essential gains for its customers. It also *relieves* real customers *pains*. For the beneficiary it offers easier access financial support rights and information regarding home care; coordinated care anticipated in case of returns at home after hospitalization, home support including assistance for the family care giver. For the sentinel referent, IsereADOM brings more value to the profession of the care giver, by taking into account their word, upgrading their skills in early detection of needs in the adjustment of the Care Assistance Plan for a better vigilance regarding the beneficiary and offering more support to their relatives.

3.3 IsereADOM Impact

The value proposition canvas is qualitative and focused on one type of customer. If they are explicit and representative of a shared customer understanding, they must nevertheless be translated into quantifiable indicators to measure real benefits.

The impact matrix makes it possible to translate the values proposition canvas into indicators giving a holistic view. It also restores a global vision of the project by specifying the stakeholders and the coherence and complementarity of the benefits for each of them. The impact matrix thus highlights the systemic approach and the circumscription of the Value Proposition of IsereADOM into the local, regional, national ecosystem.

To go further we need to apply a systemic design approach (Barbero, 2016) that integrates system thinking and human centered design) to study the externalities both positive and negative on the systems (political, economic, environmental, social, behavioral) at short, mid and long-term to be able to conceive the strategy of the scaling up of IsereADOM.

Table 1. IsereADOM Impact on Stakeholders

				STAKEHOLDERS								
				Benefi-ciary	Rela-tives	Care giver (SR)	County council	Pension Fund	Hospitals	Local authority	National Health fund	Industry start up
INDICATORS	Economic	Ind 1	Better efficiency of health & care spending	X	X	X	X	X	X		X	
		Ind 2	Improved local and national policy coordination			X	X	X		X	X	
		Ind 3	Increased no of recruitments of care givers			X	X			X		
		Ind 4	Connected device market development	X	X	X						X
	Social	Ind 5	Better quality of life	X	X							
		Ind 6	Better access to healthy aging at home info	X	X	X	X	X				
		Ind 7	Modernization of the medico social sector			X	X					
		Ind 8	Promotion of local services & activities / territorial "anchoring" and development	X	X		X			X		
	Quality of care	Ind 9	Higher efficiency of care givers	X	X	X	X					
		Ind 10	Continous integrated care vigilance by human & technological inputs			X	X	X	X		X	X
		Ind 11	Better awareness of beneficiary needs & frailty shared information	X	X	X			X			
		Ind 12	Improving ICT interfaces for health & care software	X	X	X						X
		Ind 13	Local coordination embedding			X			X	X		
		Ind 14	Responsiveness of the adaptation of health & care plans	X	X	X	X		X	X	X	

4. CONCLUSION

Two main sets of interest arise from the work of formalizing the value proposition map, in addition to its strategic support. Firstly, it is necessary to structure the values proposition in a clear manner for a systemic project with so many components and geographical levels like IsereADOM. Replacing words with meaningful and instructive pictures give a better readability of the VPC thanks to the mapping of the value created for its customers. This readability is necessary for all the actors of the project but also the stakeholders in connection with the subjects that the project addresses. Secondly, innovation projects for frail people living at home must consider not only their acceptability as beneficiary but also the acceptability of the referent sentinel (private healthcare professionals or care givers). Furthermore, the VPC of IsereADOM provides an interesting synthesis of the key points of home care for elderly people in terms of challenges, expectations, levers.

Strategy is about creating actual value so this work is a contribution to entrepreneurs and innovation project managers who need to renew their business models by calling the interest and the feasibility of the innovation into the ecosystem into question.

REFERENCES

Barbero, S., 2016. Opportunities and challenges in teaching Systemic Design. The evolution of the Open Systems master courses at Politecnico di Torino. *In Proceedings of the 6th International Forum of Design as a Process*, Universitat Politècnica de València, Valencia, pp. 57-66.

Gorskia, I., Brama, J. T., Sutermastera, S., Eckmana, M. and Mehtab K. Value propositions of mHealth projects, *In Journal of Medical Engineering & Technology*, august *2016*.

Jean, C., Duong, T., Cardinal, J., Jankovic, M., Bocquet, J., and Espinoza, P., 2016. Sharing economic value between the stakeholders of a telehealth project: methodological issues?, *In Eur. Res. Telemed.*, vol. 5, pp. 37–44.

Martin, A., Canu, C., Jean C., Breda, G., 2018. E-health deployment business model tele-expertise case study, *In MCCSIS e-Health 2018 Proc, Outstanding Paper Award 2018.*

Martin, A., Cosandier- Rimele, D., Canu C., Breda G., 2019. New system economic evaluation of ehealth innovation, *In MCCSIS e-Health 2019 Proc, Best Paper Award 2019.*

Osterwalder, A. and Pigneur, Y., 2010. *Business Model Generation.* John Wiley & Sons.

Osterwalder, A., Pigneur Y., Bernarda G., Smith A. and Papadakos T., 2015. Value Proposition Design How to create products and services customers want, *In Journal of Business Models*, vol. 3, no. 1, pp. 81-89.

Porter, M., and Kramer, M., 2011. Creating shared value - How to reinvent capitalism and unleash a wave of innovation and growth, *In Harvard Business Revue.*, vol. 39, pp. 62–77.

Zott, C., Amit, R. and Massa, L., 2011. The Business Model: Recent Developments and Future Research, *In J. Manage.*, vol. 37, no. 4, pp. 1019–1042.

VERIFICATION OF AUTOMATA MODELS FOR CLINICAL DECISION SUPPORT SYSTEMS

Grigorii Shovkoplias[1], Ivan Smirnov[1], Mark Tkachenko[1], Natalia Gusarova[1], Aleksandra Vatian[1], Anatoly Shalyto[1] and Rajdeep Niyogi[2]
[1]*ITMO University, Sankt-Petersburg 197101, Russia*
[2]*Indian Institute of Technology Roorkee. 247667, India*

ABSTRACT

In the treatment of any disease, the doctor must strictly follow the clinical protocol. However, this document cannot be fully considered as a graph diagram of the algorithm of the doctor's actions, since most of the transition rules in it cannot be implemented without the involvement of the doctor's knowledge. So, based on the clinical protocol, it is necessary to create a clinical decision support system (CDSS). Existing CDSSs do not allow describing the clinical protocol as a sequence of actions. This drawback can be eliminated by using the automata model as a CDSS inference engine, since the terminology of the treatment process itself leads to a natural thought - to model this process in states, i.e. use an automata approach. At the same time, it is vital to ensure the extensibility of the automata model, which will allow among other things for taking into account when choosing drugs any possible adverse reactions to drugs already prescribed.
The article discusses the process of development and verification of an automata model as an inference engine for CDSS. The process, based on the Automata approach and performed offline, uses a sequence of refinements, which allows you to get a more detailed model that reflects the typical options for the doctor. In addition, with each replacement or addition of the drug, online testing of the model is performed, during which compatibility with the already prescribed drugs and the patient's history is checked. The developed process is applied to the construction and verification of CDSS for the management of patients with multiple sclerosis.

KEYWORDS

Automata, Clinical Decision Support System, Verification, Multiple Sclerosis

1. INTRODUCTION

Clinical decision making is a fundamental aspect of the work of the clinician, which has a direct impact on the course of treatment of the patient. Clinical decision making is defined as a contextual, continuous, and evolving process through which data are collected, interpreted, and evaluated to make informed choices [Tiffen]. The set of factors that determine the context is very wide [Johansen]; it includes the patient's health state, available treatment options (including equipment, medicines, etc.), the clinical experience of the doctor, the patient's personal characteristics, and the level of accessible healthcare environment in general. At the same time, the guiding document for the doctor is the official clinical recommendations for the management of a specific disease. Thus, when making clinical decisions, the clinician must take into account a complex set of conditions, and to support this process, the use of clinical decision support systems (CDSS) is effective [Horsky, Bologva].

An official source [Clinica] defines CDSS as a system that provides clinicians with knowledge reasonably filtered or presented at the right time to improve health and medical care. Accordingly, these definitions cover a wide range of healthcare support systems that have been announced as projects or are already on the market (see reviews [Rawson, Reis]. A detailed analysis showed that most of CDSS implement the following functionalities individually or in combination: support for individual diagnostic procedures (for example, automatic analysis of radiographs, CT and MRI images), statistical evaluation of indicators and access to medical calculators, reference functions (for example, checking with drug capacities, structured access to health services); record indicators with remote monitoring of patients; manual recording of tasks (during treatment in the clinic). However, they do not allow describing the clinical protocol as a formalized sequence of actions.

Significant advantages in this regard are provided by the use of the automata model [Khoussainov, Khoussainov, Polikarpova] in the framework of CDSS. The concept of formal methods is widely used in healthcare (see review [Bonfanti]). A substantial analysis of the terminology used to describe the clinical process suggests that the state in the automata model and the state of the patient in the clinical process are close not only terminologically, but also conceptually. In particular, the archetype of State Machine as a basic concept is recommended for use in the openEHR medical standard [openEHR]. Note that modern means of supporting automated programming provide ample opportunities for modeling decision-making processes [Mitkin]. Many authors [Bonfanti, Banach] consider the verification of automata models (and formal modeling methods in general) a basic advantage for modeling medical processes, where the safety requirement is critically important. In [ten Teije, Groot, M´ery] models of medical protocol are proposed, but they consider the protocol as purely formal, i.e. do not imply variability in the behavior of the doctor in relation to a particular patient.

Thus, the introduction of an automated approach to the development and verification of CDSS to support the clinical process is an important and urgent task. In this article, this problem is solved by the example of the management of patients suffering from multiple sclerosis [NINDS] – the most common autoimmune disorder of the central nervous system, resulting in physical, mental, and psychiatric problems up to disability.

2. PROBLEMS OF USING AUTOMATA MODELS IN CDSS

Although studies of multiple sclerosis are conducted on a broad front, a radical cure for this disease still does not exist [NINDS]. The management scheme for a patient with multiple sclerosis [19] includes primary diagnosis, assessment of the degree of disability, treatment in the acute stage, as well as the prescription of drugs that can change the course of multiple sclerosis. Each stage is described by a set of recommendations in natural language, accompanied by a detailed commentary, and contains so called "Algorithms for the management of the patient", which are models (schemes) of the corresponding stage of the clinical process. A detailed analysis showed that this documentation structure is characteristic not only for multiple sclerosis, but also for other diseases.

Unfortunately, these models, although called algorithms, are not such in the formal sense, since they contain different types of logical discrepancies and inconsistencies [ten Teije, Groot, M´ery, Gusarova, Vatian]. Many authors [ten Teije, Groot, M´ery] propose a rigorous formalization of clinical protocols as a whole, however, such a path rarely appears to be acceptable in real clinical practice. For the practical use of clinical protocols, it is necessary by default to "load" an extensive and dynamically changing context associated with the knowledge and experience of a clinician. At the same time, the more detailed the fragment of the clinical protocol, the more formally its model and corresponding context can be set. These circumstances open up the principal possibility of constructing an automata model of the clinical process as an inference engine for CDSS.

A significant role in the clinical process is played by possible contradictions between the various types of medical activities offered to the patient. For example, the prescribed medication may worsen the course of another disease that the patient suffers from, may be contraindicated in connection with certain medical facts from the medical history, or may enter into undesirable interactions with other medicines that the patient is taking. The identification of such contradictions and their demonstration to the doctor is one of the tasks of the CDSS. In the case of multiple sclerosis, this problem is especially acute. Here, in order to achieve and maintain the desired effect, drug therapy should continue continuously and for a long time, and the existing range of drugs has a very wide range of adverse reactions up to complete intolerance.

The need for individual selection and flexible change of drugs prescribed for a particular patient poses the problem of making and verifying the current changes in the automata model used in the CDSS.

As emphasized in [Mashkoor], for modeling of medical devices, rigorous method are needed, and the notion of refinement plays the pivotal role here, supporting the correct-by-construction approach for development. Moreover, each refinement is accompanied by a verification stage, which guarantees secure operation of a medical device. It seems appropriate to extend these approaches to the development of automata models as inference engines in CDSS. To the best of our knowledge, there is no previous study that analyzes the possibilities of making and verifying the current changes in the automata model used in the CDSS.

Therefore, the article raises the following questions:

- How to define a refinement in which the clinician can make flexible changes in CDSS states without the participation of an IT developer?
- How to ensure verification of such a refinement?

The further material is organized as follows. Section 3 discusses possibilities and conditions of implementation of flexible changes in an automata model. In section 4 the ways of verification of flexible change in the automatic model are under consideration. In section 5 as a case study implementation of an automata model with flexible change as an inference engine in CDSS for multiple sclerosis is considered.

3. IMPLEMENTING OF FLEXIBLE CHANGES IN AN AUTOMATA MODEL

Consider an example. Figure 1 shows a fragment of an automata modeling the clinical protocol for the treatment of multiple sclerosis, which includes the following standard states: `Identification of exacerbation degree`, `SMT`, `Pulse therapy and result evaluation`, `Plasmophoresis (ACTH)`, `Checking the type of course`. In addition, according to the physician's opinion, in the case of a severe exacerbation, dexamethasone may be used instead of pulse therapy (the corresponding state `Dexamethasone and result evaluation` is shown in Figure 1 as optional).

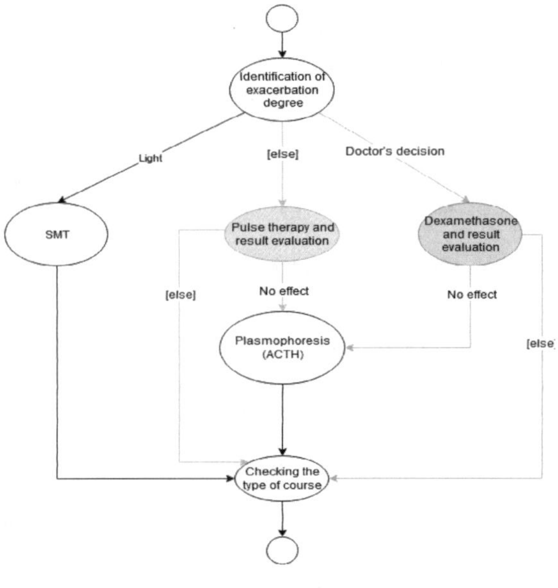

Figure 1. Fragment of a clinical protocol for the treatment of multiple sclerosis

However, in this case, the physician needs to check whether the use of dexamethasone is contraindicated in relation to all precedent and subsequent appointments and procedures, which are provided for by the further execution of the clinical protocol. A formal description of this problem is illustrated in Figure 2. Let the basic automata is in state A, state B called "standard" is responsible for fulfilling some medical activities (medicines and / or procedures), state C is the resulting one, state X can be added to basic automata optionally. Adding state X leads to the appearance of a new path from state A to state C through the new state X.

Medical activities being performed in state B may conflict with medical activities performed in previous states (they are conditionally labeled as state D in Figure 2a). However, these contradictions are revealed during the verification of the basic automata, and the history of step-by-step verification is preserved. On this basis, when state X is added, possible contradictions of its medical activities with those of previous states can be detected automatically.

Consider now some state E (Figure 2b) attainable from state C (possibly through several transitions). Prior to adding state X, the automata was checked, respectively, states A, B, C, and E did not contradict each other, however, the added state X may contradict state E, for example, due to contraindications between medicines in some states or due to some transitions non-availability. However, the possible contradictions of the medical activities of state X with the medical activities performed in subsequent state E cannot be predicted by the above method.

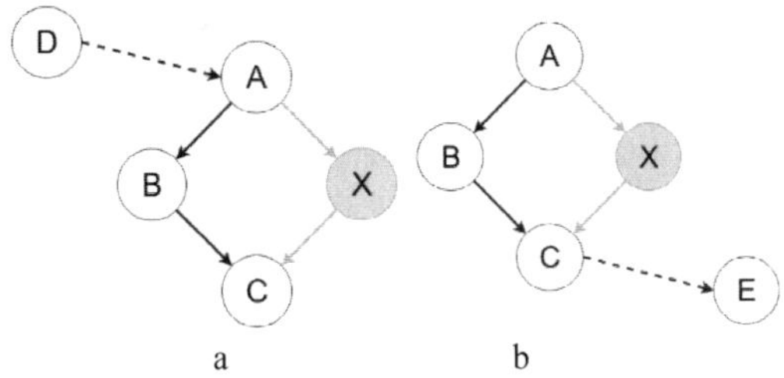

Figure 2. A formal model of a fragment of an automata with contradictions

To understand the conditions under which it is possible to add a new state X to an already existing automata model, we introduce the concepts of a "similar" state.

Let the medicine or procedure Db be used in state B with the contraindications CDb1, CDb2, ... Let state B also have an input transition from state A, and (possibly several) output transitions to states C1, C2 ...

Then let us call a "similar" state of state B, such state X that it uses medicine or procedure Dx with contraindications CDx1, CDx2, ... and:

– there is an input transition from state A to state X and output transitions C1, C2 ...;

– there are as many exit transitions from state X as there are from state B;

– there is a bijection (one-to-one correspondence) between each output transition from B and each output transition from X.

The input transitions from state A to states B and X must have different conditions (to maintain general orthogonality). Since the transition to state X is performed by the doctor's decision, the condition for the transition to X is a predicate obtained as the logical "AND" conditions for the transition to B and the doctor's decision, which will provide the necessary orthogonality.

The transformation of the base state of an automata into a "similar" state can rightfully be considered as a transition in automata modeling from ground model to refinement. Thus, in order to get the source code describing the "similar" state X, it is enough to take the code describing the "standard" state B and change the attributes that are responsible for the parameters, namely the medicines and \ or procedures and contraindications.

4. VERIFICATION OF FLEXIBLE CHANGES IN THE AUTOMATA MODEL

According to the review [Bonfanti], model checking [Emerson] dominates among verification methods for medical devices. The general limitations of the model checking method include the following: (1) Only models with a finite number of states are supported. (2) There is a state-space explosion problem characteristic of large multiprocessor systems. (3) Using the method, only a system model is checked instead of a real automatic program, and not any model properties, but only temporal ones. However, due to the specifics of the subject area of application of the automata model, namely the support of the clinical process in the presence of flexible changes, restrictions (1) and (2) are not fundamental.

The list of model-checking tools for medical devices is rather vast [Bonfanti]. However, in relation to the automata used as an inference engine in CDSS, the most effective was the use of the UniMod.Verifier, that works by integrating the UniMod tool [Gurov, Ricca] and the Bogor program verifier [Robby]. When using UniMod.Verifier, there is no need to convert the automatic program into the input language of the verifier. A significant advantage of UniMod.Verifier compared to other verifiers is that it uses the UniMod interpreter of automata systems. Therefore, UniMod.Verifier does not check the correct operation of the program model, but the automata system executed by the UniMod tool. This eliminates the above limitation (3) of the model checking method.

Program requirements of the UniMod.Verifier are written in the temporal logic language LTL using the following operators: LTL.always (G); LTL.eventually (F); LTL.next (X); LTL.until (U); LTL.weakUntil (W), p W q = (p U q) || G (p && !q); LTL.release (R, p R q = !(!p U !q); LTL.negation; LTL.equivalence; LTL.implication; LTL.conjunction; LTL.disjunction.

To verify the automata program in UniMod.Verifier, one first needs to create an XML description that is automatically executed in UniMod. Moreover, when using some other verifiers (for example, Convert), UniMod can also be used to save the automatic program in XML format.

Then one needs to write down the formula for verification. To do this, first the predicates are declared, and then the formula is written using these predicates as nested function calls. For example, a predicate with the name "no_effect" and the value AutomataModel.wasEvent (model, "CMT") is true when the last processed event is CMT.

The verifier is called by a special command, for example:

```
verifier.cmd TestHospital.connectivity AfterDegreeOfExacerbation,
```

where `TestHospital.connectivity` is the name of the description file of the automated program in UniMod format with the `connectivity` extension, and `AfterDegreeOfExacerbation` is the name of the function for verification.

Let us compare the formula for verifying a fragment of the automata model presented in figure 1 written using the described formalisms. For ground model, i.e. using only "standard" states, the formula looks as follows:

```
fun AfterDegreeOfExacerbation() returns boolean = LTL.temporalProperty(
        Property.createObservableDictionary(
                Property.createObservableKey("no_effect",
AutomataModel.wasEvent(model, "SMT")),
                Property.createObservableKey("Plasmophoresis (ACTH)",
AutomataModel.wasAction(model, "Pulse-therapy")),
                )
        ),
        LTL.weakUntil (
                LTL.negation(
                        LTL.prop("Plasmophoresis (ACTH)")
                ),
                LTL.prop("no_effect")
        )
);
```

After calling the appropriate command, standard information about the verification process will be written to standard output. The result of the verification can be either a success message or an error message after verification in the form of a counterexample that describes the steps of the state of the machine, for example:

```
Stack of transitions leading to the error:
Model [ step [0] event [null] guards [null] transitions [null] actions [null]
states [null] ] fsaState [bad$accept_init]
Model [ step [1] event [SMT; Light] guards [true->true] transitions
[Identification of exacerbation degree. Pulse therapy #true] actions [no effect]
states [(/AClient:1. [else] /AServer) - (Top); (/AClient:2. Plasmaphoresis
(ACTH)/AServer) - (Top);] ] fsaState [bad$accept_init]
Done!
```

When using a "similar" state, i.e. state `Dexamethasone and result evaluation`, the formula for verifying the automata is as follows:

```
fun AfterDegreeOfExacerbation() returns boolean = LTL.temporalProperty(
        Property.createObservableDictionary(
                Property.createObservableKey("no_effect",
AutomataModel.wasEvent(model, "SMT")),
                Property.createObservableKey("Plasmophoresis (ACTH)",
AutomataModel.wasAction(model, "Pulse-therapy")),
```

```
            Property.createObservableKey("Plasmophoresis (ACTH)",
AutomataModel.wasAction(model, "Dexametazon"))
)
        ),
        LTL.negation(
                LTL.weakUntil (
                        LTL.negation(
                                LTL.prop("Plasmophoresis (ACTH)")
                        ),
                        LTL.prop("no_effect")
                ),
                LTL.negation(
                        LTL.prop("Plasmophoresis (ACTH)")
                )
        ),
        LTL.prop("no_effect")
);
```

It is easy to see that in this case a transition is added after the `Dexametazon` state and in the automata model new conditions for the transition from previous states to the final one are added (the corresponding changes are highlighted in the code). For other possible contraindications that will appear in subsequent states, changes to the verification formula are to be made in a same way.

Thus, upon activation of a "similar" state, a pre-prepared verification code is launched that corresponds to all interactions with this state.

5. IMPLEMENTATION OF AN AUTOMATA MODEL WITH FLEXIBLE CHANGE AS AN INFERENCE ENGINE IN CDSS FOR MULTIPLE SCLEROSIS

The implemented CDSS is based on a combination of "design patterns + automata", which ensures compliance with the requirements of standardizing the clinical process both from the point of view of clinicians and from the point of view of IT developers. To work with the contradictions that arise during the clinical process, specialized design patterns are used in the architecture of the developed CDSS. The architecture diagram is shown in Figure 3.

For a flexible selection of various strategies of working with contradictions, the Strategy and Facade patterns are used. The *`IMedicalStrategy`* interface is implemented using specific strategies containing behavior for each of the scenarios of working with contradictions arising between various medical activities offered to the patient, including drugs and / or procedures. The logic of working with automata is hidden in the `StateMachine` class, which provides the `setState()`, `getState()`, and `getAllStates()` methods as an interface. To make clinical decisions, the attending physician should be provided with a quick and easy way to work with various sources of medical information. To do this, the Adapter pattern is foreseen, which allows to combine various data sources into a single interface for interaction and use them together. The `UserInterfaceControl` class represents the user interface with which the doctor will interact.

The whole complexity of working with an automata machine including resolving contradictions is hidden in ground model variant behind the `AutomataFacade` view, which is one of the points of work with the solution presented.

The presented scheme makes it easy to make refinements of the kind described above. Namely, to add a test for contraindications when introducing a new "similar" state, it is enough to add one class, called `Verifier`, whose responsibility will be the verification of a new change, described in section 4. The scheme will change as follows (the changes are highlighted in green in Figure 3): a dependency on the `Verifier` class will be added to `AutomataFacade`, and the `verify()` method will delegate verification to this class. To add a new "similar" state, the `addSimilarState()` method appears in `AutomataFacade`, which calls the `verify()` method, and, if successful, adds the passed "similar" state to the automata so that it can be used in the clinical process. An element (for example, a button) is added to the user interface, allowing the doctor to add a “similar” state.

Thus, the developed CDSS architecture based on the proposed automata model allows us to expand the scenarios of work with possible contradictions between medical activities contained in the clinical protocol without changing the code structure.

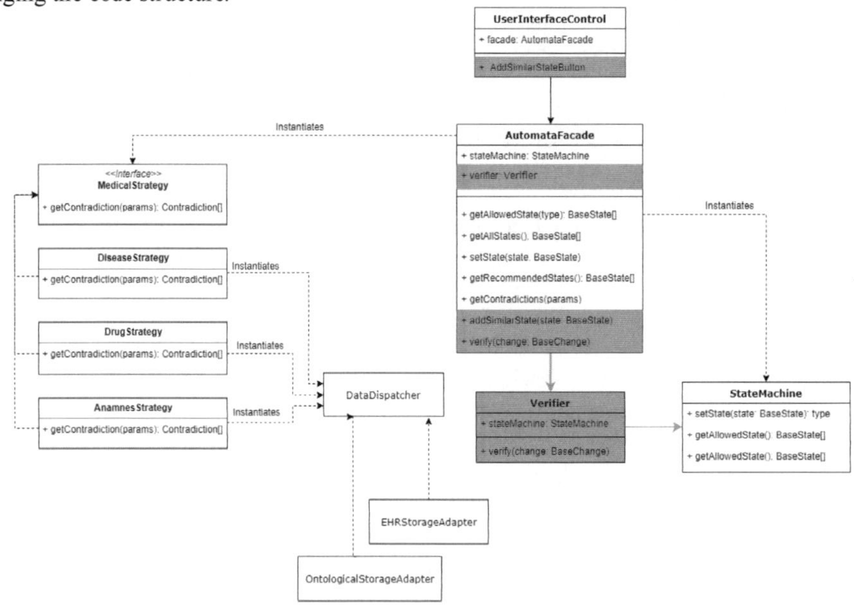

Figure 3. CDSS architecture diagram using design patterns

6. CONCLUSION

The following questions were posed in the article:

- How to define a refinement in which the clinician can make flexible changes in CDSS states without the participation of IT developer?
- How to ensure verification of such a refinement?

The answer to the first question was reduced to understanding the conditions under which it is possible to add a new state X to an already existing automata model. For this purpose the concepts of a “similar” state was introduced and defined. Thus, in order to get the source code describing the “similar” state X, it is enough to take the code describing the “standard” state B and change the attributes that are responsible for the parameters, namely the medicines and \ or procedures and contraindications.

When answering the second question, it was shown that in relation to the automata used as an inference engine in CDSS, the most effective was the use of the UniMod. Verifier. Using a specific example of a fragment of an automata model to support the clinical process of multiple sclerosis, it was demonstrated that verification of refinement associated with the introduction of a “similar” state comes down to checking for possible contradictions between the parameters of medical activities in this state and in subsequent states. Thus, upon activation of a “similar” state, a pre-prepared verification code that corresponds to all interactions with this state can be launched without the help of IT-developer.

As a case study the implementation of CDSS for multiple sclerosis using an automata model as an inference engine in was depicted. It is shown that adding a “similar” state to the automata model requires very small changes in the CDSS scheme and can be called up and verified with a single interface button, i.e. this can be done by the doctor himself without the participation of an IT developer.

The further prospects of the work are associated with the expansion of the nomenclature and characteristics of possible online changes to the automata model in accordance with the specifics of other clinical processes.

ACKNOWLEDGEMENT

This work was supported by Russian Science Foundation, Grant 19-19-00696.

REFERENCES

Banach R. Hemodialysis Machine in Hybrid Event-B //Abstract State Machines, Alloy, B, TLA, VDM, and Z. 5th Int.Conf., ABZ 2016, Linz, Austria, May 23-27, 2016. P. 376–393. (2016).

Bologva E.V., Prokusheva D.I., Krikunov A.V., Zvartau N.E., Kovalchuk S.V. Human-Computer Interaction in Electronic Medical Records: from the Perspectives of Physicians and Data Scientists // Procedia Computer Science. V.100. P. 915–920 (2016).

Bonfanti S., Gargantini A., Mashkoor A. A Systematic Literature Review of the Use of Formal Methods in Medical Software Systems. J. Softw. Evol. and Proc. 00:1–23 (2017).

Clinical Decision Support (CDS). Office of the National Coordinator for Health Information Technology. 2013. https://www.healthit.gov/policy-researchers-implementers/clinical-decision-support-cds, last access 31.08.2017.

Emerson A.E., Clarke E.M. Characterizing correctness properties of parallel programs using fixpoints. Automata, Languages and Programming, Lecture Notes in Computer Science, 85: 169–181 (1980)

Groot P., Hommersom A., Lucas P., Balser M., Schmitt J. Experiences in Quality Checking Medical Guidelines using Formal Methods. Proceedings Verification and Validation of Software Systems (VVSS 2007) March 23, 2007, Eindhoven, The Netherlands,, 2007; 164–78.

Gurov V.S., Mazin M.A. , Narvsky A.S., Shalyto A.A. UniMod: Method and Tool for Development of Reactive Object-Oriented Programs with Explicit States Emphasis /Proceedings of St. Petersburg IEEE Chapters. Year 2005. International Conference “110 Anniversary of Radio Invention”. SPb ETU "LETI". 2005. V. 2, pp. 106–110.

Gusarova N.F., Vatian A.S., Osipov S.V. Egorov N.V., Boitcov V.V., Ryngach E., Treshkur T., Tatarnikova A., Shalyto A. Retrieving and Processing Information from Clinical Algorithm via Formal Concept Analysis // Lecture Notes in Computer Science (including subseries Lecture Notes in Artificial Intelligence and Lecture Notes in Bioinformatics), 2019, Vol. 11871, pp. 525-534.

Horsky J., Schiff G.D., Johnston D., Mercincavage L., Bell D., Middleton B. Interface design principles for usable decision support: a targeted review of best practices for clinical prescribing interventions // J Biomed Inform. V. 45. P. 1202–1216 (2012).

Johansen M.L., O’Brien J.L. Decision making in nursing practice: A concept analysis. Nursing Forum. V. 51. P. 40–48 (2016).

Khoussainov B., Nerode A. Automata Theory and its Applications. Progress in Computer Science and Applied Logic, Volume 21. Birkhauser, 2001.

M´ery D, Singh N. Medical protocol diagnosis using formal methods. International Symposium on Foundations of Health Informatics Engineering and Systems - FHIES 2011, Lecture Notes in Computer Science, vol. 7151, Johannesburg, South Africa, 2012; 1–20, doi: 10.1007/978-3-642-32355-3 1.

Mashkoor A., Sametinger J. Rigorous Modeling and Analysis of Interoperable Medical Devices. MSM '16: Proceedings of the Modeling and Simulation in Medicine SymposiumApril 2016 Article No.: 5. Pages 1–8

Mitkin S. Automata-Based Programming in DRAKON Language // Programmnaya Ingeneria. 2019. V. 10. No. 1. P. 3–13. DOI: 10.17587/prin.10.3-13

NINDS Multiple Sclerosis Information Page. National Institute of Neurological Disorders and Stroke. 18 Oct 2019. http://www.ninds.nih.gov/disorders/multiple_sclerosis/multiple_sclerosis.htm, last access 12.01.2020

openEHR Architecture. Architecture Overview. Revision 1.1. Pages: 87. Availible at: https://specifications.openehr.org/releases/1.0.1/architecture/overview.pdf, ac-cessed: 08.08.2019.

Polikarpova N. I., Shalyto A. A. Automata-based programming. SPb.: Piter. 2009 (rus)

Practice Guideline Recommendations Summary: Disease-modifying Therapies for Adults with Multiple Sclerosis. Recommendations from current practice guideline. Endorsed by the Consortium of Multiple Sclerosis Centers, the Multiple Sclerosis Association of America, and the National Multiple Sclerosis Society. April 2018. https://www.aan.com/Guidelines/home/GuidelineDetail/898, last access 12.01.2020

Rawson T.M., Moore L.S.P., Hernandez B. et al. A systematic review of clinical decision support systems for antimicrobial management: are we failing to investigate these interventions appropriately? // Clinical Microbiology and Infection. Aug. 2017. V. 23. Is. 8. P. 524–532.

Reis W.C., Bonetti A.F., Bottacin W.E. et al. Impact on process results of clinical decision support systems (СПКРs) applied to medication use: overview of systematic reviews // Pharm Pract (Granada). Oct-Dec 2017. V.15 (4).

Ricca F., Leotta M., Reggio G., Tiso A., Guerrini G., Torchiano M. Using UniMod for Maintenance Tasks: An Experimental Assessment in the Context of Model Driven Development / Proceedings of 4-th International Workshop on Software Engineering (MISE), 2012. http://dx.doi.org/10.1109/MISE.2012.6226018

Robby D.M., Hatcliff J. Bogor: An Extensible and Highly-Modular Model Checking Framework, March 2003. // Proceedings of the Fourth Joint Meeting of the European Software Engineering Conference and ACM SIGSOFT Symposium on the Foundations of Software Engineering (ESEC/FSE 2003). Technical Report, SAnToS-TR2003-3.

Shalyto A.A. Software Automation Design: Algorithmization and Programming of Problems of Logical Control. Journal of Computer and Systems Sciences International. 2000. Vol.39. №6, p.899-916.

ten Teije A., Marcos M., Balser M., van Croonenborg J., Duelli C., van Harmelen F., Lucas P., Miksch S., Reif W., Rosenbrand K., Seyfang A. Improving medical protocols by formal methods. Artif Intell Med. 2006 Mar; 36(3): 193-209.

Tiffen J., Corbridge S.J., Slimmer, L. Enhancing clinical decision making: Development of a contentious definition & framework. Journal of Professional Nursing. V. 30. P. 399-405 (2014).

Vatian A., Dudorov S., Ivchenko A., Smirnov K., Chikshova E., Lobantsev A., Parfenov V., Shalyto A., Gusarova N. Design Patterns for Personalization of Healthcare Process // ACM International Conference Proceeding Series. ICGDA 2019: Proceedings of the 2nd International Conference on Geoinformatics and Data Analysis 2019, pp. 83-88.

PIPELINE FOR CONTROL OF THE DYNAMICS OF LOCALIZED BRAIN PATHOLOGIES IN MAGNETIC RESONANCE IMAGES

Artyom Lobantsev, Grigorii Shovkoplias, Mark Tkachenko, Ksenia Morokova, Roman Soldatov, Aleksey Zubanenko and Anatoly Shalyto
ITMO University
Sankt-Petersburg 197101, Russia

ABSTRACT

A reliable assessment of changes in the dynamics of brain pathologies is primordial for accurate diagnostics, treatment and predicting the course of the disease. Magnetic resonance imaging (MRI) is the method of choice for it. In the paper, we explore the possibilities of semi-automatic control of the dynamics of localized brain pathologies in MRI. Using specific clinical examples, we investigated the sources of errors that accompany various methods for assessing the dynamics of the development of brain pathologies. We built a pipeline for semi-automatic control of the dynamics of these pathologies based on the Chan-Vese algorithm. The accuracy of estimating changes in the volume of pathological zones by proposed pipeline is comparable with the results obtained under idealized conditions of laboratory experiments. The proposed pipeline provides a significant gain in processing time and labor costs of radiologists is undemanding in computing resources and the availability of training datasets and can be easily implemented in real clinical practice.

KEYWORDS

Dynamics of Neoplasm, MRI, Active Contour, Chan-Vese Algorithm

1. INTRODUCTION

Localized brain pathologies, such as brain tumors, constitute a large and increasing share of the overall picture of morbidity in the modern world. So far, the annual incidence of tumors of the central nervous system in USA is 19 cases per 100,000 person, among them 6.4 cases per 100,000 person are malignant primary brain tumors [Todd]. The emergence and development of localized pathologies in the brain is still difficult to predict. Aggressive tumors as glioblastoma show different growth dynamics. According to [Stensjøen], in a third of patients between the initial diagnosis and surgery, a significant (more than two times in volume) tumor growth was observed, while in another third the tumor remained stable or even decreased.

Therefore, a reliable assessment of changes in the dynamics of brain pathologies is primordial for accurate diagnostics, treatment and predicting the course of the disease.

To assess pathological processes in the brain, high-tech medical images are widely used, among which magnetic resonance imaging (MRI) is the method of choice. For brain imaging, the most valuable modalities of MRI images are T2 weighted images with Flair inversion and T1 weighted images with contrast enhancement [Bauer]. Regardless of the chosen modality, MRI represents the analyzed zone of interest as a set of its sections (slices), where tissues of different types have different contrasts. Combining these slices, the viewer creates a pseudo-3D image of the zone of interest, which can be visually evaluated. However, in most cases, such a visual assessment is not enough, and quantitative measurements of the lesion are required.

The obtained medical images are used to solve various problems, among which the most important are the assessment of the volume of damage to the substance of the brain and the control of its time dynamics. The size of the pathology zone can be measured by evaluating several characteristic diameters of it [Wen] or by segmenting all the slices containing it [Odland]. Currently such evaluations can be performed manually, automatically and semi-automatically, and each of these options is not free from disadvantages.

For example, manual segmentation is so far considered the gold standard, but it is very time-consuming and requires experience [Gordillo]. Moreover, the detailed investigations [Odland] show a considerable inter-observer variability among experts defining tumor volume, which reduces the reliability and reproducibility of estimates, especially important in assessing the dynamics of pathological processes. Automatic measurements based on machine learning methods require large amounts of source data (datasets), which are not always available due to the high variability of controlled pathologies. When using semi-automatic methods (like active contour methods), the accuracy of estimating the volume and dynamics of pathological sites is greatly influenced by the configuration of heterogeneities of the lesion zone and neighboring tissues.

In the paper, we explore the possibilities of semi-automatic control of the dynamics of localized brain pathologies in magnetic resonance images. We set ourselves the following tasks:

- to investigate the sources of errors that accompany various methods for assessing the dynamics of the development of brain pathologies;
- to build a pipeline for semi-automatic control of the dynamics of these pathologies, which allows us to eliminate and / or compensate for the identified errors;
- to investigate the effectiveness of the proposed pipeline using examples from clinical practice.

Further material of the article is organized as follows. In section 2, we consider the specifics of the manifestations of brain lesions in MRI images, as well as generally accepted metrics and existing methods for quantifying the dynamics of such lesions. In section 3, we substantiate the choice of the Chan-Vese algorithm as the basis for our experiments, and present the scheme and characteristics of the developed pipeline. In Section 4, we present the results of an experimental assessment of the accuracy and speed of the developed pipeline on specific clinical examples in comparison with manual methods, as well as provide an interpretation of the results and outline possible ways to improve them.

2. BACKGROUND AND RELATED WORKS

2.1 Manifestations of Localized Brain Lesions in MRI Images

Challenges associated with the segmentation of a brain pathology are widely discussed (see, for example, [Havaei]). In this article, we highlight those manifestations that are most significant when assessing changes in the volume of pathological lesions.

As a rule, localized brain lesions have an irregular, complex shape; the boundaries of formation often become blurred and layered on another tissue, which makes it difficult for radiologists to interpret them [Goetz].

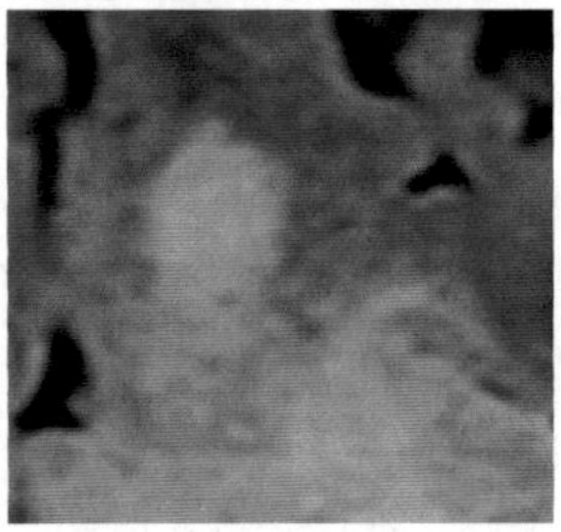

a

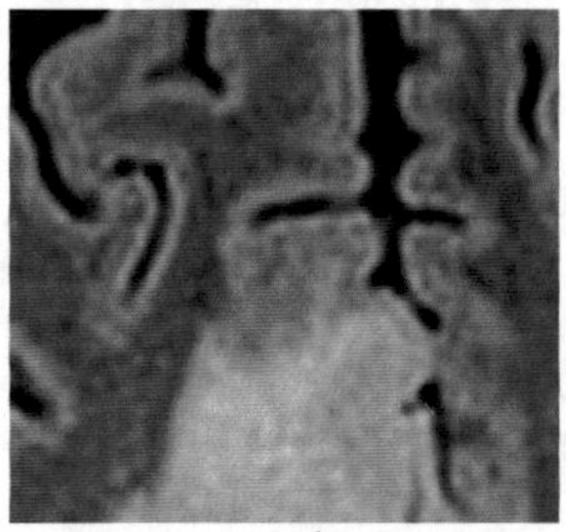

b

Figure 1. Sources of errors in the measurement of brain tumors (gliomas): a, b – blurred border between the tumor and the swelling zone

Tumors (e.g. gliomas) in most cases have a heterogeneous structure and are surrounded by an edema zone, while tumor cells can infiltrate into peritumoral edema [Claes]. This circumstance is an important source of error when segmenting the edema zone, since the boundaries between the tumor and the edema become blurred (Figure 1 a, b). As a result, the radiologist is forced to label borderline voxels subjectively, based on his experience and a visual assessment of the brightness characteristics of adjacent tissues. In addition, as shown in Figure 2a, b, the zones of edema, as well as some gliomas, can form a signal response similar to normal tissue. which increases the variability of changes between radiologists, in the case of low-grade gliomas up to 13.61%. [Kaus]

Significant influence on the manifestation of the lesions have the characteristics of MR tomograph [Dance]. The impact factors include the equipment parameters (like configuration of coils and scanner drifts) as well as the experiment parameters (like slice thickness and flip angle). As a result, voxel volume may vary from image to image. Besides, the signal fluctuations in a given voxel are influenced by physiological noise (like cerebral metabolism).

Since the head of the patient is not rigidly fixed during MR imaging, this causes angular and linear displacements of the region of interest in successive images (Figure 2). It is fundamentally possible to bind the coordinates of the region of interest to the characteristic points of the skull, however, most methods for processing MRI images of the brain involve the operation of skull-stripping [Valvano], and that is, images with missing skull bones are being presented for comparison.

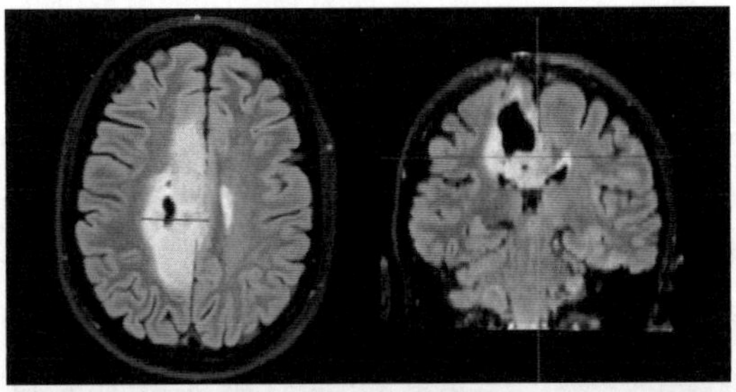

a

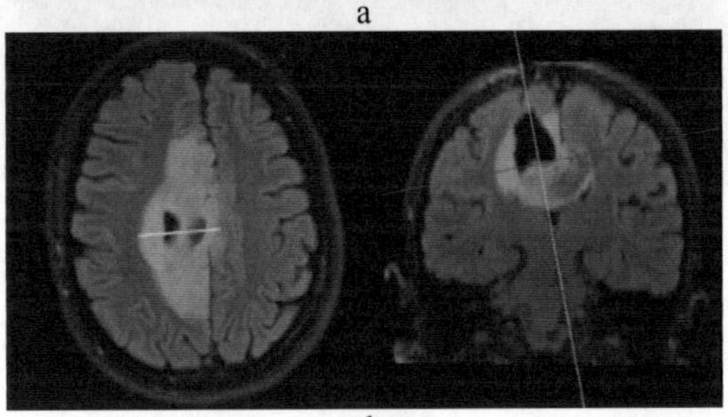

b

Figure 2. Illustration of geometric errors when comparing MRI images: a - the study was performed when setting the slices in the midline; b - the study was performed without alignment in the midline

All of the above complicates the pixel-by-pixel comparison of successive MRI images, which is important when assessing the dynamics of development of localized brain lesions.

2.2 Methods for Evaluations of Localized Brain Lesions in MRI Images

As mentioned above, the evaluations of localized brain lesions can be performed manually, automatically and semi-automatically. In the first case, the radiologist "by eye", guided by his experience, draws the pixel-by-pixel boundaries of the zone of interest on each slice, thereby segmenting it; the sum of the number of voxels corresponding to the zone of interest on all slices serves as an estimate of the volume of this zone. The process is very laborious: according to our estimates, the average time for such processing of one MRI exam is 40 minutes.

A simplified version of the evaluations of the size of gliomas offer RANO recommendations [Wen]. In this case, the radiologist selects the most distant points on the lesion contour and measures the linear distance between them. Obviously, the relevance of such estimates substantially depends on the shape and size of the controlled area. For example, [Kim] reports good correlation between diameter-based and segmentation-based measurements of lesion approximated by an ellipsoid. However, our studies have revealed a number of clinical cases when such a measurement is unlawful and can lead to false estimates of the volume and dynamics of the affected area (Figure 3, a, b). Note that similar studies for breast cancer [Franca] also showed a large percentage of deviations: good agreement between the tumor size measured by MRI and pathology in 69.0%, MRI overestimated tumor size in 24.1% and underestimated in 6.9% cases.

The progress of automatic methods of lesion segmentation in recent years has been associated with the use of machine learning and, above all, deep neural networks (see reviews [Havaei], [Mehrabian], [Nadeem]). Nevertheless, all the reviews show that the lack of large training datasets is a main obstacle for deep learning methods, this obstacle being fundamental in the case of brain lesions manifesting a great variety of configurations. Besides, as already mentioned above, the labeling of MRI brain images for forming training datasets is extremely time- and experience-consuming.

Semi-automatic methods include region merging, graph-based and active contour methods. The first ones consider the homogeneity characteristics of single pixels within a region; thereby an image is partitioned into a collection of regions and objects [Peng], [Zhuo]. Graph-based methods [Santle], [Lai] consider image pixels as the set of graph vertices connected by edges, and the more the pixels are “similar” to each other, the less weight the corresponding edge has. Both of these methods use "greedy" algorithms and are very noise-sensitive, which puts them out of the brackets when organizing the control of the dynamics of such low-contrast and noisy images as MRI brain images.

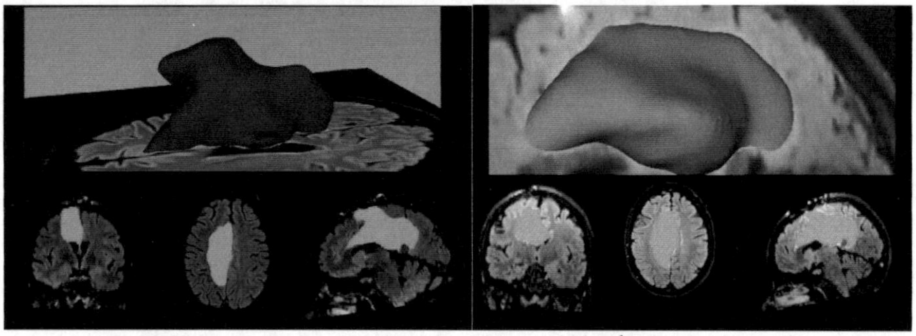

a b

Figure 3. An example of the growth dynamics of an irregularly shaped tumor: a - the first examination; b - the second examination; on top - 3D tumor models; bottom - sections of the tumor in 3-plane

Active contour model (ACM)-based methods [Hemalata], in turn, are categorized as edge-based and region-based methods. Edge-based methods are built on on information of image gradient in the border area. In this regard, they are unstable to noise and do not work well on images with weakly expressed and blurry borders without serious algorithmic and semantic tricks [Khvostikov]. Region based methods are integral, that is, based on an assessment of the distribution of energy in a segmented image, and thereby are in principle devoid of these drawbacks.

One of the most popular global region method for medical applications is Chan-Vese method [Chan], designed especially for segmenting objects without clearly defined boundaries. The Chan-Vese method minimizes the following energy functional:

$$\arg\min_{c_1,c_2,C} \mu \text{Length}(C) + \nu \text{Area}(inside(C)) + \lambda_1 \int_{inside(C)} \|f(x) - c_1\|^2 dx + \lambda_2 \int_{outside(C)} \|f(x) - c_2\|^2 dx, \quad (1)$$

where $f(x)$ is the original image, C is the desired segmentation contour, λ_1, $\lambda_2 \geq 0$ are scaling constants, constant $\mu \geq 0$ scales the Euclidian length of the curve C and constant $v \geq 0$ scales the area term inside the contour C. The disadvantage of the method is its sensitivity to heterogeneity within the segmented contour, and its various modifications [Zawish], [Xiangrui Meng], [Soomro], [Mostaar] are aimed firstly at eliminating this drawback. However, as clinical practice shows, when monitoring the dynamics of the development of localized brain lesions, the clinician is primarily interested in changes in their external contour. Therefore, to solve the problem posed in the article, the Chan-Vese algorithm was used as the basic algorithm.

In contrast to articles dealing with various aspects of medical image segmentation, works devoted directly to monitoring the dynamics of regions of interest in medical images are much scarcer in the literature. Basically, studies have been reported [Holzgreve], [Souza], [da Silva] on experimental animals (C6 mice), in which brain tumors, according to a number of indicators, correspond to human ones. The tumor volume was calculated by the sum of tumor area in each slice multiplied by the thickness and gap per slice. In all studies when recording MRI, the animals were anesthetized and immobilized, which should reduce measurement errors. Nevertheless, even in such idealized conditions, the accuracy of measuring the volume of the region of interest was not high (from 9 to 26%), which allows one to adequately evaluate only its significant changes. In articles connected with the assessment of the dynamics of neoplasms in humans, the regions of interest were evaluated using the diameter-based method, and the statistical metric of overall survival (OS) rates was used.

Thus, the analysis performed in Section 2 confirmed the relevance of the problem statement and the prospects of the selected approaches to solve it.

3. PROPOSED SOLUTION

In developing the pipeline for processing MRI images of the brain in order to assess the dynamics of the localized brain lesions, we sought to eliminate or compensate for the sources of error described in Section 2.1 above. The developed pipeline is shown in Figure 4.

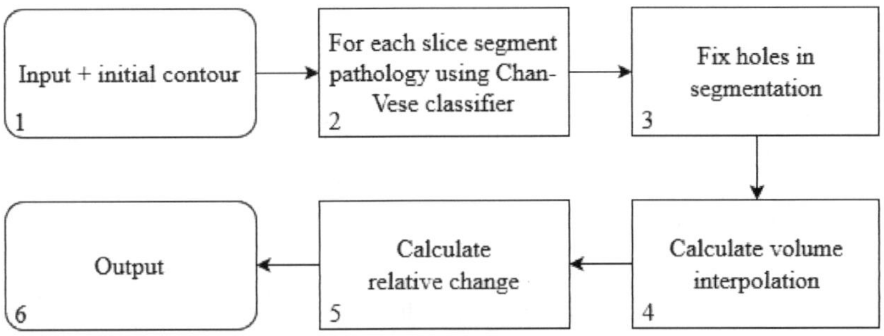

Figure 4. Proposed pipeline

The input data is a set of slices on which region of pathology is present. The region of location of the pathology is marked manually, forming the initial contour (Figure 5, a). At step 2 each slice is segmented using the Chan-Vese algorithm. We used the algorithm implemented in Python [Chan-Vese] with the following parameters: μ= 0.15, λ_1 = 1.7, λ_2 = 1, tol = 1e-5, max_iter = 3000, dt = 0.01. The values of the settings μ, λ_1, λ_2 of (1) as well as iteration parameters were empirically selected and result in stable work on all experimental data, but can be easily adjusted for other data if necessary. At step 3, the region highlighted by the algorithm is checked for internal inhomogeneities ("hole fixing"); further consideration includes the volume of the entire region inside the external contour.

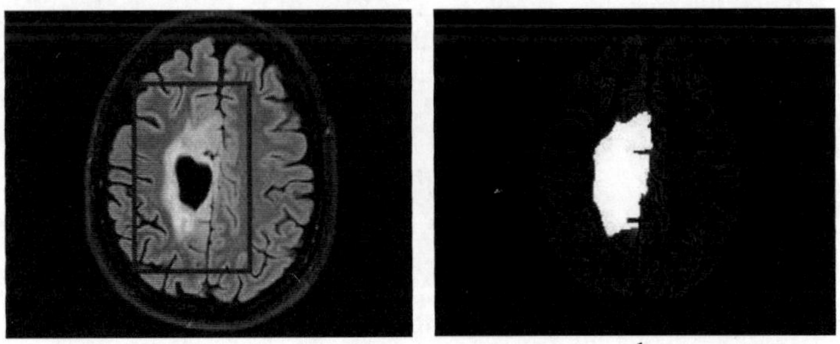

a b

Figure 5. Slice with a manually marked area of interest (a) and the result of its segmentation by proposed pipeline (b)

At step 4, in order to eliminate the errors described in section 2, the interpolation procedure is used to calculate the pathology volume. We compare the values of voxels in an identical position on the current and adjacent slices; if both of them are True (that is, they belong to the pathology identified in step 3), then we add the full voxel (1) to the volume value, if both are False, then nothing (0); in other cases we add half the voxel volume (0.5). Thus, each voxel with coordinates x, y on slice s is assigned a weight $W(s, x, y) \in \{0,0.5,1\}$.

Converting the volume of pathology in voxels to the absolute values, we take into account the possible difference in the size of the slices in the compared MRI images. That is, the pathology volume in absolute units is calculated as follows:

$$V_{abs} = \left(\sum_{s \in Slices} \sum_{x,y} W(s,x,y) \right) \times \left(\frac{OriginalSize}{Pixels} \right)^2 \times ThicknessPerSlice \quad (2)$$

where $W(s, x, y) \in \{0, 0.5, 1\}$ is weight of voxel (x, y) at slice s, *OriginalSize* (mm) is original length and width of slices, *Pixels* is length and width of slices in pixels, the value of *ThicknessPerSlice* includes the thickness of the slice itself increased by the gap per slice.

At step 5, the value of relative changes in the volume of pathology is calculated:

$$RelativeChange = \frac{V_{end} - V_{init}}{V_{init}} \quad (3)$$

where V_{init} and V_{end} are calculated in step 4 by the formula (2) for the compared images, respectively.

4. RESULTS AND DISCUSSION

We investigated the effectiveness of the proposed pipeline using examples from clinical practice. As a dataset, we selected pairs of MRI images characterizing the dynamics of diseases in 9 patients with gloiomas. To calculate the relative changes according to formula (3), we used three methods: A - manual segmentation of pathologies and per-pixel volumetry; B - manual segmentation of pathologies and diameter-based volumetry; C - volume estimation using the proposed pipeline (see section 2.2 for a detailed description of methods A and B).

The experimental results are presented in the table 1 and are graphically depicted in Figure 6. In the table 1: V_{init} and V_{end} [mm^3] is absolute volume of the pathological zone in the compared MR images, *RC* [%] is relative change of the pathological zone calculated according to (3), *T* [min] is total evaluation time.

Table 1. Experimental results

	A				B				C			
	V_{init},	V_{end},	*RC*	*T*	V_{init},	V_{end},	*RC*	*T*	V_{init},	V_{end},	*RC*	*T*
1	66377	76333	15.00	23	73621	77302	5,00	5	56070	68501	22.17	4.71
2	137563	165075	20.00	37	170549	189309	11,00	7	108604	133893	23.29	4.96
3	97348	153697	57.88	42	110921	149998	35,23	10	64710	108378	67.48	5.07
4	158437	168457	6.32	32	175469	180118	2,65	6	111101	122606	10.36	5.29
5	124567	147389	18.32	45	149553	172898	15,61	11	84186	117314	39.35	4.38
6	72893	60598	-16.87	25	79563	69218	-13	6	54367	43579	-19.84	4.3
7	98549	108463	10.06	39	103546	112865	9	9	71563	89702	25.35	5.06
8	45690	60954	33.41	31	51793	67035	29,43	7	30633	48090	56.99	4.67
9	137459	108385	-21.15	42	147954	119842	-19	8	103725	71819	-30.76	5.2

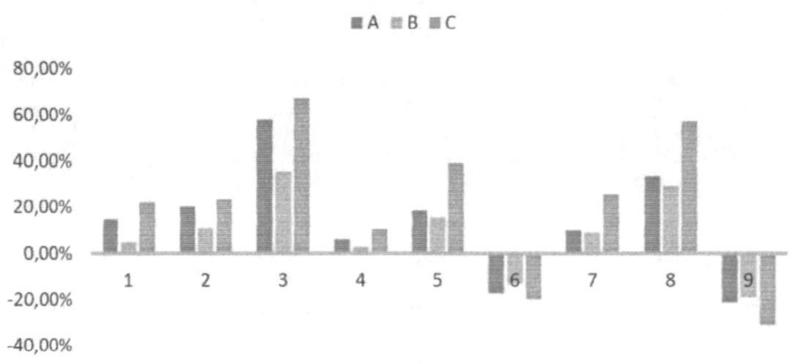

Figure 6. Comparison of relative changes obtained by: A - manual segmentation of pathologies and per-pixel volumetry; B - manual segmentation of pathologies and diameter-based volumetry; C - volume estimation using the proposed pipeline

To analyze the results of experiments, the measurement method A was chosen as a baseline.

The data show that proposed method C evaluates relative changes in the volume of pathology (RC) much more accurately than method B (see Figure 6). Note that method B can give an assessment bias in both directions, both overstating and underestimating the real value depending on the configuration of the pathology zone (see Figure 3). Meanwhile the proposed method C always overestimates the estimates, which can be used in the future for correcting them. It is also worth noting that the error values of method C compared with method A are comparable with the errors in evaluating MRI images obtained under idealized conditions of laboratory experiments on mice (see section 2), i.e., the proposed method quite effectively eliminates the sources of errors stated in section 2.The Table 1 also shows that the proposed method C exceeds methods A and B in the total evaluation time *T* and, most importantly, in radiologists' required time. Note that the presented results were obtained without optimizing the Chan-Vese algorithm in terms of speed and resources, using a standard laptop (CPU Intel(R) Core(TM) i7-8550U CPU @ 1.80GHz).

The analysis also revealed the limitations of the proposed solution. Chan-Vese algorithm showed a relatively high and non-linear sensitivity to the contrast of the analyzed image. The search for optimal contrast requires a separate study. In addition, a separate study requires the use of the proposed pipeline for such a significant pathology as multiple sclerosis. In this case, the pathology is not concentrated in one area, but consists of many small foci for which the error in determining the volumes and increments will be obviously higher.

5. CONCLUSION

In the paper, we explored the possibilities of semi-automatic control of the dynamics of localized brain pathologies in magnetic resonance images.

Using specific clinical examples, we investigated the sources of errors that accompany various methods for assessing the dynamics of the development of brain pathologies.

We built a pipeline for semi-automatic control of the dynamics of these pathologies based on the Chan-Vese algorithm, which allows us to eliminate and / or compensate for the identified errors.

We have shown the effectiveness of the proposed pipeline using examples from clinical practice. The accuracy of estimating changes in the volume of pathological zones by proposed pipeline is comparable with the results obtained under idealized conditions of laboratory experiments. The proposed pipeline provides a significant gain in processing time and labor costs of radiologists is undemanding in computing resources and the availability of training datasets and can be easily implemented in real clinical practice.

ACKNOWLEDGEMENT

This work was supported by Russian Science Foundation, Grant №19-19-00696

REFERENCES

Bauer S., Wiest R., Nolte L.-P., Reyes M. (2013). A survey of MRI-based medical image analysis for brain tumor studies. In. *Phys. Med. Biol*. 58: R97. 10.1088 / 0031-9155 / 58/13 / R97

Chan T. and Vese L. An Active Contour Model without Edges. *Scale-Space Theories in Computer Vision*, 1999, DOI:10.1007/3-540-48236-9_13

Chan-Vese Segmentation. Skikit-image documentation. https://scikit-image.org/docs/dev/auto_examples/segmentation/plot_chan_vese.html. Last access 01.05.2020.

Claes A, Idema AJ, Wesseling P. Diffuse glioma growth: a guerilla war. Acta Neuropathol. 2007. November; 114(5): 443–58. 10.1007/s00401-007-0293-7

da Silva A.C. et al. Tumor growth analysis by magnetic resonance imaging of the C6 glioblastoma model with prospects for the assessment of magnetohyperthermia therapy. *Einstein (São Paulo)* vol.10 no.1 São Paulo Jan./Mar. 2012 http://dx.doi.org/10.1590/S1679-45082012000100004

Dance D.R. et al. *Diagnostic Radiology Physics.* A Handbook for Teachers and Students. IAEA, 2014. 682 p. ISBN 978–92–0–131010–1

Franca L.K.L. et al. (2018) Tumor size assessment of invasive breast cancers: which pathological features affect MRI-pathology agreement? *Appl Cancer Res* **38,** 2 (2018). https://doi.org/10.1186/s41241-018-0055-y

Goetz M. et al. (2016). DALSA: Domain adaptation for supervised learning from sparsely annotated MR images. *IEEE Trans Med Imaging.* 2016 Jan;35(1):184-96. doi: 10.1109/TMI.2015.2463078.

Gordillo N., Montseny E., Sobrevilla P. (2013). State of the art survey on MRI brain tumor segmentation *Magn Reson Imaging.* 2013 Oct;31(8):1426-38. doi: 10.1016/j.mri.2013.05.002.

Havaei M. et al. (2016). Deep learning trends for focal brain pathology segmentation in MRI. In: Holzinger A. (eds) *Machine Learning for Health Informatics*. Lecture Notes in Computer Science, vol 9605. Springer, Cham, pp 125-148.

Hemalata R.J. et al. Active Contour Based Segmentation Techniques for Medical Image Analysis. In: R. Koprowski (Ed.) *Medical and Biological Image Analysis* (2018). ISBN: 978-1-78923-331-5

Holzgreve A. et al. Monitoring of Tumor Growth with [18F]-FET PET in a Mouse Model of Glioblastoma: SUV Measurements and Volumetric Approaches. *Front. Neurosci.,* 14 June 2016 | https://doi.org/10.3389/fnins.2016.00260

Kaus M.R. Automated Segmentation of M et.al. RI of Brain Tumors Article in Radiology 218(2):586-91 · March 2001

Khvostikov A.V. et al. Trainable Active Contour Model for Histological Image Segmentation. *Scientific Visualization*, 2019, Vol. 11, № 3, pp. 64 - 75, DOI: 10.26583/sv.11.3.06

Kim H.J., Kim V. Method of tumor volume evaluation using magnetic resonance imaging for outcome prediction in cervical cancer treated with concurrent chemotherapy and radiotherapy. *Radiation Oncol J.* 2012 Jun; 30(2): 70–77.

Lai Z, Qu X, Liu Y, Guo D, Ye J, Zhan Z, et al. Image Reconstruction of Compressed Sensing MRI Using Graph-based Redundant Wavelet Transform. *Medical Image Analysis.* 2015:1–27. doi: 10.1016/j.media.2015.05.012.

Mehrabian H. et al. Advanced Magnetic Resonance Imaging Techniques in Management of Brain Metastases. *Front. Oncol.,* 04 June 2019 | https://doi.org/10.3389/fonc.2019.00440

Mostaar A., Houshyari M., and Badieyan S. A Novel Active Contour Model for MRI Brain Segmentation used in Radiotherapy Treatment Planning. *Electron Physician*. 2016 May; 8(5): 2443–2451.

Nadeem M.W. et al. (2020) Brain Tumor Analysis Empowered with Deep Learning: A Review, Taxonomy, and Future Challenges. *Brain Sci.* 2020, Vol. 10, 118, 33 pp.

Odland A. Volumetric glioma quantification: comparison of manual and semi-automatic tumor segmentation for the quantification of tumor growth. In *Acta Radiol.* 2015 Nov;56 (11):1396-403. doi: 10.1177/0284185114554822

Peng B, Zhang L, Zhang D. (2012) Automatic image segmentation by dynamic region merging. IEEE Trans Image Process. 2011;20(12):3592–605. doi: 10.1109/TIP.2011.2157512.

Santle Camilus K., Govindan V.K. A Review on Graph Based Segmentation. *I.J. Image, Graphics and Signal Processing*, 2012, 5, 1-13. DOI: 10.5815/ijigsp.2012.05.01

Soomro S, Munir A, Choi KN (2018). Hybrid two-stage active contour method with region and edge information for intensity inhomogeneous image segmentation. *PLoS ONE* 13(1): e0191827. https://doi.org/10.1371/journal.pone.0191827

Souza T.K.F. et al. Image and motor behavior for monitoring tumor growth in C6 glioma model. *PLoS One* v.13 (7); 2018PMC6062126

Stensjøen A.L. et al. Growth dynamics of untreated glioblastomas in vivo. In *Neuro-Oncology*, Volume 17, Issue 10, October 2015, Pages 1402–1411, https://doi.org/10.1093/neuonc/nov029.

Todd R.F. et.al. (2012) *Tumor Board Review: Guideline and Case Reviews in Oncology.* Demos Medical Publishing, 448 p.

Valvano G. et al. Training of a Skull-Stripping Neural Network with efficient data augmentation. arXiv:1810.10853v1 [cs.CV] 25 Oct 2018

Wen PY et al. Updated response assessment criteria for high-grade gliomas: response assessment in neuro-oncology working group. *J Clin Oncol.* 2010 Apr 10;28(11):1963-72. doi: 10.1200/JCO.2009.26.3541.

Xiangrui Meng et al. Brain MR image segmentation based on an improved active contour model. *PLoS ONE* 12 (8): e0183943. https://doi.org/10.1371/journal.pone.0183943

Zawish M. Brain Tumor Segmentation in MRI images using Chan-Vese Technique in MATLAB. 2018 International conference on Computing, Electronic and Electrical Engineering (ICE Cube). DOI: 10.1109/ICECUBE.2018.8610987

Zhuo Zhao et al. (2020) Hybrid image segmentation method based on anisotropic Gaussian kernels and adjacent graph region merging. *Review of Scientific Instruments*, Volume 91, Issue 1, 015104 (2020)

ICT SUPPORTED OCCUPATIONAL STRESS MANAGEMENT: THE VIEWS OF HR MANAGERS IN SRI LANKAN SOFTWARE COMPANIES

Manoja Weerasekara[1,2], Åsa Smedberg[2] and Hélène Sandmark[3]
[1]*School of Computing, NSBM Green University Town, Pitipana, Homagama, Sri Lanka*
[2]*The Department of Computer and Systems Sciences, Stockholm University, Kista, Sweden*
[3]*Ramazzini Research, Lidingö, Sweden*

ABSTRACT

The software industry is considered to be a growing industry worldwide but with a high propensity to cause stress reactions among employees due to its demanding and technology-driven nature. As a human capital-intensive industry, too much of stress on employees can also create a negative impact on the quality of work. Thus, organizations take numerous initiatives at both individual and organizational levels to address this problem. One area of concern is the use of ICT supported occupational stress management interventions. In Sri Lanka, the dominant ICT occupation today is software engineer, and there is a national awareness of mental health issues due to occupational stress. Thus this study aims to investigate current practices used to frame employees' well-being inside software organizations in Sri Lanka, and to understand Sri Lankan software companies' viewpoints of using ICT supported occupational stress management interventions through the lens of human resource (HR) managers. Ten HR managers were interviewed covering eight software companies. Through thematic analysis, we identified three main themes and multiple subthemes. Results revealed various types of individual level and organizational level strategies, policies and practices implemented to create a healthy work setting for the employees. Findings also show that the HR managers had limited knowledge and experiences of ICT supported occupational stress management interventions. Nevertheless, they appreciated the immediate possible benefits of ICT interventions and demonstrated the willingness to adopt them in their organizations. Moreover, they revealed several functional requirements and feature sets they would like to see in such applications to better support their employees. Therefore, this study can possibly provide insights for researchers and practitioners who are to design and develop ICT supported interventions to manage occupational stress, to promote well-being and healthy organizations.

KEYWORDS

Well-Being, Digital Health Intervention, Occupational Stress Management, Sri Lanka, Software Industry, Human Resource Managers

1. INTRODUCTION

Occupational work is inherently stressful, and workplace stress is more on the rise than ever. According to the world health organization, stress has been dubbed as the "Health Epidemic of the 21st Century" (Fink, 2017). A recent global survey commissioned by The Regus Group, which included 1,000 corporations across 15 countries, also shows similar indications. The study found that six in ten workers in significant global economies experienced increased workplace stress, which is a significant rise compared to the previous two years (Esmond, no date). Occupational stress is a kind of stress associated with an individual's work environment and job conditions. The gap arises from the individual's knowledge, skills and capacity, and job demands causing stress reactions (Mustafa *et al.*, 2015). However, occupational stress, or stress in general, is not uniformly negative. A desirable amount of stress motivates and encourages employees to learn and face challenging situations where an excessive amount of stress could cause negative impact on individuals and harm health and well-being of employees (Lazarus, 1993). Overcoming such stressful conditions mainly depends on the individual's capacity to adjust, and provision of adjustments in the work environment (Swanson and Fouad, 1999). Therefore, every organization promotes diversified approaches at both individual and organizational levels to reduce stress and to improve employee well-being. Though most all occupations are subjected to a series of stressors, comparatively, the software industry is considered a stressful workplace due to its fast-moving and demanding nature. This rapidly growing nature has created enormous opportunities and demands in the software industry. It often leads to a lot of health problems that come with continuous physical and mental stress (Padma *et al.*, 2015). Thus, it is essential to explore the current context and future possibilities of stress management practices within a software house.

Sri Lanka is a multi-cultural and multi-ethnic country lying southeast of India with a population of 21.07 million in 2019. Recent statistics at the Mental Health Unit of Sri Lanka's Health Ministry mirror the global stress conditions; the statistics show that one in five public sector employees suffers from stress, and this has become a real issue that needs to be addressed at the workplace (Rosayro, 2017). However, as a country with a national awareness of mental health concerns due to occupational stress, Sri Lanka clarifies the vision for comprehensive mental health service for the citizens of Sri Lanka based on a community care model. It also promotes the treatment of mental illness holistically and discusses the role of a multidisciplinary team (Kathriarachchi, Seneviratne and Amarakoon, 2019). Currently, mental healthcare is provided through a combination of public, private and non-government services. Mental Health ATLAS 2017 Member State Profile of Sri Lanka recorded that there are only 1,480 mental healthcare professionals in both government and non-government sector, which gives 7.14% mental healthcare workers per 100,000 population. Because of the current shortage of mental healthcare human resources, there has been particular interest in introducing innovations to deal with the situation (Minas, Mendis and Hall, 2017). Along with the Government of Sri Lanka's (GOSL) visionary journey of transforming Sri Lanka into a "Knowledge Economy' the Information and Communication Technology (ICT) has a significant role to play in this journey. According to recent statistics published by the Information and Communication Technology Agency of Sri Lanka (ICTA), there is a dominance of software engineering followed by software quality assurance. In parallel to the GOSL vision, the SLASSCOM (Sri Lanka Association of Software and Service Companies - the apex body of Sri Lankan software and service companies) plan to create 200,000 direct jobs, enable 1,000 start-ups and to grow the industry revenue up to US $ 5 billion by 2022 (ICTA, 2019). With the fast-paced industry and service environment, the competitive run to reach the next level requires employees to work long hours and to strive for continuous skill and capacity updates. This has resulted in lack of work-life balance and caused mental stress and mental disorders among workers in the software industry (Padma *et al.*, 2015).

Thus, this study specifically aims to understand the Sri Lankan software companies' viewpoints of using ICT supported occupational stress management interventions through the lens of human resource (HR) managers working in Sri Lankan software companies. The outcome of this research could feed in as knowledge to the design and development of ICT tools for occupational stress management, to yield optimum efficacy levels.

2. RESEARCH AIM AND OBJECTIVES

As discussed in the introduction, this study aims to explore current HR and administrative practices of the Sri Lankan software companies to create a healthy work environment for the employees, and to investigate the companies' perceptions of the use of ICT for stress management interventions from the viewpoint of HR managers. The specific objectives of this research are: (1) to identify current HR management practices to enhance employee health and well-being inside a software company, (2) to learn about managers' experience of dealing with situations involving employee mental distress, and (3) to ascertain the managers' perspectives of the use of ICT supported occupational stress management interventions.

3. LITERATURE REVIEW

3.1 Occupational Stress

Occupational stress is associated with one's work environment or job conditions. Studies exploring the impact of work-related stress reactions show that it does not only impact on an individual level but also at the organizational and national level. At an individual level, high levels of stress can cause adverse cognitive difficulties (e.g. concentration problem, poor judgment, constant nervousness), behavioural issues (e.g. sleeping disorder, poor eating habits, use of drugs and alcohol), emotional problems (e.g. short temper, restlessness, impatient, depression) and physical problems (e.g. back and neck pains, heart problems, blood pressure, nausea, anxiety) (Burman and Goswami, 2018). It impacts on the organizational level with the increased absenteeism and presenteeism, as well as reduced commitment, motivation and satisfaction, along with a higher rate of staff turnover and intention to quit (Quick, 2013). Moreover, employees' stressful behaviours influence the company's overall competitiveness, productivity, and public image. At a broader level, stress-related disorders affect the economic and social environment of both individuals and organizations (Narváez, Tobar and López, 2014).

3.2 Well-Being in the Workplace: Human Resource (HR) Management

Employee health and well-being is considered the main focus area of HR management as it directly links with employee satisfaction and performance (Martin, Woods and Dawkins, 2018). The organizations consider employee health "not out of altruism, but as a key part of their business strategy" (Burke, 2014, p. 4). Within a rapidly changing economic and technological environment, a healthy workforce becomes a competitive advantage for organizations (Biron, Karanika-Murray and Cooper, 2012). The reduced performance due to sickness, absenteeism and presentism incur a substantial cost for employers. It affects the employee's innovative nature, communication and problem-solving skills, and it causes overall productivity loss (DeFrank, 2012). Thus the companies invest more and more to keep their workers present, healthy, and well (Black, 2009).

3.3 Stress Management Interventions

There are different job-related stress reduction strategies suggested in the literature to adjust individuals and work environments to create more desirable employee and work settings. Such adjustments are two-fold: 1. Individual adjustment strategies; 2. Environmental adjustment strategies (Black, J *et al.*, 2019). Individual strategies include (1). Developing Self-Awareness- knowing the job requirement and prepare yourself to accept challenges, (2). Promoting Outside Interests- select suitable activities based on individual preferences, e.g. sports, music to take mindset off work, etc., (3). Finding a Personal or Unique Solution– individuals may identify suitable practices by their own in order to cope with stressful situations, e.g. guided imaginations, listen to music, etc., (4). Physical Exercise- engage in regular exercises to keep energy up, (5). Cognitive Perspective- framing the situations positively by distinguishing factors that are within as well as outside your control and influence, and (6). Leaving the organization- if an employee cannot adjust to the environment and the environment is also not changing in a positive way for the individual then the employee may investigate alternative employment opportunities (Black, J *et al.*, 2019). The organizational level adjustment strategies are viewed as more desirable as they bring more control over the organization's employees and have more opportunity to contribute to creating a healthy work setting. Such organization strategies involve (1). Personnel Selection and Placement- pay more attention to the selection and placement process to recruit suitable job applicants for the job and the specific work environment, (2). Skills Training- provide numerous training programs to improve job-related skills, interpersonal skills, etc., (3). Job Redesign– job enrichment, job rotation, etc., to enhance job-person fit, (4). Company-Sponsored Counseling and Health Programs- provide awareness activities, programs and tools to cope with stress, (5). Increased Participation and Personal Control- allow employees greater participation and personal control in decisions affecting their work, (6). Work Group Cohesiveness- attempt to build workgroup cohesiveness through team building activities, and (7). Improved Communication- open communication channels, so that employees are more informed about what is happening in the organization (Black, J *et al.*, 2019).

3.4 ICT Supported Occupational Stress Interventions

In the current context, digital health interventions are considered a growing phenomenon with promising outcomes. With the technological advancements and availability of digital resources, numerous initiatives have been explored to cater to human needs using eHealth and mHealth strategies. It is possible to carry them out at a larger scale, regardless of time, place or group of people (Cohen, Kessler, & Gordon, 1995), and the user can access information in a quick, easy and confidential manner (McKenna & Bargh, 2000). Looking at the ICT supported stress management interventions (ICTSMI), there are diversified approaches practiced using the web, mobile, sensor and hybrid platforms (Weerasekara and Smedberg, 2019). Such interventions provide support and interaction using various modalities; guided through human experts, guided through machines, guided and supported by peers, moderately guided, unguided, etc. (Weerasekara and Smedberg, 2019). It is not only the personalized and individualized care such interventions facilitate; they can also create a collaborative and interactive platform. These digital platforms allow getting connected with professional healthcare providers and people with similar health conditions. Such interactions enable the sharing of knowledge and experience (Brough, Dollard and Tuckey, 2014), providing a sense of belongingness and comfort leading to patient empowerment (LLUCH, 2011).

3.5 Software Industry

The growing needs of the organizations and industry have given birth to software houses (Rajeswari and Anantharaman, 2003). These software houses intend to fulfil the demands of their clients by using the latest technology and skills in the market (Nayak, 2014) and are considered as a human capital intensive industry with knowledge workers in a technology-intensive environment (Rajeswari and Anantharaman, 2003). Hence, the software industry is regarded as a stressful workplace due to its fast-moving and demanding nature. With the rapid technological advancements and changing nature of business requirements, software professionals need to abreast with them. Software engineers must confront different sorts of pressures, including (1). global pressures: market and rivalry, (2). technology pressures: rapid tech-advancements, and (3). local pressures: working hours and changing work culture due to its global software development nature (Amin *et al.*, 2011).

4. METHODOLOGY

This study involves an exploratory and qualitative approach to understand and gain more knowledge on the current HR and administrative processes used to create and maintain healthy work settings for the employees. The first phase of the data collection process was designated to identify suitable companies for the investigation. To assure the uniformity of the selection of companies, only the SLASSCOM registered companies located in the Colombo district were selected. An online questionnaire was disseminated through email to the software and service companies on the list to gather demographic profiles of the companies. This mainly focused on identifying the company's core business area, operational status, number of employees, gender distribution support services, etc. The companies were selected based on the scale of the company in terms of the number of employees working in the organization. This scale was formed based on the enterprise classification defined by the World Bank for Sri Lankan companies; those with fewer than 49 employees were categorized as small; those with 50-99 employees were classified as medium-sized; and those with 100 employees or more as large companies (Ponnamperuma, 2000). Thus, the complete response list comprised of 11 small scale companies, five medium scale companies and 11 large scale companies. Based on the 27 responses received, a combined convenience and purposeful criterion sampling method was used to select respondents for the next phase. The selected eight companies consisted of three small scale companies, two medium scale companies and three large scale companies. The selection of companies was also based on the core business area (comprised of software development, quality assurance, networking, IT consultancy) and operation status (serves an only local market, serves both domestic and international market and serves only the global market). This purposive selection of participants provided an opportunity to receive different views and ideas from different perspectives to enrich the understanding of experience (Polkinghorne, 2005). Demographic data of the selected companies is detailed below in Table 1.

For the selected companies, an email was sent to the HR managers (in COMP2 there was no separate HR team, and thus the principal administration officer was contacted) requesting available date and time to have the interviews. Two email reminders and one text message were sent. Eight interviews were conducted with ten interviewees according to their convenience and preferences. Among them, four interviews were conducted as video conference meetings using either Skype or Zoom software applications. Three audio conferences were held using Skype, and one audio conversation took place through a mobile phone call.

Table 1. Company and Respondent Demographics Data

Company No.	Core Business Area*	Operational Status	No. of Employees	Respondent No.	Gender	Age	Education Qualification	Experience (no. of years)
COMP1	SWD, QAE, NET	Local and international	45 (Small)	HR1	Female	40	Postgraduate Diploma in HR	8
COMP2	SWD, QAE	Local	19 (Small)	HR2	Male	31	MBA	6
				HR3	Male	33	MBA	6
COMP3	SWD, QAE, ITC	International	22 (Small)	HR4	Female	25	BSc in IT	2
COMP4	SWD	International	55 (Medium)	HR5	Female	33	MBA	9
COMP5	SWD	Local and international	90 (Medium)	HR6	Female	34	MBA	8
COMP6	SWD	International	900+ (Large)	HR7	Male	43	MBA	17
COMP7	SWD	Local and international	200 (Large)	HR8	Female	28	MBA	4
				HR9	Male	30	Postgraduate Diploma in HR	6
COMP8	SWD, QAE, NET, ITC	International	200+ (Large)	HR10	Female	31	MBA	6

*SWD: Software Development, QAE: QA Engineering, NET: Networking, ITC: IT Consultancy

Since this study was carried out during the outbreak of the coronavirus disease 2019 (COVID-19), data collection was conducted online. Thus, several initiatives were taken to conduct an ethically sound investigation to mitigate possible ethical challenges of internet-based research. Data were collected through semi-structured interviews, which allowed the researcher to focus on specific themes while adhering to the conversational style of interviewing (Polkinghorne, 2005). The interviews were conducted using English or in the participant's native language (Sinhala), and each meeting took 15-35min. The interview guide focused on background data and current HR practices, support services, stress management programs, the interviewee's awareness and experience of ICT in stress management practices, and perception of ICT supported stress management interventions. The interviews were recorded digitally upon receiving verbal consent from the participants. All the interviews were transcribed to written text for analysis purpose. The verbatim transcribed interviews were analyzed using inductive thematic analysis. The inductive thematic analysis supports identifying, analyzing, and reporting patterns from responses or data, which allows the researcher to organize and describe the data in detail (Braun and Clarke, 2006).

5. RESULTS

As a result of the inductive thematic analysis of the participants' responses, three major themes emerged. In this section, the themes and their subthemes (see Table 2) are explained in detail. The descriptions are supported with direct quotations from the participants.

Table 2. Summary of Themes and Subthemes

Themes	Subthemes	Categories	Remark
Healthy Workplace and Happy Employee	Workplace ergonomics	Seating arrangement, Floor plan	Identifies different organizational strategies placed to create a healthy work environment
	Availability of support services	Insurance schemes, Leave facilities, Transport, Food and beverages, Daycare	
	Workplace strategies	Company structure and environment, Cohesive Team and Team Work, Recruitment & Job Design, Training and Evaluations, Inculcating outside interests	

The HR Role and Stressed Employee	Awareness and acceptance	Stressful periods, Work planning, support and guide, Human skills development training	Explores the manager's experience of interacting with stressed employees
	Communication	Channels, Trust, Openness, Availability	
	Challenges	Age barrier, Stigma, Stressors: family or work Early identification, size of the company	
ICTSMI: Accept or Reject	Positive attitudes	Awareness, Willingness, Acceptance	Discusses managers' perspectives on ICT supported stress-management intervention
	Doubtful attitudes	Lack of awareness and experience, Negative experience, human to machine replacement	
	Feature requirements	Conversational agent, Early detection, Alerts, Differentiate workplace stressors and family matter, Emoji usage	

5.1 Theme 01: Healthy Workplace and Happy Employee

The theme "Healthy workplace and happy employee" elaborates the various types of strategies implemented by the organizations to assure the employee well-being. This includes the setup of the physical environment, various support services and facilities provided, and company level strategies implemented in the organizations to create a healthier work setting for its employees.

5.1.1 Workplace Ergonomics

This subtheme discusses the various approaches taken to set up a healthy physical environment. The participants strongly believed, since employees spend a lot of time inside the building, that the physical environment influences employee well-being and has direct impact on the employees' work performance and productivity. Therefore, they had taken different measures to create a favourable layout of the office space. This included the work cabin arrangements, workspace design with numerous physical facilities including pantry, lobby areas, private meeting rooms, prayer rooms, sick rooms, etc. Such arrangements became evident through:

HR1: "This place is small and cosy. Currently, our administration is trying to renovate the environment to make this place more relaxing."

HR5: "Our office space is designed to provide a convenient and friendly atmosphere for our employees."

HR7: "There are pool tables and relaxation areas, the pantry area is also quite large so they can use it to gather and relax, the sick room is there, if you feel uncomfortable you can go there to relax, we have prayer rooms separately, conference rooms (…)."

5.1.2 Availability of Support Services

The second subtheme explores the different benefit packages offered by the companies above the salaries and wages to their employees to make them benefited and satisfied. These employee benefits packages include health insurance, vacation and holiday facilities, transport facilities, a daycare facility for children, free food and beverages, etc.

HR10: "We offer hotel vouchers so they can go on vacation with their families utilizing their annual leaves."

HR4: "If they stay late at night, we reimburse transport and food expenses. All the employees have a medical insurance policy as well."

HR7: "Lunch is provided, the medical facility is there, the daycare facility is there, more than 50% is paid by the company (…)."

HR6: "We have a health insurance policy for all the employees, we provide meals and transport facilities when they work late nights."

5.1.3 Workplace Strategies

The final subtheme of Theme 01 consists of explicit and implicit strategies created, or evolved, and maintained in an organization to make its culture healthier for its employees. There are five main categories identified in this subtheme: Company structure and environment, Cohesive Team and Teamwork, Recruitment & Job Design, Training and Evaluations, and Inculcating outside interests.

The companies claimed to follow a flat structure and to have a friendly atmosphere inside the organization. They followed the open-door policy, to have the employees feel easy and welcomed to meet any other employee or manager without any prior appointment.

HR8: "Here we practice open-door policy, any employee can visit any employee's workspace without getting any prior approval. Even this is the same with the top management."

HR10: "We practice open-door policy, even for the CEO or chairpersons there's no separate room, and everyone is sitting on the same floor, if you want to talk to the CEO you can go directly and talk."

HR3: "We have no hierarchy; we have a flat structure here."

They also provided flexible working hours and working from home facility to let the employees feel more relaxed at work.

HR10: "We provide working from home facilities, so they don't have travelling problems."

HR1: "We follow flexi hours. Normal working hours are from 8.30-5.30 pm, but normally no one comes at 8.30 am."

A shared practice identified in all the companies was the "teamwork." All the companies formed their workflows into 'team' mode. The team was given a task, and members worked as a team to reach the set targets. The team was managed by a team lead/s who took on a supervisor role. The respondents figured that the team members were more open to the supervisor and their peers, and that such structure tended to alleviate communication regarding work as well as obstacles they were facing.

HR2: "We practice teamwork, employees recognize (company name) as a family; it's not an image that we created but created by themselves and maintained by themselves."

HR7: "We try to maintain good communication, collaboration and a kind of bonding with the manager and the member."

The "Buddy" concept had been introduced in some companies, along with the team concept. The buddy was described as a peer working in the same project and assigned to co-work with new employees. The buddy was responsible for and supported the work assignments, and this person helped others to adjust to the work setting and work culture.

HR9: "We assign a buddy who may not be a supervisor. Buddy is like a senior fellow who will help him to understand the context. We provide a budget for the buddy and the newcomer to go for lunch together (buddy lunch out)"

Companies also provided team allowances to let teams go out for team outings. This was intended to strengthen the teams and to avoid conflicts and create a more friendly environment.

HR10:" We provide leisure allowance for the teams where they can go outing as a team."

The participants explained that, while using internal strategies to create the cohesive teams and well-performing employees, they also strived to identify the right candidates for job positions.

HR3: "When it comes to recruitment, our interviews are a little bit tough because we need to select the best ones. We recruit only people who can match the level of the existing ones. (…) We look at whether they have the potential to learn from others, if we think they can learn, yes, we recruit them."

HR10: "We hire them based on their talent whenever there's a vacancy in a team, we recruit the right talent."

Next, the HR managers explained that the companies paid attention to training and evaluation process. They revealed that they tried to have a transparent evaluation process throughout the year, to let employees become aware of the areas that they should improve.

HR4: "We have surveys from time to time to check employee's satisfaction and happiness."

HR7: "We try to have one-to-one meetings, and we strive to give feedback, both good and bad, continuously, so they are not surprised at the annual evaluations because they know what's going on and they feel comfortable in their work positions."

It was also explained that the companies provided various workshops and training to improve the required knowledge and skills of their employees.

HR4: "We have technical training workshops for employees to improve their knowledge and skills. We have frequent knowledge sharing sessions."

Inculcating outside interest is another aspect the addressed companies focused on. Besides the work-related activities, companies introduced various extracurricular activities; annual trips, team outings, sports, musical events, etc., for their employees to increase work-life balance.

HR1: "We arrange different activities. Once a month we arrange a movie out, every Tuesday and Thursday evening we go for badminton."

HR8: "Very recently, we had Valentine's day party, which they enjoyed a lot. We also have carom tournaments, movie nights and paduru party (musical events) to change their work environments and to make them enjoy and relax."

5.2 Theme 02: The HR Role and Stressed Employee

The second theme discusses the managers' experience with stressed employees. This will give details on their awareness of stressful conditions within the work setting, how they were managing and dealing with such situations and employees, and the perceived challenges.

5.2.1 Awareness and Acceptance

It was evident from the interviews that the managers, particularly in large and medium scale organizations, had noticed and worked with stressed employees. In such cases, they positively accepted and attempted to take necessary actions to reduce such stressful situations and to improve work-life balance.

HR10: "When there's delivery/deployment, the stress level goes high because they have to put their full strength, work in full time, work overnight, deal with customers and managers. Those days their stress level is high, and that we know. That is why we create a space for them to relax, provide flexible working hours and flexibility in everything (...)."

HR1: "There are few specialized resources for those we have no replacements, so they are bit stressed sometimes. But as a team, they try to manage their workload."

HR5: "Yes, we have recognized employees who suffered from stress, and we are aware that there are periods which are stressful for our employees."

However, participants from two of the small scale organizations believed that the employees were not stressed nor that the company environment needed to take any special stress management measures.

HR2: "Here they are not stressed. In our collaborative work environment they support each other."

HR4: "I believe our culture is very positive and friendly and not that stressful."

5.2.2 Communication

All the participants revealed that effective communication is the tool they used to reach their employees and to create a culture for them to bring up their problems. They tried to make themselves available and open to any matter. They believed trust and friendliness geared towards more open discussions.

HR8: "Sometimes we are busy, but when an employee reaches us with a grievance or problem or just for a chat, we make sure that we are available for them."

HR3: "We use a communication channel called slack. So, it helps to have improved efficient communication among employees."

HR6: "We are very close to our employees, so they tend to raise any concerns they have."

5.2.3 Challenges

HR managers said that they had observed age, stigma and company size as main problems averting the communication between HR and employees. They emphasized that with a young HR panel in the same age as their employees, they tended to communicate all the problems more openly, but this could change with more aged HR personnel.

HR2: "I think age matters in this. We are also not that aged, and employees are also the same, so they all cooperate very well. If there is an age gap, they might not easily open up like this. They come and talk to us if there's a problem at home still they talk about it if they want a salary increment, they talk to us, not as a dispute but just as a friendly talk."

HR1: "What I have noticed is that some employees if they have a problem, they are not communicating with HR. They just communicate when they are about to leave the organization."

HR9: "We know that sometimes with different attitudes, some employees are not willing to open up and communicate their problems."

Size of the company also acted as a challenge; in small scale organization, they explained that they could provide one-to-one attention, but in large organizations, it was regarded challenging to manage early detection of stress symptoms and stressors.

HR3: "If one person is getting late to work for a couple of days or leave work early frequently or if he's not in a good mood at a meeting we figure out that there's a problem. Since team size is small, we can manage and give one-to-one attention, because of this practice, many issues are resolved easily at an early stage."

5.3 Theme 03: ICTSMI: Accept or Reject

The final theme "ICTSMI: accept or reject" addresses the company perspectives on ICTSMI through the lens of HR managers. This theme comprises of three subthemes: "Positive attitudes", "Doubtful attitudes" and "Feature requirements".

5.3.1 Positive Attitudes

Participants elaborated on their awareness of ICTSMI. Though some managers had only minimal awareness and experience with ICTSMI, still they perceived ICT supported interventions as an interactive and efficient means to identify and manage stress among employees.

HR4: "Yes, I think it's good to have a platform or system that we could use to support employees who are suffering from stress (…). "

HR10: "Yes, we believe that ICT supported interventions could support our employees to manage their stress levels. As an organization, we are happy to use them and promote them among our employees.

5.3.2 Doubtful Attitudes

Though the positive perceptions outweighed the doubtful opinions, some managers did not have confidence in ICTSMI due to negative experiences, lack of awareness, and knowledge and expertise. Participants appreciated the value of face-to-face communication in the counselling process and doubted that machine-based communication settings could replace the effectiveness of the human conversation.

HR10: "When it comes to stress management, rather letting people read, listen and watch videos, I think stress management should be related to discussions, conversations, the sort of things involving experts."

HR7: "I have one experience where he was stressed, and he had a family issue, and he was trying to use different applications available online, and he failed, then he came to us and talked about his problem. I know that there are useful applications, but when it comes to stress management, I think counselling is a must."

HR5: "I haven't used or are aware of such interventions. If they could offer the required functions, we may use them in our organization."

5.3.3 Feature Requirements

Conversations with the HR managers revealed several feature requirements that they expected from ICTSMI. These mainly included creating an effective platform for communication, stress level and stressors identification and enhancing employee awareness on stress management.

HR9: "I think there should be an initial screening process where individuals could assess themselves for stress levels so they could decide when to seek support."

HR4: "I prefer to have a platform that employees can use to communicate with us directly."

HR1: "It'll be great if we can have a system to get to know about how they feel and if they can use it to communicate with the HR directly. Especially if we can recognize whether it's a problem related to their personal life or work life."

HR2: "If we can have a system feature to express their feelings using emojis or images, then they could easily express their feelings rather than typing."

6. DISCUSSION

The general objective of this study was to explore the software company perspective of ICT supported interventions through the views of HR managers. The three main themes and nine subthemes extracted from the interviews provided valuable insights into the current strategies implemented in software organizations to create a healthier work setting for their employees. The findings did highlight not only managers' positive perception of ICTSMI but also doubtful thoughts which might hinder the possible use or implementation of ICTSMI within an organization.

According to literature, reducing stress and improving employee well-being is facilitated via diversified approaches at both individual and organizational level (DeFrank & Cooper, 1987). The individual level interventions target individual employees to develop their skills to manage stress. In contrast, the corporate level approaches focus on the systematic changes to organizational practices that either target all the employees or a specific group of employees (Holman, Johnson and O' Connor, 2018). Considering the focus of the stress management interventions, they could be categorized as primary, secondary and tertiary (DeFrank and Cooper, 1987). The primary interventions focus on preventing stress by eliminating stressors and enhancing cases of well-being. Secondary interventions aim to reduce the severity of the stress and to avoid the level of stress becoming problematic. Tertiary interventions target rehabilitation and maximizing the functionality of employees who are already suffering from mental ill-health. Accordingly, the current findings also show similar mechanisms and strategies adopted in software companies to assure well-being of the employees.

Healthy physical setup of the organization, flat hierarchy and the open-door policy concept mentioned in the interviews are valuable to create a friendly atmosphere within the organization supporting more open conversations among employees and administration (Black, J *et al.*, 2019). Regardless of the size of the company, these types of structures were present in all the eight companies. Another factor revealed by all the companies and mostly highlighted by the respondents in the large scale companies was how they arranged the physical environment. It was noted that the large companies strived to setup more convenient physical work settings and make a more cohesive environment to bridge distance and provide more opportunities to work closely with colleagues (Aries, Veitch and Newsham, 2010). This type of installation aims to mirror the one-to-one attention provided by the small and medium companies to their employees. Literature also shows that characteristics of the physical office environment can have a significant impact on behaviour, perceptions and productivity of employees (Kamarulzaman et al., 2011).

All the organizations offered different types of benefits packages to their employees. It was shown that the large scale software companies in the study had rich and more established benefit schemes over the small and medium scale companies. Benefits schemes, including health insurance, annual leaves, daycare facility, free meals transport services, etc., could provide a sense of comfort and help to attract and retain employees (Bernhard, 2018), and could support to alleviate stressful conditions which are more common in larger scale companies (Dekker and Barling, 1995). Such schemes could be considered as a way of responding to employee stress issues and labelled as an employee benefit for employer's benefit (Wheeler and Lyon, 1992).

It is noteworthy to mention that flexible working hours and working from home facility provided by the software companies act as a family-friendly policy which supports employees to balance their work-life (Shagvaliyeva and Yazdanifard, 2014).

In all the interviews, it was explained how the companies facilitated a team-based working culture where peers could help each other. This kind of initiative is in line with a study by Peterson et al. (2008), suggesting that peer-support groups could be a useful and comparatively inexpensive tool in alleviating work-related stress and burnout. However, organizations must pay careful attention to such implementation as the same initiatives might create unintended pressure on trained employees (Agarwal, Brooks and Greenberg, 2019). Secondary-individual strategies like health and relaxation, promoting programs investing in company trips, teams outings, sports activities, and parties provide employees with opportunities to engage in stress-reducing activities (Kondo, Jacoby and South, 2018). Acknowledging different personalities is important still since direct or indirect forceful participation or availability of preferred options might hinder the involvement and effectiveness of such initiatives (Edwards, 2012).

The selection and assessment procedures implemented inside an organization are considered as a primary-individual level strategy which ensures the selection of a candidate who has the right set of skills and attitudes to manage the demand from the job (Pulakos, 2005). However, this might implicitly force the employees to be in a continuous struggle to achieve their best to secure their job positions, imposing stressful mentality on them (World Health Organization, 2002).

Results revealed that the HR managers in the small scale software companies believed that their work settings were stress-free environments. However, over expectations and misbeliefs about the current situation might impact negatively on the employees (World Health Organization, 2002). Some employees might feel that if they suffer from stress in such a situation, company culture will characterize them as personal mismatches. Thus, they might remain unobserved and unspoken regarding their stressful conditions.

Therefore it is essential to identify a more personalized and customized solution to manage employee stress within an organization.

Despite the HR managers' awareness and willingness to accept ICTSMI, the study also revealed challenges and drawbacks of using ICT in occupational stress management practices. They expressed how they doubted the possibility of replacing the human with a machine in a stress counselling situation. Such doubtful attitudes were based on negative experiences and also limited knowledge of ICTSMI. Since the successful implementation of ICTSMI in an organization tends to accelerate by the positive perception of managers (Ariani, Koesoema and Soegijoko, 2017), it is essential to take necessary measures to increase their awareness and positive experience of ICTSMI.

7. CONCLUSION

Over time, the research literature and experience with ICTSMIs within organizations will grow, and more consensus will be reached regarding which ICT interventions have a greater benefit to employees who need stress management. Thus, the results of this study could contribute to the already existing knowledge base, through increased understanding of current organizational practices and strategies implemented to create healthy workplaces and of the possibilities and challenges related to using ICTSMI in a technology-intensive work setting. It also elaborates features suggested by HR managers that, integrated into ICTSMIs, could help to better target stress-related problems at work. The findings could contribute to future design and development of ICT tools for stress management not only in the Sri Lankan context but also at the international level. However, it is also relevant to investigate employees' experiences and views on ICT supported stress management in future studies.

REFERENCES

Agarwal, B., Brooks, S. K. and Greenberg, N. (2019) 'The Role of Peer Support in Managing Occupational Stress: A Qualitative Study of the Sustaining Resilience at Work Intervention', *Workplace Health & Safety*. SAGE Publications Inc, 68(2), pp. 57–64. doi: 10.1177/2165079919873934.

Amin, A. *et al.* (2011) 'Occupational stress, knowledge sharing and GSD communication barriers as predictors of software engineer's creativity', *IEEE International Conference on Industrial Engineering and Engineering Management*, pp. 394–398. doi: 10.1109/IEEM.2011.6117946.

Ariani, A., Koesoema, A. P. and Soegijoko, S. (2017) *Innovative Healthcare Applications of ICT for Developing Countries*. doi: 10.1007/978-3-319-55774-8.

Aries, M. B. C., Veitch, J. A. and Newsham, G. R. (2010) 'Windows, view, and office characteristics predict physical and psychological discomfort', *Journal of Environmental Psychology*, 30(4), pp. 533–541.

Bernhard, K. (2018) *To attract and retain key employees, offer benefits that they value.* Available at: https://www.bizjournals.com/bizjournals/news/2018/09/11/to-attract-and-retain-key-employees-offer-benefits.html (Accessed: 15 May 2020).

Biron, C., Karanika-Murray, M. and Cooper, C. L. (2012) 'Organizational interventions for stress and well-being – An overview', *Improving Organizational Interventions for Stress and Well-Being: Addressing Process and Context*, (2012), pp. 1–18. doi: 10.4324/9780203723494.

Black, J, S. *et al.* (2019) *Organizational Behavior*. Texas: OpenStax. Available at: https://openstax.org/details/books/organizational-behavior.

Black, C. (2009) 'Working for a healthier tomorrow', *Occupational and Environmental Medicine*, 66(1), pp. 1–2.

Braun, V. and Clarke, V. (2006) 'Using thematic analysis in psychology', *Qualitative Research in Psychology*, 3(2), pp. 77–101. doi: 10.1191/1478088706qp063oa.

Brough, P., Dollard, M. F. and Tuckey, M. R. (2014) 'Theory and methods to prevent and manage occupational stress: Innovations from around the globe', *International Journal of Stress Management*, 21(1), pp. 1–6. doi: 10.1037/a0035903.

Burke, R. J. (2014) 'Corporate wellness programs: an overview', *Corporate Wellness Programs*, pp. 3–27.

Burman, R. and Goswami, T. G. (2018) 'A Systematic Literature Review of Work Stress', *International Journal of Management Studies*, V(3(9)), p. 112. doi: 10.18843/ijms/v5i3(9)/15.

DeFrank, R. S. (2012) 'Teaching a Comprehensive Course on Stress and Work', *Journal of Management Education*, 36(2), pp. 143–165. doi: 10.1177/1052562911427834.

DeFrank, R. S. and Cooper, C. L. (1987) 'Worksite Stress Management Interventions':, *Journal of Managerial Psychology*. MCB UP Ltd, 2(1), pp. 4–10. doi: 10.1108/eb043385.

Dekker, I. and Barling, J. (1995) 'Workforce Size and Work-Related Role Stress', *Work & Stress*, 9, pp. 45–54.

Edwards, J. (2012) 'Understanding the predictors of participation and the barriers to employee involvement in workplace health promotion programmes', pp. 1–81. Available https://mro.massey.ac.nz/bitstream/handle/10179/4078/02_whole.pdf.

Esmond, J. (no date) *STRESS FACTS*. Available at: http://www.gostress.com/stress-facts/ (Accessed: 17-09-2019).

Fink, G. (2017) 'Stress: Concepts, Definition and History', *Reference Module in Neuroscience and Biobehavioral Psychology*, (January), pp. 0–9. doi: 10.1016/b978-0-12-809324-5.02208-2.

Holman, D., Johnson, S. and O' Connor, E. (2018) 'Stress Management Interventions : Improving Subjective Psychological Well-Being in the Workplace', *Handbook of well-being*, pp. 1–13. doi: nobascholar.com.

ICTA (2019) *NATIONAL IT - BPM WORKFORCE SURVEY 2019*.

Kamarulzaman, N. *et al.* (2011) 'An overview of the influence of physical office environments towards employees', *Procedia Engineering*, 20, pp. 262–268. doi: 10.1016/j.proeng.2011.11.164.

Kathriarachchi, S., Seneviratne, Vl. and Amarakoon, L. (2019) 'Development of mental health care in Sri Lanka: Lessons learned', *Taiwanese Journal of Psychiatry*, 33(2), p. 55. doi: 10.4103/tpsy.tpsy_15_19.

Kondo, M. C., Jacoby, S. F. and South, E. C. (2018) 'Does spending time outdoors reduce stress? A review of real-time stress response to outdoor environments', *Health and Place*. Elsevier Ltd, 51(March), pp. 136–150.

Lazarus, R. S. (1993) 'FROM PSYCHOLOGICAL STRESS TO THE EMOTIONS " A History of Changing Outlooks', *Annual Review of Psychology*, 44, pp. 1–21.

LLUCH, M. (2011) 'Empowering Patients through ICT Organisational Impact on Healthcare Systems in England and Scotland', *COMMUNICATIONS & STRATEGIES*, 83(3), pp. 37–58.

Martin, A., Woods, M. and Dawkins, S. (2018) 'How managers experience situations involving employee mental ill-health', *International Journal of Workplace Health Management*, 11(6), pp. 442–463.

Minas, H., Mendis, J. and Hall, T. (2017) 'Mental Health System Development in Sri Lanka'.

Mustafa, M. *et al.* (2015) 'Causes and Prevention of Occupational Stress', *IOSR Journal of Dental and Medical Sciences*, 14(November 2015), pp. 98--104. doi: 10.9790/0853-1411898104.

Narváez, S., Tobar, Á. M. and López, D. M. (2014) 'Systematic review of interventions supported by ICT for the prevention treatment of occupational stress', in *Studies in Health Technology and Informatics*, pp. 71–80.

Nayak, R. D. (2014) 'Anxiety and Mental Health of Software Professionals and Mechanical Professionals', *International Journal of Humanities and Social Science Invention*, 3(2), pp. 52–56.

Padma, V. *et al.* (2015) 'Health problems and stress in Information Technology and Business Process Outsourcing employees', *Journal of Pharmacy & Bioallied Sciences*. India, pp. S9–S13. doi: 10.4103/0975-7406.155764.

Polkinghorne, D. E. (2005) 'Language and meaning: Data collection in qualitative research.', *Journal of Counseling Psychology*. American Psychological Association, 52(2), pp. 137–145. doi: 10.1037/0022-0167.52.2.137.

Ponnamperuma, E. N. (2000) 'SMEs in Competitive Markets, Country Paper', *Asian ProductivityOrganization Tokyo*, pp. 295–313.

Pulakos, E. D. (2005) 'Selection Assessment Methods: A guide to implementing formal assessments to build a high-quality workforce', *SHRM Foundation*, pp. 1–54. Available at: https://www.shrm.org/hr-today/trends-and-forecasting/special-reports-and-expert-views/documents/selection-assessment-methods.pdf.

Quick, J. C. (2013) *Preventive stress management in organizations*. 2nd ed. Washington, District of Columbia: American Psychological Association.

Rajeswari, K. S. and Anantharaman, R. N. (2003) 'Development of an Instrument to Measure Stress Among Software Professionals: Factor Analytic Study', *Proceedings of the ACM SIGMIS CPR Conference*, pp. 34–43.

Rosayro, G. de (2017) 'Society must help in battle with depression and anxiety', *dailynews*, 9 October. Available at: http://www.dailynews.lk/2017/10/09/features/130539/society-must-help-battle-depression-and-anxiety?page=29.

Shagvaliyeva, S. and Yazdanifard, R. (2014) 'Impact of Flexible Working Hours on Work-Life Balance', *American Journal of Industrial and Business Management*, 04(01), pp. 20–23. doi: 10.4236/ajibm.2014.41004.

Swanson, J. L. and Fouad, N. A. (1999) 'Applying theories of person-environment fit to the transition from school to work.', *The Career Development Quarterly*. US: American Counseling Assn, pp. 337–347.

Weerasekara, M. and Smedberg, A. (2019) 'Design practices and implications in information and communication technology supported occupational stress management interventions', *2019 International Conference on Advanced Computer Science and Information Systems, ICACSIS 2019*, pp. 285–294. doi: 10.1109/ICACSIS47736.2019.8979813.

Wheeler, S. and Lyon, D. (1992) 'Employee Benefits for the Employer's Benefit: How Companies Respond to Employee Stress', *Personnel review*. Bradford, England : doi: 10.1108/00483489210021053.

World Health Organization (2002) *Mental health and work: Impact, issues and good practices*.

APPLICATION OF AUTOMATON MODEL TO SUPPORT REHABILITATION OF MOTOR SKILLS OF PATIENTS

Grigorii Shovkoplias, Eduard Glekler, Vladislav Povyshev, Jaafar Hammoud, Natalia Gusarova, Aleksandra Vatian and Anatoly Shalyto
ITMO University
Sankt-Petersburg 197101, Russia

ABSTRACT

Rehabilitation of patients during illnesses associated with the loss of motor skills, such as multiple sclerosis, is an important task of medicine. Effective in these cases are gaming approaches that increase patient motivation, which in the field of e-Health form the direction of "Serious Games". The essential requirements in the development of Serious Games are the continuity of training, which should be carried out not only in a specialized center, but al-so by the patient himself at home, as well as the accuracy of the movements performed by the patient; otherwise, the outcome of the rehabilitation will be doubtful or even negative. As the basis for satisfying the rigorous requirements for the correctness of the movements performed the automatic simulation of the game process is chosen.

Serious Games differ from regular games in a significantly smaller number of possible states, which makes them convenient for applying this approach. At the same time, in contrast to the previously considered applications of au-tomaton models in medical systems, such as, for example, systems for sup-porting hemodialysis, in the case of a game, the parameters of each state can vary over a fairly wide range.

In the article, this problem is considered in the example of rehabilitation of pinch movements of hands by performing physical exercises in a specialized video game form using motion capture of fingers on a smartphone. The Ab-stract State Machine method is used to build the automaton model applicable to the Serious Game for pinch movement, which provides opportunities and motivation of the patient to do independent exercises. To build the resulting model, the method of successive refinements is used, on each of which veri-fication is performed, which provides the necessary control over the accuracy of patient movements.

KEYWORDS

Abstract State Machine, Serious Games, Rehabilitation Systems, Pinch Movements

1. INTRODUCTION

Rehabilitation of patients during illnesses associated with the loss of motor skills, such as multiple sclerosis, is an important task of medicine. Multiple sclerosis is a disease of the central nervous system, which may be accompanied by various neurological signs, including loss of motor skills. Radical treatments for multiple sclerosis have not yet been identified, and one of the most important factors in the rehabilitation of patients after exacerbations, as well as preventing or delaying the development of new exacerbations, is the regular exercise [NINDS, Halabchi]. But the patient's independent implementation of the complex of exercises is associated with motivational difficulties, as well as with the difficulty of assessing the correctness of the exercises without direct supervision of a specialist.

To increase the patient's motivation for regular exercise, a game approach is widely used, which is called in medical applications "serious games". In particular, in order to support the rehabilitation of patients with multiple sclerosis specialized video games using various motion capture technologies are offered [Haase, Jonsdottir, Midaglia]. However, most often the proposed video games require the use of special game consoles or specialized motion controllers, which can be difficult for the patient in his daily life. In addition, during rehabilitation through physical exercises, the correct execution of individual movement elements is important, which should be combined with a minimum restriction of patient freedom of movement. Otherwise, the exercises may not give the desired effect or even harm the patient. These contradictory requirements get critical when training in fine motor skills. The article discusses the possibilities of satisfying them by the example of developing serious games for training the pinch movement of the fingers.

2. BACKGROUND

As the analysis showed, there are several categories of hand and finger motion controllers on the market that implement various motion capture technologies, including gloves with motion sensors like [CaptoGlove], Leap Motion controllers like [Leap], full-scale game controllers with six degrees of freedom like [Nintendo] and smartphones like [Xiaomi]. The first three categories allow you to fully capture all types of movements of the hands (the first two categories also allow you to fully capture all types of movements of the fingers), taking into account all degrees of freedom, while smartphones have the ability to fully capture the movements of the distal phalanges of 10 fingers simultaneously only in the plane of the smartphone screen.

However, smartphones, unlike other categories of devices, are gadgets of everyday use, they are familiar and convenient for the vast majority of patients. According to [Haase], approximately 85.6% of a sample of 570 patients with MS regularly use smartphones. Thus, the use of smartphones as a means of organizing regular rehabilitation exercises, inscribed in the daily life of the patient, and for motivational purposes is quite justified.

Consider the need and the possibility of organizing serious games when using a smartphone as a means of motion capture. From the point of view of modeling, most smartphone games for widespread use involve evaluation of the player's finger getting into a certain predetermined area of the screen, which, for example, can be a target or terminal state; however, tracking the current position of the finger outside this area is not required. At the same time, as mentioned above, when implementing rehabilitation exercises, the correct execution of individual movement elements is important. This opposition is confirmed in practice. According to [Jonsdottir], improvements in clinical indicators were detected in patients undergoing rehabilitation using specialized video games after 12 gaming sessions, while in patients playing active video games for a wide audience, no improvement was noted, although the patients themselves noted moral satisfaction from the process and subjective improvement in health status. This means that from the patient's point of view, the scenario of specialized games should be as close as possible to regular games, but at the same time keep all the advantages described above.

Thus, for the sterling rehabilitation of motor skills in patients with multiple sclerosis, it is necessary to develop specialized "serious games", and special requirements must be imposed on them:

- Automatic control and current monitoring of the correct execution of the patient's movements during the game, which will allow, on the one hand, to increase the effectiveness of the exercises, and on the other hand, to meaningfully interpret the results demonstrated by the patient without regular personal examination [Midaglia].
- Possibility of individual adaptation of a set of exercises and game scenario. For example, movements that are obviously impossible for a particular patient should be easily excluded from the scenario; otherwise, a deterioration in the state of health and demoralization of the patient may follow [Thirumalai].

To implement these requirements, it seems advisable to use the automaton model [Khoussainov, Shalyto, Polikarpova]. The concept of formal methods is widely used in healthcare (see review [Bonfanti]). As a rule, such applications as hemodialysis machine [Butler], infusion pump [Zhang], electrocardiography [Sobrinho], blood separator machine [Tolvanen] and other medical devices with a rigid and uniquely defined operation algorithm are considered. Respectively, the authors focus mainly on modeling, verification and validation of medical devices in accordance with pre-developed requirements. At the same time, the works related to the application of the automaton approach to the development of medical game applications ("serious games") were not found in the literature accessible to the authors.

Thus, the article sets the following tasks:

- Modeling hand and fingers movements in order to formulate requirements and constraints for the automaton model.
- Modeling the automaton that implements the identified requirements using abstract state machine tools.
- Offering an example of a scenario of a "serious game" for rehabilitation of motor skills restricted after multiple sclerosis based on a developed automaton model.

3. MODELING HAND AND FINGERS MOVEMENTS FOR REHABILITATION

Pinch movement is the movement of the thumb and forefinger, the result of which is a pinch grip [Kapandji]. Pinch movement is the movement of the thumb and forefinger, the result of which is a pinch grip. From the point of view of patient rehabilitation, pinch grip, carried out by the pads of the thumb and index finger and allowing to hold relatively large objects, is of the greatest importance.

As a basis for assessing the correctness of pinch movement on a smartphone, the "Pinching Test" test was used from the set of active application tests for smartphones "FLOODLIGHT" [Midaglia]. This test consists in the fact that large round objects appear on the smartphone screen in front of the patient, which must be captured with a pinch movement. However, success is considered to be any closure of the thumb and forefinger, which does not always correspond to the anatomically correct pinch movement. In particular, our experiments showed that when the distal phalanges of the thumb and forefinger came together at an angle of 90°, the FLOODLIGHT application counted this movement as "success", although in this case the capture of a real object is fundamentally impossible.

In this regard, based on the anatomical description of the movements of the phalanges of the fingers during pinch grip, we built its mathematical (skeleton) model, which is illustrated in Figure 1.

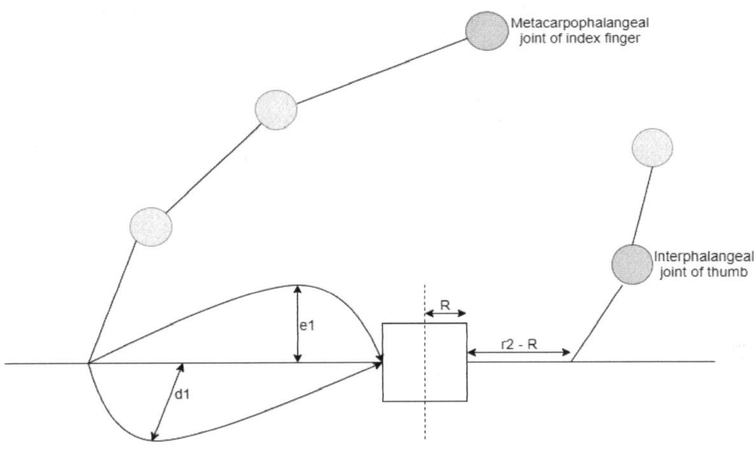

Figure 1. A skeleton model of pinch movement

The pinch movement begins with the location of the distal phalanges of the thumb and forefinger so that the center of the captured object is on a straight line with the corresponding phalanges. Moreover, hereafter, the center of the captured object is understood to be not necessarily the geometric center of the object, but a certain point near which the person wants to grip the given object. Next, the flexion movement of the metacarpophalangeal joint of the index finger occurs. The remaining joints at this time are almost motionless. A slight flexion movement of the distal interphalangeal joint of the thumb is acceptable. As a result of the bending of these joints, the distal phalanges of the index and thumb are approached at distances to the center of the object r1 and r2. The condition for the end of the movement is $\max(r1, r2) \leq R$, where R is the distance from the center of the griped object to its edge. The distal phalanges of the thumb and forefinger approach each other in a straight line on which lies the center of the captured object, while the orthogonal components of the phalangeal shift vector d and e orthogonal to this straight line should not exceed D (in the plane orthogonal to the distal phalanx of the thumb) and E (in the plane, in which lies the distal phalanx of the thumb).

From the point of view of the effectiveness of rehabilitation, it is also necessary to introduce a restriction T on the duration t of the movement, i.e. $t \leq T$, as well as the number of repetitions of the pinch movement in one game session m with the restriction M, i.e. $m \leq M$.

The control of pinch movement begins only from the moment the distal phalanges of the fingers are placed in a position on a straight line with the captured object (with a deviation b not exceeding the value of B, otherwise the movement is not considered erroneous, but it does not begin to be measured). The parameter

$n = r1 / r2$ is measured at the moment of the beginning of the movement and must be outside the norm N ($N \geq 1$), otherwise the movement is immediately recognized as erroneous. Parameters d and e are measured throughout the movement; they must be within the limits of the norms D and E ($d \leq D$, $e \leq E$, for both distal phalanges) and remain within these limits at any stage of the movement, otherwise it is immediately recognized as erroneous. The movement is considered complete when the parameters r1 and r2 are within the norm R, if the time limit T has not been exceeded.

Thus, going beyond any of the seven parameters described above (n, d and e for the thumb and forefinger, t, m) is considered as an error in the pinch movement. Let us explain the anatomical meaning of these parameters. In Figure 1, the joints that should be active or accordingly inactive during the pinch movement are marked in red and blue. Going beyond the parameters d and e means that there has been a flexion of the joints, which should be inactive, or the movement of the hand or arm as a whole. It may also mean that there was a flexion of the joints involved in the movement, but not in that plane. Going beyond the parameter n means that the distal phalanges were initially set incorrectly (for the correct movement, the distal phalanx of the thumb should be closer to the captured object).

In a meaningful sense, going beyond the specified parameters means errors in the implementation of rehabilitation movements and requires a statistical analysis with the subsequent involvement of a doctor for their interpretation and correction.

When implementing pinch movement on the smartphone screen, the distal phalanges of both fingers should be placed on the screen and then during the movement not to tear themselves away from it, i.e. in the above model, $E = 0$. Figure 2 shows the red trajectories of the distal phalanges on the screen in red, the black circle indicates the captured round object on the screen. After both distal phalanges have been placed on the screen, the distance b is measured from the line between them to the center of the captured circle on the screen. If it is less than B, the movement is considered to have begun. This limitation prevents accidental clicks. Next, there is a parallel transfer of the line between the distal phalanges to the point of the center of the circle. Subsequently, the deviation d for both phalanges (for each separately) will be calculated from this straight line. r1 and r2 are also counted from the projection of the coordinates of the phalanges onto this line.

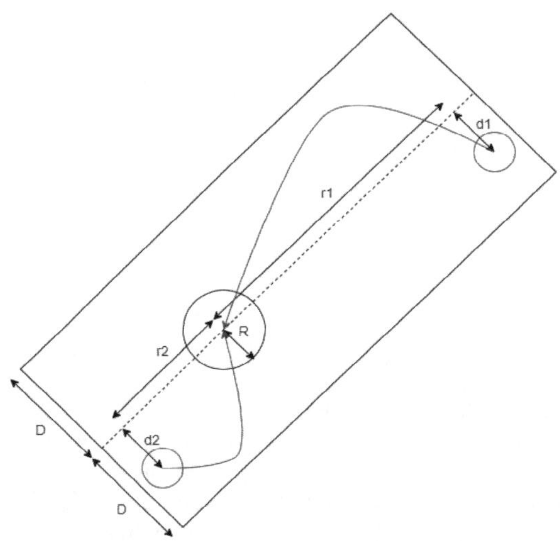

Figure 2. Trajectories of distal phalanges and their parameters

Based on the developed description of hand and fingers movements, we conclude that the game scenario can be a single sequence of states, but with variable parameters that are controls. Consequently, we can formulate requirements for the automaton model of "Serious Game" for pinch movement rehabilitation:

- it is required to provide the ability to change parameters directly during the game, i.e. after several iterations;
- statistics are required for each parameter separately.

4. AUTOMATA MODELING

In order to build the automaton model we use the Abstract State Machine (ASM) method. The ASM for the game is built using the refining method as a sequential forming of the ground model, and then two levels of refinement. The ground model contains the basis of the test, not taking into account the errors that the patient may make during the capture procedure. The first refining includes the processing and collection of statistics for errors. The second refining adds the ability to the doctor to adapt the game scenario for a specific patient, based on the collected data. A detailed description of each stage of refinement is given below.

Ground model (Fig. 3). After the "Start of the session" (Initial state), the procedure for capturing the item occurs, according to the scenario. After M successful capture attempts, the game ends in the "End of the session" state. In the ground model, parameter b is measured and used to prevent accidental clicks. The r1 and r2 parameters are used to determine the condition for successful completion of the attempt. The parameters n, d1, d2, and t are not used at this stage, since there is no separate collection of statistics for these parameters.

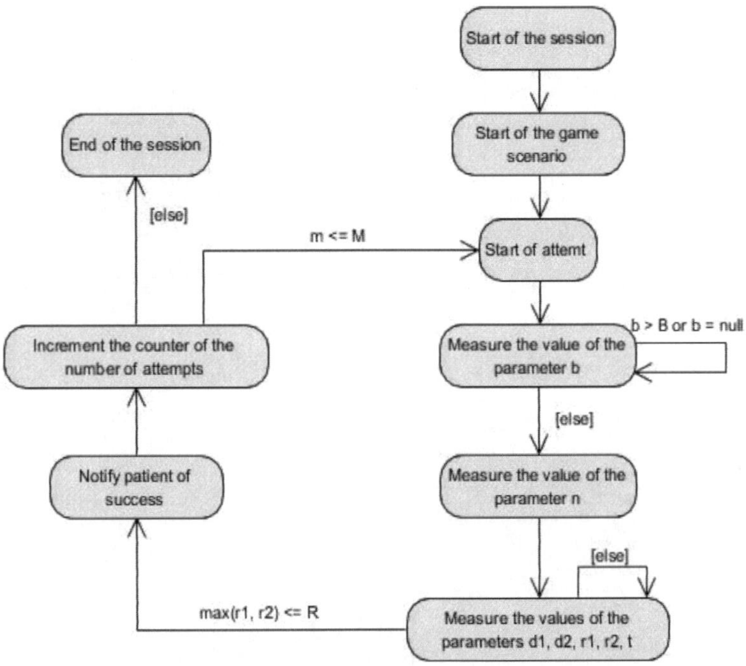

Figure 3. Ground model of automata

First refinement (Figure 4). As described earlier, when making a capture, various errors in hand movements can be made, which will be reflected in the parameters n, d1, d2 and t. All the described deviations are orthogonal and independent of each other, therefore, they can be added to the model in the same way except for the parameter n, which is checked at an earlier stage - when placing fingers on the screen. In the framework of the same refining, the ability to collect statistics was added so that the doctor could later analyze it; for this, several states were included in the model that allow counting the number of successful and erroneous attempts (errors are calculated separately for each parameter).

Second refinement (Figure 5). In order for the doctor to adapt the game scenario, it is necessary to transfer information from the device to him and give him the opportunity to change the constants R, D, N, T and M, then, by the decision of the doctor, the game either continues or is stopped by transitions to the start one, or to final state.

Thus, the automaton modeling process for a serious game under consideration is described. The automaton model satisfies the above requirements, providing the ability to control and adapt constants affecting the game scenario. After each step, the automaton can be verified by standard means.

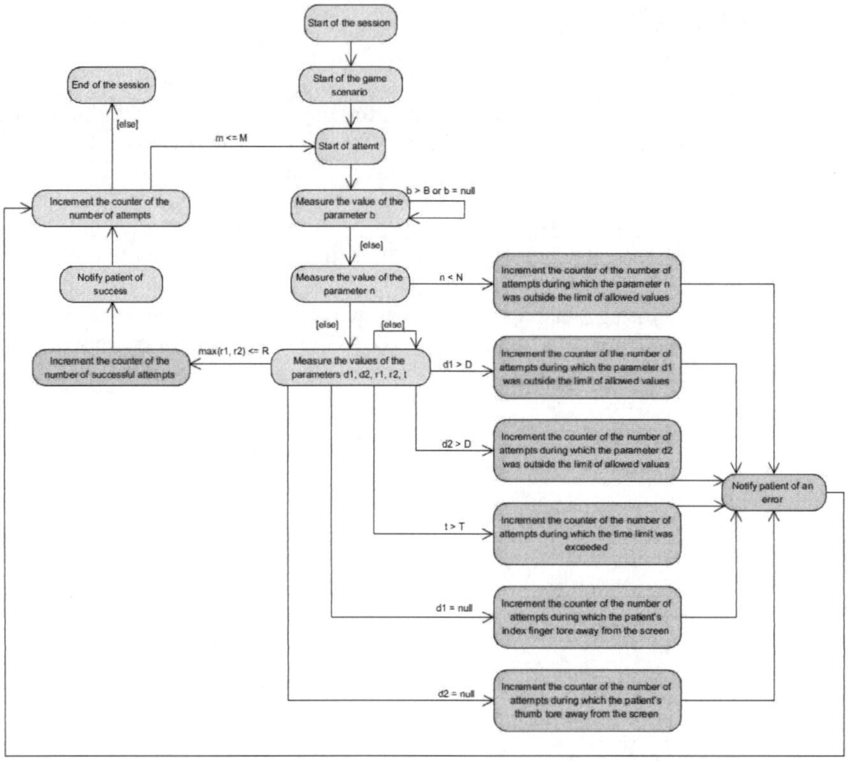

Figure 4. First refinement of automata

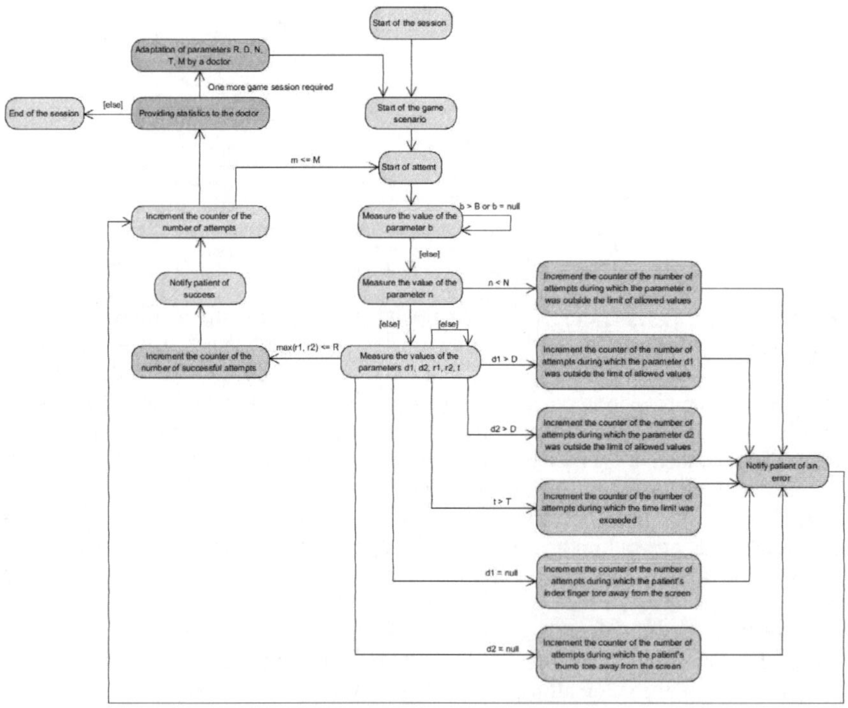

Figure 5. Second refinement of automata

5. "SERIOUS GAME" SCENARIO FOR REHABILITATION OF PINCH MOVEMENT

The game should provide the patient with several attempts to pinch movement in an intuitive game form, in accordance with the previously established requirement for the involvement and motivation of patients.

An example of a possible test scenario is given below.

Before starting the game, the patient is explained his task, namely: to squeeze round objects appearing on the smartphone screen with the thumb and forefinger at a pace accessible to him, but preferably as soon as possible.

Next, the first round object appears on the screen in a random position. After the patient has set the distal phalanges in the correct starting position and started pinching movements, the object on the screen visually shrinks until the movement completes successfully - in this case, the object disappears. If the movement is made with an error, then the item on the screen restores its original shape. Thus, the patient is provided with intuitive visual information about whether he is moving correctly.

If the plucking attempt was completed with an error, the patient is offered a new attempt to capture the same object. Upon successful completion, the captured item disappears and a new one appears on the screen in a random position. The result of any attempt (successful or erroneous, taking into account the type of error) is recorded. The patient is offered M attempts in one session of the game. After the session is completed, the final results are generated, which counts the number of successes and mistakes made by the patient during the session, with each type of error taken into account separately. The results can be provided to a specialist doctor who, on their basis, can adjust the limitations (R, D, N, T and M) taking into account the characteristics of this patient for the next session.

It is easy to see that the developed game scenario satisfies the requirements formulated in Section 2: it adapts to the capabilities and needs of the patient, and at the same time, it contains variable parameters that are controls.

6. CONCLUSION

The article discusses the problem of rehabilitation of patients during illnesses associated with the loss of motor skills, such as multiple sclerosis. Effective in these cases are gaming approaches that increase patient motivation – so called "Serious Games".

Serious Games differ from regular games in a significantly smaller number of possible states, which makes them convenient for applying this approach. At the same time, in contrast to the previously considered applications of automaton models in medical systems, such as, for example, systems for supporting hemodialysis, in the case of a game, the parameters of each state can vary over a fairly wide range.

In the article, this problem is considered in the example of rehabilitation of pinch movements of hands by performing physical exercises in a specialized video game form using motion capture of fingers on a smartphone. The Abstract State Machine method is used to build the automaton model applicable to the Serious Game for pinch movement, which provides opportunities and motivation of the patient to do independent exercises. To build the resulting model, the method of successive refinements is used, on each of which verification is performed, which provides the necessary control over the accuracy of patient movements.

Thus, all the tasks formulated by the authors in the article are solved, namely:

- The model of hand and fingers movements is developed, that allowed us to formulate requirements and constraints for the automaton model.
- Using abstract state machine tools modeling the automaton was fulfilled that implements the identified requirements.
- Based on a developed automaton model an example of a scenario of a "serious game" for rehabilitation of motor skills restricted after multiple sclerosis is offered.

The developed approach seems quite general and can be extended to more complex and promising motion controllers, including those that support 3D movements, as well as to other types of hand movements. The authors consider such extensions as directions for their future work.

ACKNOWLEDGEMENT

This work was financially supported by Russian Science Foundation, Grant 19-19-00696.

REFERENCES

Bonfanti S., Gargantini A., Mashkoor A. A Systematic Literature Review of the Use of Formal Methods in Medical Software Systems. J. Softw. Evol. and Proc. 00:1–23 (2017).

Butler MJ, Schewe K, Mashkoor A, Bir´o M ((eds.)). Abstract State Machines, Alloy, B, TLA, VDM, and Z - 5th International Conference, ABZ 2016, Linz, Austria, May 23-27, 2016, Proceedings, Lecture Notes in Computer Science, vol. 9675, Springer, 2016, doi: 10.1007/978-3-319-33600-8. URL https://doi.org/10.1007/ 978-3-319-33600-8.

CaptoGlove applications. https://www.captoglove.com/features/, last access 2020/12/01

Haase R, Schultheiss T, Kempcke R, Thomas K, Ziemssen T. Use and Acceptance of Electronic Communication by Patients With Multiple Sclerosis: A Multicenter Questionnaire Study. J Med Internet Res 2012;14(5): e135

Halabchi F., Alizadeh Z., Sahraian M. A. & Abolhasani M. Exercise prescription for patients with multiple sclerosis; potential benefits and practical recommendations. BMC Neurology, V.17, Article number: 185 (2017)

Jonsdottir J., Bertoni R., Lawo M., Montesano A. Serious Games for Arm Rehabilitation of Persons with Multiple Sclerosis. A randomized controlled pilot study. Multiple Sclerosis and Related Disorders 19, 25–29 (2018).

Kapandji I.A. The Physiology of the Joints, Volume 1: Upper Limb. 6th ed. Churchill Livingstone (2007).

Khoussainov B., Nerode A. Automata Theory and its Applications. Progress in Computer Science and Applied Logic, Volume 21. Birkhauser, 2001.

Leap Motion Controller. https://www.leapmotion.com/technology/, last access 2020/12/01

Midaglia L., Mulero P., Montalban X. et al. Adherence and Satisfaction of Smartphone- and Smartwatch-Based Remote Active Testing and Passive Monitoring in People With Multiple Sclerosis: Nonrandomized Interventional Feasibility Study. J Med Internet Res.;21(8): e14863 (2019). DOI: 10.2196/14863.

NINDS Multiple Sclerosis Information Page. National Institute of Neurological Disorders and Stroke. 18 Oct 2019. http://www.ninds.nih.gov/disorders/multiple_sclerosis/multiple_sclerosis.htm, last access 12.01.2020

Nintendo Wii. https://www.nintendo.ru/Wii/Wii-94559.html, last access 2020/12/01

Polikarpova N. I., Shalyto A. A. Automata-based programming. SPb.: Piter. 2009 (rus)

Shalyto A.A. Software Automation Design: Algorithmization and Programming of Problems of Logical Control. Journal of Computer and Systems Sciences International. 2000. Vol.39. №6, p.899-916.

Sobrinho, Cunha P, Da Silva L, Perkusich A, Cordeiro T, R^ego J. A simulation approach to certify electrocardiography devices. International Conference on E-health Networking, Application Services (HealthCom), Boston, MA, USA, 2015; 86–90, doi: 10.1109/HealthCom.2015.7454478.

Thirumalai M, Rimmer JH, Johnson G, Wilroy J, Young HJ, Mehta T, Lai B. TEAMS (Tele-Exercise and Multiple Sclerosis), a Tailored Telerehabilitation mHealth App: Participant-Centered Development and Usability Study. JMIR Mhealth Uhealth. 6(5): e10181(2018). DOI: 10.2196/10181

Tolvanen JP, Djuki´c V, Popovic A. Metamodeling for medical devices: Code generation, model-debugging and run-time synchronization. Procedia Computer Science, vol. 63, 2015; 539–544, doi: 10.1016/j.procs.2015.08.382.

Xiaomi Redmi Note 8 Pro. https://www.mi.com/ru/redmi-note-8-pro/, last access 2020/12/01

Zhang Y, Jetley R, Jones PL, Ray A. Generic Safety Requirements for Developing Safe Insulin Pump Software. Journal of Diabetes Science and Technology Nov 2011; 5(6):1403–1419. URL http://www.ncbi.nlm.nih.gov/pmc/articles/PMC3262707/, 00008.

AN EFFICIENT DATA IMPUTATION TECHNIQUE FOR HUMAN ACTIVITY RECOGNITION

Ivan Miguel Pires[1, 2], Faisal Hussain[3], Nuno M. Garcia[1] and Eftim Zdravevski[4]
[1]*Instituto de Telecomunicações, Universidade da Beira Interior, Covilhã, Portugal*
[2]*Department of Computer Science, Polytechnic Institute of Viseu, Viseu, Portugal*
[3]*Department of Computer Engineering, University of Engineering and Technology (UET), Taxila, Pakistan*
[4]*Faculty of Computer Science and Engineering, University Ss Cyril and Methodius, Skopje, North Macedonia*

ABSTRACT

The tremendous applications of human activity recognition are surging its span from health monitoring systems to virtual reality applications. Thus, the automatic recognition of daily life activities has become significant for numerous applications. In recent years, many datasets have been proposed to train the machine learning models for efficient monitoring and recognition of human daily living activities. However, the performance of machine learning models in activity recognition is crucially affected when there are incomplete activities in a dataset, i.e., having missing samples in dataset captures. Therefore, in this work, we propose a methodology for extrapolating the missing samples of a dataset to better recognize the human daily living activities. The proposed method efficiently pre-processes the data captures and utilizes the k-Nearest Neighbors (KNN) imputation technique to extrapolate the missing samples in dataset captures. The proposed methodology elegantly extrapolated a similar pattern of activities as they were in the real dataset.

KEYWORDS

Daily Activities, Data Imputation, Sensors, Mobile Devices, Missing Data

1. INTRODUCTION

The recent advancements of the technology (Cardinale and Varley, 2017; Kim et al., 2016; Ni Scanaill et al., 2011; Patel et al., 2012; Sousa et al., 2015) and the presence of sensors in the commonly used off-the-shelf mobile devices (Shahriyar et al., 2010; Stankevich et al., 2012; Steele, 2011; Tian et al., 2019; Ventola, 2014) allow the development of solutions for the identification of the human daily living activities in order to monitor its lifestyles, *e.g.*, the creation of a Personal Digital Life Coach (Garcia, 2016). However, the accuracy of these systems and their resilience of fails is essential for the recognition of activities in different environments (Dimitrievski et al., 2016a; Pires et al., 2017, 2018a; Zdravevski et al., 2015). In general, sportspeople, older adults, and other persons with special needs are living in conditions with bad network connection, but the development of solutions for this type of people is vital to improve their quality of life (Dimitrievski et al., 2016b; Sendra et al., 2012; Seneviratne et al., 2017).

There are different types of daily activities, and most of these are detected with inertial sensors, e.g., walking, running, moving upstairs, moving downstairs, and standing (Ferreira et al., 2020; Pires et al., 2018b, 2018c, 2018d, 2019, 2020). These are simple activities with different types of motion that can be distinguished with low-cost sensors. The combination of artificial intelligence methods and the data acquired sensors available in the off-the-shelf mobile devices may empower the monitoring of older adults, and the development of intelligent solutions for sports and medicine (Costa et al., 2015; E. Zdravevski et al., 2017; Kumar and Venkatesan, 2014).

Currently, the devices used for the data acquisition may fail due to the memory, battery, and power processing constraints (Pires et al., 2017, 2018a) which cause missing samples/values while data acquisition. Thus, the activities are wrongly recognized. However, the correct recognition of activities is essential to develop solutions that support various types of people in different daily activities. The implementation of data imputation minimizes these constraints measuring the data in faults. As the proposal uses three sensors (*i.e.,* accelerometer, magnetometer, and gyroscope), they have the same number of records and the same frequency of data acquisition to be fused and comparable. Before this study, we used different statistical features for the identification of activities and environments. Still, the data imputation allows the implementation of different techniques for the recognition of the activities, which may improve the accuracy of the recognition performed by our framework (Ferreira et al., 2020; Pires et al., 2018b, 2018d, 2019, 2020).

This paper's primary motivation is to increase the reliability of the previously proposed framework for the recognition of daily activities (Ferreira et al., 2020; Pires et al., 2020, 2019, 2018b, 2018d) with the imputation of the missing data to fill the data and correctly identify the daily activities.

This paper proposes implementing the k-Nearest Neighbors (KNN) imputation algorithm for the estimation of the values of the different datasets to fulfill the number of outputs correctly. The inertial sensors, including accelerometer, magnetometer, and gyroscope sensors, have three axes (*i.e.,* x, y, and z), where the measurement should be related to the different axis. The proposed method first identifies the missing data, then performs data segmentation, and, finally, imputes the missing data.

In the literature, the KNN imputation is one of the most used methods to support the recognition of activities and environments. With our implementation, it is verified that the pattern of the imputed data is similar but with higher amplitude than the original data.

This paper is organized as follows: Section 2 presents the methodology, including the definition of data acquisition, identification of missing samples, and implementation of data segmentation and imputation algorithms. The discussion and results are presented in Section 3. Finally, Section 4 presents the conclusions of the study.

2. METHODOLOGY

Figure 1 shows the flow diagram of the proposed methodology to extrapolate the missing samples in a human activity recognition dataset. The proposed method consists of four major stages, *i.e.,* data acquisition (section 2.1), missing samples identification (section 2.2), data segmentation (section 2.3), and data imputation (section 2.4). These stages are explained in the subsequent sections.

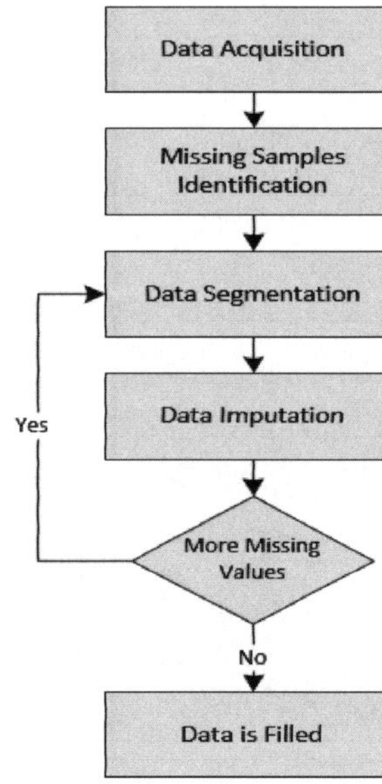

Figure 1. The workflow of the data imputation scheme

2.1 Data Acquisition

Data acquisition is the premier stage for performing the proceeding steps. In this stage, we first acquire the dataset for which we have to extrapolate the missing samples. Thus, we used a publicly available dataset (Pires, 2018). The dataset (Pires, 2018) includes five daily living activities, i.e., walking, running, standing, moving upstairs, and moving downstairs. The dataset (Pires, 2018) is acquired using three motion sensors, *i.e.,* accelerometer, gyroscope, and magnetometer. The five daily living activities included in the dataset (Pires, 2018) are performed by 25 subjects age ranging from (20-60) with sedentary and active lifestyles.

This dataset contains several captures with 5 seconds of sensors' data acquired every 5 minutes. While having a smartphone in the pocket, each activity was recorded by using an Android application. The data was obtained by hundreds of hours to implement methods for the identification of the activities. Due to the memory, battery, and power processing constraints, sometimes the data acquisition may fail (Pires et al., 2017, 2018a), and the imputation of the mission data is needed.

2.2 Missing Samples Identification

Once the dataset (Pires, 2018) is acquired, the next step is to check if there are some missing samples in each recorded activity. The missing samples in a dataset have a lousy impact while training the machine learning (ML) models. It occurs because due to missing samples, the ML models do not learn the activity patterns properly. The missing samples in the dataset are mainly due to two reasons, *i.e.,* the user did not perform an activity for a complete defined activity duration, or there could have some issue in the device that is being used for recording the activity.

To identify the missing samples in the dataset, we first analyzed each activity time duration and frequency rate. In our case, each activity was performed for a time duration of 5 seconds. The frequency rate for accelerometer and gyroscope was 100 Hz (*i.e.,* 100 samples/s), while for magnetometer, it was 10 Hz (*i.e.,* 10 samples/s). So, for a 5sec activity, there should be 5 x 100 = 500 samples for the accelerometer. On the other hand, there should be 5 x 10 = 50 samples for the gyroscope in an activity of 5 seconds.

Once we figured out the number of samples for each activity across the three sensors, we then write a python script to identify the missing samples in each activity to perform further steps. While recognizing the missing samples in an activity, we ignored the activities having missing samples for duration more than 1sec, i.e., ignored the captures that have more than 100 missing samples in case of accelerometer or more than 10 missing samples in case of gyroscope. It is done to be closer to the originality of the data than filling all synthetic samples to fulfill the space of missing samples.

2.3 Data Segmentation

After identifying the missing samples, we first inserted Null values to fulfill the space of missing samples. Then, we segmented the samples based upon the missing samples count. If the missing samples count was greater than 10, we segmented data into a chunk of 100 samples to include 90 original samples and the first 10 samples with Null values. If the missing samples count is less than or equal to 10 samples, we made a chunk of the last 100 samples, including the recently inserted and the Null value samples and original samples.

2.4 Data Imputation

Once the data is segmented, we then applied the KNN imputation algorithm (Beretta and Santaniello, 2016) to extrapolate the missing values. The KNN impuation technique is based upon the KNN algorithm. In KNN imputation, we firstly find k-closest neighbors to the missing data and then impute these missing values based upon known k-closest neighbors.

We noticed the count of missing samples every time, before applying the KNN imputation. If the missing samples were less than 10 samples, then the missing data is filled in the first iteration. However, if the missing samples are more than 10, we need to perform segmentation steps again to make another chunk of data and apply the KNN imputer algorithm still to extrapolate the missing values. As shown in Figure 1, this process continues until all the missing samples are extrapolated.

3. RESULTS AND DISCUSSION

The proposed data imputation technique is applied to a publicly available dataset (Pires, 2018) of human activity recognition. Based on the time of 5 seconds (500 samples per file in case of accelerometer and gyroscope and 50 samples per file in case of magnetometer), the dataset had many missing samples. Table 1 shows the missing samples in dataset (Pires, 2018) for the accelerometer data across each activity. In Table 1, the second column shows the total number of records available in the dataset (Pires, 2018) across each activity. The proceeding columns shows the number of missing samples in the dataset (Pires, 2018) for each corresponding activity. It can be verified that the most missing samples are related to moving upstairs activity.

Table 1. Accelerometer Missing Samples Statistics in Dataset

Activity	Total Records	Records having missing samples		
		<=10	>10 & <=100	>100
Walking	783	317	327	139
Running	487	403	72	12
Standing	399	360	37	2
Moving upstairs	707	173	115	419
Moving downstairs	364	111	75	178

To cope with this challenge, we proposed an imputation technique for extrapolating the missing samples by following the steps as described in the methodology section. Figure 2 presents an excerpt of accelerometer data for moving downstairs activity, taken from the dataset. The selected excerpt had 50 missing samples. We first applied the proposed methodology to identify the missing samples to validate the actual missing count. The proposed method accurately identified the correct missing samples and found 50 missing samples in the given excerpt.

After discovering that 50 samples are missing in the given excerpt, we then filled the missing samples by NULL values, as shown in Figure 3, for further implementation of the KNN imputation algorithm. Afterwards, we performed data segmentation, as explained in Section 2.3. Figure 4 shows how the data is segmented for the given excerpt.

Finally, we applied the KNN imputation algorithm to extrapolate the unknown samples, as explained in the previous sections. Figure 5 shows the missing calculated values extrapolated after using the KNN imputation algorithm.

Sr.#	Timestamp	Ax	Ay	Az
1	1493996698893	-2.145	-9.174	3.802
2	1493996698902	-0.612	-9.625	3.984
3	1493996698914	-0.641	-10.678	3.84
		- - -		
448	1493996702662	-0.641	-8.533	3.84
449	1493996702663	-0.632	-8.399	3.61
450	1493996702672	-0.526	-8.322	3.39

Figure 2. Excerpt of accelerometer data for walking activity

450	1493996702672	-0.526	-8.322	3.39
451	1493996702682	Null	Null	Null
452	1493996702692	Null	Null	Null
		- - -		
499	1493996703162	Null	Null	Null
500	1493996703172	Null	Null	Null

Figure 3. Missing samples filled as NULL values

441	1493996702582	-0.201	-8.169	3.352
442	1493996702593	-0.392	-8.102	3.208
443	1493996702604	-0.507	-8.207	3.112
		- - -		
449	1493996702663	-0.632	-8.399	3.61
450	1493996702672	-0.526	-8.322	3.39
451	1493996702682	Null	Null	Null
452	1493996702692	Null	Null	Null

Figure 4. Data segmentation before applying the KNN imputation technique

449	1493996702663	-0.632	-8.399	3.61
450	1493996702672	-0.526	-8.322	3.39
451	1493996702682	-0.329	-9.866	-1.05
452	1493996702692	-0.332	-9.197	2.405

Figure 5. Missing samples filled after applying the KNN imputation technique

This technique is implemented for the files that have less than 100 missing samples in case of accelerometer and gyroscope and 10 missing samples in case of magnetometer. If more than 100 samples are missed, this capture should be discarded. In short, we ignored the activities having missing samples for duration more than 1sec, i.e., ignored the captures that have more than 100 missing samples in case of accelerometer or more than 10 missing samples in case of gyroscope. It is done to be closer to the originality of the data than filling all synthetic samples to fulfill the space of missing samples as explained in (section 2.2).

Figure 6 shows an example of an activity recorded while moving downstairs, where it is possible to verify that the file has 375 samples. As the file must have 500 samples, 125 samples are missing. By applying KNN Imputation methods, we found data with the same format, but it has more high amplitude. Thus, the peaks for the downstairs activity are similar, as presented in Figure 7.

For the future work, the data classification should be implemented with different machine learning methods, where we can now use different variables of previous works, *e.g.,* the mean, standard deviation, variation, and median of the values acquired and imputed from each axis. Thus, we can discover different patterns to improve the development of a method for the recognition of daily activities and environments.

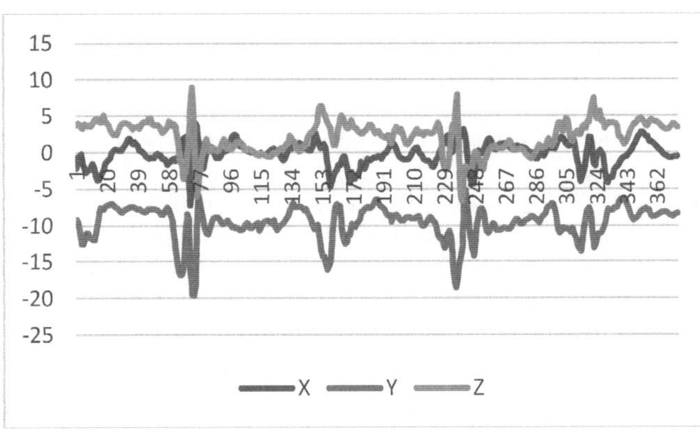

Figure 6. Accelerometer sample related to moving downstairs

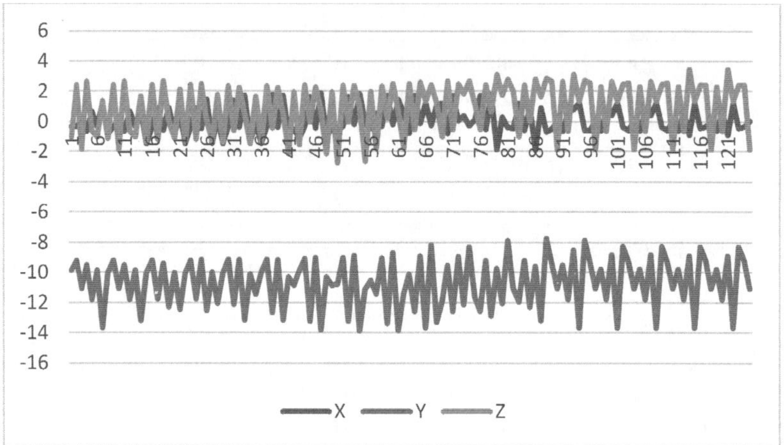

Figure 7. Imputed data for the accelerometer sample related to moving downstairs

4. CONCLUSION

The missing samples in a dataset capture affect the performance of machine learning models. Therefore, in this work, we proposed a methodology to extrapolate the missing samples of every dataset capture related to motion sensors data recorded for human activity recognition. The proposed method first identifies the missing samples in each excerpt with respect to time duration and frequency of the dataset. Secondly, it inserts Null values to fil the gap of missing samples. Thirdly, it performs the segmentation of the data, and finally, it applies the KNN imputation technique to extrapolate the missing samples. The proposed methodology extrapolated the missing samples with the same pattern as found in the original dataset but with some high amplitude.

As future work, we aim to improve the data amplitude to make it closer to the reality and train different machine learning models to better detect human activity patterns.

ACKNOWLEDGEMENT

This work is funded by FCT/MEC through national funds and when applicable co-funded by FEDER-PT2020 partnership agreement under the project **UIDB/EEA/50008/2020**. (*Este trabalho é financiado pela FCT/MEC através de fundos nacionais e cofinanciado pelo FEDER, no âmbito do Acordo de Parceria PT2020 no âmbito do projeto UIDB/EEA/50008/2020*).

This article is based upon work from COST Action IC1303-AAPELE—Architectures, Algorithms and Protocols for Enhanced Living Environments and COST Action CA16226–SHELD-ON—Indoor living space improvement: Smart Habitat for the Elderly, supported by COST (European Cooperation in Science and Technology). COST is a funding agency for research and innovation networks. Our Actions help connect research initiatives across Europe and enable scientists to grow their ideas by sharing them with their peers. It boosts their research, career and innovation. More information in www.cost.eu.

REFERENCES

Beretta, L., Santaniello, A., 2016. Nearest neighbor imputation algorithms: a critical evaluation. BMC Med Inform Decis Mak 16. https://doi.org/10.1186/s12911-016-0318-z

Cardinale, M., Varley, M.C., 2017. Wearable Training-Monitoring Technology: Applications, Challenges, and Opportunities. International Journal of Sports Physiology and Performance 12, S2-S2-62. https://doi.org/10.1123/ijspp.2016-0423

Costa, S.E.P., Rodrigues, J.J.P.C., Silva, B.M.C., Isento, J.N., Corchado, J.M., 2015. Integration of Wearable Solutions in AAL Environments with Mobility Support. J Med Syst 39, 184. https://doi.org/10.1007/s10916-015-0342-z

Dimitrievski, A., Zdravevski, E., Lameski, P., Trajkovik, V., 2016a. Towards application of non-invasive environmental sensors for risks and activity detection, in: 2016 IEEE 12th International Conference on Intelligent Computer Communication and Processing (ICCP). Presented at the 2016 IEEE 12th International Conference on Intelligent Computer Communication and Processing (ICCP), pp. 27–33. https://doi.org/10.1109/ICCP.2016.7737117

Dimitrievski, A., Zdravevski, E., Lameski, P., Trajkovik, V., 2016b. A survey of Ambient Assisted Living systems: Challenges and opportunities, in: 2016 IEEE 12th International Conference on Intelligent Computer Communication and Processing (ICCP). Presented at the 2016 IEEE 12th International Conference on Intelligent Computer Communication and Processing (ICCP), pp. 49–53. https://doi.org/10.1109/ICCP.2016.7737121

E. Zdravevski, P. Lameski, V. Trajkovik, A. Kulakov, I. Chorbev, R. Goleva, N. Pombo, N. Garcia, 2017. Improving Activity Recognition Accuracy in Ambient-Assisted Living Systems by Automated Feature Engineering. IEEE Access 5, 5262–5280. https://doi.org/10.1109/ACCESS.2017.2684913

Ferreira, J.M., Pires, I.M., Marques, G., García, N.M., Zdravevski, E., Lameski, P., Flórez-Revuelta, F., Spinsante, S., Xu, L., 2020. Activities of Daily Living and Environment Recognition Using Mobile Devices: A Comparative Study. Electronics 9, 180. https://doi.org/10.3390/electronics9010180

Garcia, N.M., 2016. A Roadmap to the Design of a Personal Digital Life Coach, in: Loshkovska, S., Koceski, S. (Eds.), ICT Innovations 2015, Advances in Intelligent Systems and Computing. Springer International Publishing, Cham, pp. 21–27. https://doi.org/10.1007/978-3-319-25733-4_3

Kim, S., Gajos, K.Z., Muller, M., Grosz, B.J., 2016. Acceptance of mobile technology by older adults: a preliminary study, in: Proceedings of the 18th International Conference on Human-Computer Interaction with Mobile Devices and Services. Presented at the MobileHCI '16: 18th International Conference on Human-Computer Interaction with Mobile Devices and Services, ACM, Florence Italy, pp. 147–157. https://doi.org/10.1145/2935334.2935380

Kumar, K.M., Venkatesan, R.S., 2014. A design approach to smart health monitoring using android mobile devices, in: 2014 IEEE International Conference on Advanced Communications, Control and Computing Technologies. Presented at the 2014 IEEE International Conference on Advanced Communications, Control and Computing Technologies, pp. 1740–1744. https://doi.org/10.1109/ICACCCT.2014.7019406

Ni Scanaill, C., Garattini, C., Greene, B.R., McGrath, M.J., 2011. Technology Innovation Enabling Falls Risk Assessment in a Community Setting. Ageing Int 36, 217–231. https://doi.org/10.1007/s12126-010-9087-7

Patel, S., Park, H., Bonato, P., Chan, L., Rodgers, M., 2012. A review of wearable sensors and systems with application in rehabilitation. Journal of NeuroEngineering and Rehabilitation 9, 21. https://doi.org/10.1186/1743-0003-9-21

Pires, I., Felizardo, V., Pombo, N., Garcia, N.M., 2017. Limitations of energy expenditure calculation based on a mobile phone accelerometer, in: 2017 International Conference on High Performance Computing & Simulation (HPCS). IEEE, pp. 124–127.

Pires, I.M., 2018. impires/August_2017-_Multi-sensor_data_fusion_in_mobile_devices_for_the_identification_of_activities_of_dail [WWW Document]. GitHub. URL https://github.com/impires/August_2017-_Multi-sensor_data_fusion_in_mobile_devices_for_the_identification_of_activities_of_dail (accessed 5.8.20).

Pires, I.M., Garcia, N.M., Pombo, N., Flórez-Revuelta, F., 2018a. Limitations of the Use of Mobile Devices and Smart Environments for the Monitoring of Ageing People, in: HSP 2018.

Pires, I.M., Garcia, N.M., Pombo, N., Flórez-Revuelta, F., 2018b. Framework for the Recognition of Activities of Daily Living and their Environments in the Development of a Personal Digital Life Coach.

Pires, I.M., Garcia, N.M., Pombo, N., Flórez-Revuelta, F., Spinsante, S., Teixeira, M.C., 2018c. Identification of activities of daily living through data fusion on motion and magnetic sensors embedded on mobile devices. Pervasive and Mobile Computing 47, 78–93.

Pires, I.M., Marques, G., Garcia, N.M., Flórez-Revuelta, F., Canavarro Teixeira, M., Zdravevski, E., Spinsante, S., Coimbra, M., 2020. Pattern Recognition Techniques for the Identification of Activities of Daily Living Using a Mobile Device Accelerometer. Electronics 9, 509. https://doi.org/10.3390/electronics9030509

Pires, I.M., Marques, G., Garcia, N.M., Pombo, N., Flórez-Revuelta, F., Spinsante, S., Teixeira, M.C., Zdravevski, E., 2019. Recognition of Activities of Daily Living and Environments Using Acoustic Sensors Embedded on Mobile Devices. Electronics 8, 1499.

Pires, I.M., Teixeira, M.C., Pombo, N., Garcia, N.M., Flórez-Revuelta, F., Spinsante, S., Goleva, R., Zdravevski, E., others, 2018d. Android Library for Recognition of Activities of Daily Living: Implementation Considerations, Challenges, and Solutions.

Sendra, S., Granell, E., Lloret, J., Rodrigues, J.J.P.C., 2012. Smart collaborative system using the sensors of mobile devices for monitoring disabled and elderly people, in: 2012 IEEE International Conference on Communications (ICC). Presented at the 2012 IEEE International Conference on Communications (ICC), pp. 6479–6483. https://doi.org/10.1109/ICC.2012.6364935

Seneviratne, S., Hu, Y., Nguyen, T., Lan, G., Khalifa, S., Thilakarathna, K., Hassan, M., Seneviratne, A., 2017. A Survey of Wearable Devices and Challenges. IEEE Communications Surveys Tutorials 19, 2573–2620. https://doi.org/10.1109/COMST.2017.2731979

Shahriyar, R., Bari, Md.F., Kundu, G., Ahamed, S.I., Akbar, Md.M., 2010. Intelligent Mobile Health Monitoring System (IMHMS), in: Kostkova, P. (Ed.), Electronic Healthcare, Lecture Notes of the Institute for Computer Sciences, Social Informatics and Telecommunications Engineering. Springer Berlin Heidelberg, Berlin, Heidelberg, pp. 5–12. https://doi.org/10.1007/978-3-642-11745-9_2

Sousa, P.S., Sabugueiro, D., Felizardo, V., Couto, R., Pires, I., Garcia, N.M., 2015. mHealth Sensors and Applications for Personal Aid, in: Adibi, S. (Ed.), Mobile Health, Springer Series in Bio-/Neuroinformatics. Springer International Publishing, Cham, pp. 265–281. https://doi.org/10.1007/978-3-319-12817-7_12

Stankevich, E., Paramonov, I., Timofeev, I., 2012. Mobile phone sensors in health applications, in: 2012 12th Conference of Open Innovations Association (FRUCT). Presented at the 2012 12th Conference of Open Innovations Association (FRUCT) and Seminar on e-Travel, IEEE, Oulu, pp. 1–6. https://doi.org/10.23919/FRUCT.2012.8122097

Steele, R., 2011. Social media, mobile devices and sensors: Categorizing new techniques for health communication, in: 2011 Fifth International Conference on Sensing Technology. Presented at the 2011 Fifth International Conference on Sensing Technology, pp. 187–192. https://doi.org/10.1109/ICSensT.2011.6136960

Tian, S., Yang, W., Grange, J.M.L., Wang, P., Huang, W., Ye, Z., 2019. Smart healthcare: making medical care more intelligent. Global Health Journal 3, 62–65. https://doi.org/10.1016/j.glohj.2019.07.001

Ventola, C.L., 2014. Mobile Devices and Apps for Health Care Professionals: Uses and Benefits. P T 39, 356–364.

Zdravevski, E., Lameski, P., Kulakov, A., Kalajdziski, S., 2015. Transformation of nominal features into numeric in supervised multi-class problems based on the weight of evidence parameter, in: 2015 Federated Conference on Computer Science and Information Systems (FedCSIS). Presented at the 2015 Federated Conference on Computer Science and Information Systems (FedCSIS), pp. 169–179. https://doi.org/10.15439/2015F90

ELECTRONIC REGISTRY SYSTEM FOR DIABETIC PATIENTS UNDERGOING HYPERBARIC OXYGEN THERAPY: DETERMINATION OF COMMON DATA ELEMENTS

Shahram Oliaei[a], Nahid Mehrabi[b], Tayebeh Noori[c], Solmaz Saeidi[d] and Esmaeil Mehraeen[e, 1]

[a]HBOT Research Center, Golestan Hospital, Islamic Republic of Iran, Navy and AJA Medical University, Tehran, Iran
[b]Assistant Professor of Health Information Management, Aja University of Medical Sciences, Tehran, Iran
[c]Department of Health Information Technology, School of Allied Medical Sciences, Zabol University of Medical Sciences, Zabol, Iran
[d]Department of Nursing, Khalkhal University of Medical Sciences; Khalkhal, Iran
[e]Department of Health Information Technologies, Khalkhal University of Medical Sciences, Khalkhal, Iran

ABSTRACT

Background: The advent of electronic registries has increasingly provided an opportunity to improve healthcare services and advanced applied research. However, the delivery of effective hyperbaric technology services requires well-functioning electronic registry systems and patient information integration to support national and global analyses.
Objectives: this study aimed to determine the most common data elements of an electronic registry for diabetic patients undergoing hyperbaric oxygen therapy (HBOT).
Methods: This was a cross-sectional study that conducted in two main stages. The first phase involves a review of the current available evidence on the study aim. In this stage, we searched related articles in PubMed, Web of Science, Up To Date, Science Direct, and Scopus databases. In the next phase, using the researcher-made questionnaire, the essential data elements and technical features of an electronic registry system were validated. The statistical population of the present study consisted of 32 endocrinologists working in Golestan Hospital. The collected data was analyzed using descriptive statistics in SPSS version 26.
Results: Through a full-text review of 12 related articles, the data elements (N=95) of an electronic registry for diabetic patients undergoing HBOT were identified and classified into three categories including: administrative (n=30), clinical (n=54), and technical (n=11). According to findings, among the 30 data elements of administrative category, 25 data elements and in clinical category, 50 data elements were selected by statistical population. Moreover, all identified technical features were selected by experts, except the "decision making support".
Conclusions: In this study we identified and validated the data elements and technical features of an electronic registry system for HBOT-treated diabetic patients. These requirements make it possible to standardize data collection and improve health care services.

KEYWORDS

Data Elements, Diabetes, Registry System, Hyperbaric Oxygen Therapy, Diving Medicine

1. INTRODUCTION

Diabetes mellitus is one of the culprits of public health problems that have led to considerably widespread economic and social consequences worldwide due to its high prevalence (Irawan et al., 2018). Complications can include kidney failure, diabetic foot ulcers, and even lower limb amputation. Oxygen plays an important role in wound healing. However, the main components of standard treatment are glycemic control, antibiotic medication, wound debridement (depletion of dead tissues), wound care, offloading and improvement of blood flow / revascularization (Saputri et al., 2018).

[1] Corresponding Author: Esmaeil Mehraeen, Assistant Professor, Department of Health Information Technology, Khalkhal University of Medical Sciences, Khalkhal, Iran. Postal code:5681761351, tel:+98-4532426801, fax: +98-4532422305,
Email: es.mehraeen@gmail.com

Hyperbaric Oxygen Therapy (HBOT) causes 100% breathing of oxygen under increasing atmospheric pressure conditions. HBOT involves breathing of pure oxygen into a room or enclosure that is used in a variety of medical settings (Hunt, 1972). During high pressure oxygen therapy, the blood hemoglobin in the vein is completely saturated with oxygen and results in increased oxygen pressure in the blood vessels (Hunt, 1972, Ruthenborg et al., 2014). HBOT treats several medical conditions such as severe infections, air bubbles in the blood vessels, and wounds caused by diabetes. Standard care for diabetes includes condition control; diabetes wound care and infection treatment. Diabetic foot is the most commonly diagnosed condition for HBOT control in diabetics. In addition to reducing insulin requirement in diabetic patients, other factors such as decreased oxygenation (hypoxia), edema, and HBOT-induced infection are reduced (Fife and Eckert, 2018).

Hyperbaric specialists are widely monitored for compliance with health insurance coverage policies and adherence to clinical practice guidelines. The best way to achieve this is registering clinical information; clinical data registration is nowadays performed using computerized registry systems (Mathieu et al., 2017). Registry systems are currently available for a wide range of clinical conditions such as infectious diseases, cancers, congenital diseases and rare diseases such as cystic fibrosis. Chronic diseases such as AIDS, diabetes, heart failure and kidney disease have been the target of registry systems in many countries in the final stages (Al-Rubeaan et al., 2013). A survey in 1040 physician organizations showed that diabetes registries were used in 40.3%, asthma registries in 31.2%, congestive heart failure registries in 34.8%, and depression registrations in 15.7% of health care organizations (Casalino et al., 2003).

In recent years, the creation of registry systems in this area is essential for data management due to the increase in oxygen hyperbaric services (Andersen et al., 2016, Noori et al., 2019). Patients undergoing HBOT treatment should have their information recorded in a system known as an electronic data recording system (such as a HBOT-treated wound information system). The data recorded in these systems may be used for benchmarking, service quality reporting, health policy or specialized data projects. Data may be requested for criteria, quality reporting, public policies or specialized data projects. The Wound Information Record System can provide benchmarking information for many wound care interventions and blood glucose medications by using the Health Record documentation for each HBOT-treated patient. This will provide the infrastructure needed to implement similar technologies such as Electronic Health Record (EHR) and distant patient information access (Fife and Eckert, 2018).

One of the key components of health registry systems is availability of standard data elements for proper collection and timely access to health information (Andersen et al., 2016). Therefore, gathering demographic and clinical data using standard data elements can be useful in providing good health care to patients with hyperbaric diabetes (Hosseini et al., 2010). In other words, standard data elements will avail detailed data to service providers. Standard and accurate data require access to valid patient information, infections, new therapies, and their impact (Ahmadi et al., 2014).

An electronic registry with standard data elements is significantly needed to enable accurate data collection in developing countries as HBOT progresses (Fife et al., 2016). There are few studies on the data elements of the registry system for HBOT-treated diabetic patients. As a result, unclear and incomprehensible data are used in HBOT-based care services (George et al., 2017). Hence, if we lack standard data elements to collect and access accurate and valid data, it may be impossible to pot for an empirical approach to improve the delivery of health services, monitoring and ongoing evaluation of diabetic patients at HBOT centers. This study aims to determine common data elements of a registry system for diabetic patients treated with HBOT.

2. MATERIALS AND METHODS

This was a cross-sectional study that conducted in two main stages. The first phase involves a review of the current available evidence to determine the data elements of an electronic registry system for diabetic patients undergoing hyperbaric oxygen therapy. In this stage, we searched related articles in PubMed, Web of Science, Up To Date, Science Direct, and Scopus databases. Hybrid keywords including hyperbaric oxygen therapy, HBOT, electronic registry, e-registry, registry system, information system, database, diabetes, technical functionality/features, data element and minimum data set were used to come up with the search strategy.

The inclusion criteria included full text of English-written original or review papers which had been published from 2010 to 2020. Abstracts, reports, letters to the editor, and news articles were excluded from the study. First, the title and abstract of articles were evaluated according to inclusion and exclusion criteria. The full text of selected articles was then studied for further evaluation of final selection.

To evaluate the quality of the final selected articles, a checklist containing 15 items (Table 1) was developed based on the related studies (BinDhim et al., 2015, Zapata et al., 2015, McKay et al., 2018). The quality of articles was then rated on three scales: low quality (0-5), medium quality (6-10) and high quality (11-15).

Table 1. Quality assessment checklist

NO.	*Question*
1	Does the study address any research question(s) or objective(s)?
2	Does the study provide any theoretical framework for the evaluation method?
3	Does the theoretical framework of the study include any health promotion theory?
4	Does the study provide a timeframe of the data collection?
5	Does the study identify the country where the search was conducted?
6	Does the study mention that the reviewed registry system (RS) was downloaded for evaluation?
7	Does the study discuss the selection criteria for HBOT to be included or excluded for review?
8	Does the study provide a clear description of the evaluation method?
9	Are there at least two independent data extractors with a consensus procedure in place in case of disagreement?
10	Is a list of the reviews HBOT registry system provided?
11	Does the study discuss the findings of the evaluation?
12	Does the study look at the reviewed HBOT registry system to promote healthcare services?
13	Does the study discuss any limitations?
14	Does the study provide any future recommendations in general?
15	Does the study state any conflict of interest?

In the second phase, a questionnaire containing the identified data elements was designed. The questionnaire was completed within ten days first by four health information management specialists; then, four endocrinologists examined it to ensure its reliability. The collected data were analyzed by SPSS software version 26, and the reliability of the questionnaire was calculated through Cronbach's alpha (=0.87). Eight experts evaluated the content validity of the questionnaire. In order to increase the number of participants in the study, in-person referral method was used.

Finally, a researcher-made questionnaire was administered to the statistical population. The statistical population of the present study consisted of 32 endocrinologists working in Golestan Hospital (The only center equipped with HBOT services in Iran). To analyze the data, the questionnaire items ranged from 1 to 3 (1 for disagreement, 2 for no idea and 3 for agreement). The collected data were analyzed through SPSS software using descriptive statistics (frequency distribution and mean reports). Each data element with a mean score of at least 1.5 and more was selected as the data element of the electronic registry system for diabetic patients treated with HBOT.

3. RESULTS

Using the aforementioned strategies, 980 sources were retrieved.

3.1 General Specifications

After reviewing the titles and abstracts of the articles, 370 duplicates were found. The 610 titles of the remaining articles were reviewed and then 91 studies with the main selection criteria were extracted. After reviewing the full text of the articles, 12 articles met the study selection criteria and were included in the final review (Figure 1). These 12 studies were published between 2011 and 2019. The mean quality score of the selected articles reviewed was 13 (in the range of 11 to 15) indicating high quality of the articles.

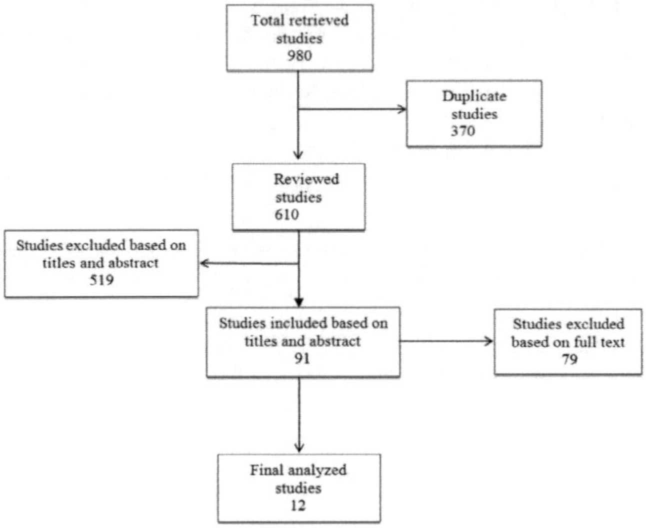

Figure 1. Flow diagram of selection process

3.2 Evaluating Classifications

In this study, a set of data elements and features (N=95) of the electronic registry system for HBOT diabetic patients were identified in three categories: administrative data elements, clinical data elements and technical features. Table 2 shows each of the classifications associated with the reviewed articles. Based on the findings, more than half of the studies (seven studies) described clinical data elements and five studies investigated the technical features of the registry system related to diabetic patients who had been treated with HBOT.

Table 2. Identified categories of data elements and features and final surveyed articles

Reference	*Data Elements/Capabilities*		
	Administrative	Clinical	Technical
(Sundh et al., 2019)			√
(Khorasani et al., 2019)	√		
(Kawecki et al., 2018)			√
(Fife and Eckert, 2018)		√	
(Irawan et al., 2018)	√	√	
(Lakshminarayanan et al., 2017)	√	√	
(George et al., 2017)		√	
(Fife et al., 2016)			√
(Andersen et al., 2016)	√	√	
(Stoekenbroek et al., 2014)		√	√
(Al-Rubeaan et al., 2013)			√
(Richesson, 2011)		√	

3.3 Sub-Categories

According to the findings, the administrative category consisted of five subcategories and 30 data elements, the clinical category consisted of six subcategories and 54 data elements, and the technical category consisted of one subcategory and 11 features. The identified data elements and features were scored by 30 samples of attended specialists (response rate was 93.8%). The results showed that among the 30 data elements of administrative category, 25 data elements and in clinical category, 50 data elements were selected by statistical population (Table 3 and 4). Moreover, all technical features except "decision making support" were selected by the experts (Table 5).

Table 3. The mean of given scores and selected administrative data elements

Category	*Subcategory*	*No*	*Data element*	*Mean*	*√/✶*
Administrative data element	Demographic	1	Patient medical record number	3	√
		2	National identity number	3	√
		3	Patient name	3	√
		4	Patient surname	3	√
		5	Father name	2.8	√
		6	Sex	2.9	√
		7	Date of birth	3	√
		8	Country of birth	2.9	√
		9	Province of birth	2.5	√
		10	City of birth	2.9	√
		11	Occupation	2.1	√
		12	Marital status	1.3	✶
		13	Education level	2	√
		14	Economic status	1	✶
		15	Patient's address	2.5	√
		16	Phone/mobile number	2.4	√
	Patient encounter	17	Date of admission or referral to the facility	3	√
		18	Time of admission or referral to the facility	2.1	√
		19	Name of institution or referral center	2.9	√
		20	Name of referrer person/ physician	2.5	√
		21	Informed consent	3	√
	Provider ID	22	Name of the service provider person	1.1	✶
		23	Family name of the service provider person	1.2	√
		24	Identifier code of the service provider person	2	√
	Facility identifier	25	Facility Name	2.3	√
		26	Identification code of the facility	2.1	√
		27	Full facility address	1.4	✶
		28	Telephone of facility	1.3	✶
	Financial data	29	Primary insurer company	2	√
		30	Insurance number	3	√

√=Selected data element; ✶= Removed data element

Table 4. The mean of given scores and selected Clinical data elements

Category	*Subcategory*	*N*	*Data element*	*Average*	*√/✶*
Clinical data element	Patient CC	1	Chief Complaint of patient/referral	2.7	√
		2	Current condition/disease sign	3	√
	Personal medical history	3	Diabetes type	3	√
		4	Diabetes duration	2.5	√
		5	Diabetic retinopathy	3	√
		6	Family history of diabetes	2.9	√
		7	Chronic Disease and Comorbidity	2.6	√
		8	Alcohol consumption/addiction-Opium abuse/addiction	2.8	√
		9	Type of opium	2.1	√
		10	The name of the medications used	2.6	√
		11	Name of the allergicsubstance or drug	2.7	√
		12	Description of specific nutritional habits	3	√
		13	Surgical procedure history	2.5	√
		14	Surgical procedure	1.3	✶
		15	surgical procedure names	1.9	√
	Clinical Evaluation	16	Weight	3	√
		17	Height	2.5	√
		18	BMIa	2.8	√
		19	Function of heart	3	√

		20	Blood pressure	3	√
		21	Respiration	2.9	√
		22	Asthma	3	√
		23	Smoking	3	√
		24	Epileptic seizure	2.8	√
		25	Sinusitis	3	√
		26	Foot ulcer	3	√
		27	Wagner grade	3	√
		28	Neuropathy	2.6	√
		29	Ischemic ulcers	2.8	√
		30	Fever	1.4	✶
		31	Diarrhea	1.1	✶
		32	Laboratory test (Name of test,Laboratory test results…)	1.9	√
		33	Radiography (Name of radiography, radiography report…)	2	√
		34	Hemoglobin A1c	2.3	√
		35	HDL b	3	√
		36	LDL c	3	√
		37	ACR d	3	√
		38	Description of consultation (Consultation type, Consult orderingPhysician, Consult referring physician, Comments, diagnosis and advice of referring physician, …)	2.6	√
	Procedures	39	Number of HBOT sessions	2.2	√
		40	Physician order description	2	√
		41	Nursing observation and assessment	1.9	√
		42	Date and time Start of HBOT	3	√
		43	Vital signs before&after HBOT(BP,RR,PR)	2.8	√
		44	HBOT duration (minutes)	3	√
		45	HBOT pressure (ATA)f	2.8	√
		46	TcpO2 e after HBOT	3	√
		47	Time end of HBOT	3	√
	Discharge condition	48	Discharge date	2.1	√
		49	Discharge time	1	✶
		50	Discharge status	3	√
		51	Medication and educational instructions at discharge	2.3	√
	Follow up	52	Description of the follow up order	2.1	√
		53	Follow up date	2	√
		54	Follow up outcome	2.9	√

[a] BMI= Body Mass Index;[b] HDL= High-density lipoprotein; [c] LDL= Low-density lipoprotein;[d]ACR= Albumin to Creatinine Ratio;[e]TcpO2=Transcutaneous oxygen pressure;[f] ATA=Atmospheres Absolute

Table 5. The mean of given scores and selected technical features

Category	Subcategory	No	Data element	Mean	√/✶
Technical features	N/A	1	Internet access	2.4	√
		2	Medication reminder	2.5	√
		3	Decision making support	1.4	✶
		4	Provide instructions	2.8	√
		5	Being user-friendly	2.5	√
		6	Provide HBOT regimens	2.7	√
		7	Online scheduling	3	√
		8	Providing management dashboard	2.9	√
		9	Advanced search	2.5	√
		10	Provide report	3	√
		11	Register of date, place and causes of death	2.9	√

4. DISCUSSION

Health care providers need accurate and up-to-date information for their professional activities. In other words, data aggregated in health care information systems should be extensively accessible to all users (Kohli and Tan, 2016). Therefore, improving the quality of information available for health care decision-making is one of the important national efforts that should be made through understanding and implementing appropriate registration systems (Tursunbayeva et al., 2016). The purpose of this study was to determine the requirements of the registry system for HBOT-treated diabetic patients.

The findings of the present study showed that the statistical population selected 25 data elements for administrative category. Most of the data registry systems consist of clinical and administrative groups. The administrative data elements are collected with the aim of identifying and communicating with patients, and including the data necessary for identification, recall and follow up of patients(Khorasani et al., 2019, Lakshminarayanan et al., 2017). Therefore, determining the age and gender of patients, the degree of illness and mortality rate, and identifying the extent of service utilization is required as the basis of management and service delivery (Casalino et al., 2003). As a result, the development of electronic registry systems provides opportunities for improving health services and enhancing applied research. However, providing electronic registries with appropriate functions and standard aggregation of administrative information to support national or global analyses necessities will need the standard requirements (Richesson, 2011).

Based on the findings of the study, 54 data elements were identified for the clinical category, out of which 50 data elements were validated by the statistical population. Standardizing health data collection is an important step in the proper processing of information and getting access to effective health care (Fife et al., 2016). In other words, the crucial factor in the prevention and control of diseases is the availability of accurate and sufficient clinical information about the prognosis, symptoms, type of treatment and the context of each disease occurrence (George et al., 2017, Stoekenbroek et al., 2014). Therefore, gathering diabetes data as a first step in managing diabetes information requires a comprehensive information registry system to record clinical data elements. By analyzing the various indicators obtained from the data recording system, we are able to measure and compare the activities performed for diabetic patients (Richesson, 2011).

The results showed that the experts selected all technical features except the "decision making support". Registry systems provide a set of technical features needed to achieve appropriate clinical results. Planning on these features supports a facility to meet the needs of its environment through the use of registry systems that meet specific health needs. Kawecki et al. in a related study found that technology-based HBOT can heal diabetic foot ulcers, reduce tissue damage, and help reduce the complications of bone and soft tissue infections (Kawecki et al., 2018).

5. CONCLUSION

The outcome of the present study was the data elements and technical features of an electronic registry system for HBOT-treated diabetic patients. These requirements make it possible to standardize data collection and improve health care services. Designing a follow-up study to implement and evaluate the electronic registry system for other diseases in HBOT can be a good recommendation for future studies in this field. Clinical data collection using the standard electronic registry system can integrate accurate and detailed data. It is beneficial for all health care providers to provide optimal health services to patients. Standard and accurate clinical data are prerequisites for a convenient access to valid patient information, new treatment modalities and their effects.

ACKNOWLEDGEMENT

The present study was extracted from the PhD thesis with code 9321458001 in the field of health information management entitled "Design, development, evaluation an electronic registry system for patients undergoing hyperbaric oxygen therapy" in 2019. We thank the statistical population and all the participants for taking time to contribute to the study.

REFERENCES

Ahmadi, M., Mohammadi, A., Chraghbaigi, R., Fathi, T. and Baghini, M.S., 2014. Developing a minimum data set of the information management system for orthopedic injuries in iran. Iranian Red Crescent Medical Journal, Vol. 16, No. 7, pp e17020.

Al-Rubeaan, K.A., Youssef, A.M., Subhani, S.N., Ahmad, N.A., Al-Sharqawi, A.H. and Ibrahim, H.M., 2013. A Web-based interactive diabetes registry for health care management and planning in Saudi Arabia. Journal of medical Internet research, Vol. 15, No. 9, pp e202.

Andersen, N., Hjortdal, J.Ø., Schielke, K.C., Bek, T., Grauslund, J., Laugesen, C.S., Lund-Andersen, H., Cerqueira, C. and Andresen, J., 2016. The Danish registry of diabetic retinopathy. Clinical epidemiology, Vol. 8, No. 1, pp 613-619.

Casalino, L., Gillies, R.R., Shortell, S.M., Schmittdiel, J.A., Bodenheimer, T., Robinson, J.C., Rundall, T., Oswald, N., Schauffler, H. and Wang, M.C., 2003. External incentives, information technology, and organized processes to improve health care quality for patients with chronic diseases. Jama, Vol. 289, No. 4, pp 434-441.

Fife, C.E. and Eckert, K.A., 2018. The Hyperbaric Oxygen Therapy Registry: Driving quality and demonstrating compliance. Undersea & hyperbaric medicine: journal of the Undersea and Hyperbaric Medical Society, Inc, Vol. 45, No. 1, pp 1-8.

Fife, C.E., Gelly, H., Walker, D. and Eckert, K.A., 2016. Rapid analysis of hyperbaric oxygen therapy registry data for reimbursement purposes: Technical communication. Undersea & hyperbaric medicine: journal of the Undersea and Hyperbaric Medical Society, Inc, Vol. 43, No. 6, pp 633-639.

George, K., Ross, D. and Rowe, L., 2017. Integration of Data to Establish a Standard Operating Procedure for the Diabetic Patient Undergoing Hyperbaric Oxygen Therapy. Journal of Wound Ostomy & Continence Nursing, Vol. 44, No. 6, pp 546-549.

Hosseini, A., Moghaddasi, H. and Jahanbakhsh, M., 2010. Designing minimum data sets of diabetes mellitus: Basis of effectiveness indicators of diabetes management. Health Information Management, Vol. 7, No. 3, pp 330-340.

Hunt, T. 1972. The effect of varying ambient oxygen tensions on wound metabolism and collagen synthesis. Surg Gynecol Obstet, Vol. 135, No. 1, p 561-567.

Irawan, H., Semadi, I.N. and Widiana, I., 2018. A Pilot Study of Short-Duration Hyperbaric Oxygen Therapy to Improve HbA1c, Leukocyte, and Serum Creatinine in Patients with Diabetic Foot Ulcer Wagner 3-4. The Scientific World Journal, Vol. 16, No. 1.

Kawecki, M., Pasek, J., Cieślar, G., Sieroń, A., Knefel, G., Nowak, M. and Glik, J., 2018. Computerized planimetry evaluation of hyperbaric oxygen therapy in the treatment of diabetic foot. Adv Clin Exp Med, 27(1), pp.39-44.

Khorasani, Z.M., Tara, M., Ebnehoseini, Z., Etminani, K. and Moosavi, Z., 2019. Quantitative Evaluation of Diabetic Medical Record Documentation. Current diabetes reviews, Vol. 15, No. 2, pp 158-163.

Kohli, R. and Tan, S.S.L., 2016. Electronic health records: how can IS researchers contribute to transforming healthcare?. Mis Quarterly, Vol. 40, No. 3, pp 553-573.

Lakshminarayanan, S., Kar, S.S., Gupta, R., Xavier, D. and Reddy, S.V.B., 2017. Primary healthcare-based diabetes registry in Puducherry: Design and methods. Indian journal of endocrinology and metabolism, Vol. 21, No. 3, pp 373-377.

Mathieu, D., Marroni, A. and Kot, J., 2017. Tenth European Consensus Conference on Hyperbaric Medicine: recommendations for accepted and non-accepted clinical indications and practice of hyperbaric oxygen treatment. Diving and hyperbaric medicine, Vol. 47, No. 1, p 24.

Noori, T., Ghazisaeedi, M., Aliabad, G.M., Mehdipour, Y., Mehraeen, E., Conte, R. and Safdari, R., 2019. International comparison of thalassemia registries: Challenges and opportunities. Acta Informatica Medica, Vol. 27, No. 1, p 58.

Richardsson, R. L. 2011. Data standards in diabetes patient registries. Journal of Diabetes Science and Technology, Vol. 5, No. 1, pp 476-485.

Ruthenborg, R.J., Ban, J.J., Wazir, A., Takeda, N. and Kim, J.W., 2014. Regulation of wound healing and fibrosis by hypoxia and hypoxia-inducible factor-1. Molecules and cells, Vol. 37, No. 9, pp 637-643.

Saputri, E.G., Setiani, O. and Dewanti, N.A.Y., 2018. Hubungan riwayat pajanan pestisida dengan kejadian diabetes melitus tipe 2 pada petani penyemprot di kecamatan Ngablak kabupaten Magelang. Jurnal Kesehatan Masyarakat (e-Journal), Vol. 16, No. 1, pp 645-653.

Stoekenbroek, R.M., Santema, T.B., Legemate, D.A., Ubbink, D.T., Van Den Brink, A. and Koelemay, M.J.W., 2014. Hyperbaric oxygen for the treatment of diabetic foot ulcers: a systematic review. European journal of vascular and endovascular surgery, Vol. 47, No. 6, pp 647-655.

Sundh, J., Bornefalk-Hermansson, A., Ahmadi, Z., Blomberg, A., Janson, C., Currow, D.C., McDonald, C.F., McCaffrey, N. and Ekström, M., 2019. REgistry-based randomized controlled trial of treatment and Duration and mortality in long-term OXygen therapy (REDOX) study protocol. BMC pulmonary medicine, Vol. 19, No. 1, pp 50-58.

Tursunbayeva, A., Bunduchi, R., Franco, M. and Pagliari, C., 2017. Human resource information systems in health care: a systematic evidence review. Journal of the American Medical Informatics Association, Vol. 24, No. 3, pp 633-654.

IMPLEMENTATION OF ELECTRONIC PRESCRIPTIONS: THE PHYSICIANS' VIEW

Kristina Drusany Starič[1], Myra Tilney[2], Ioanna Chouvarda[3], Petra Bukovec[4] and Hrvoje Belani[5]
[1]*University medical centre Ljubljana, Division of gynaecology and obstetrics, Ljubljana, Slovenia*
[2]*Faculty of Medicine and Surgery, Medical School, Mater Dei Hospital, Msida, Malta*
[3]*Aristotle University of Thessaloniki, Faculty of Medicine, Medical Informatics and Biomedical Imaging Technologies Thessaloniki, Greece*
[4]*University medical centre Ljubljana, Division of surgery, Ljubljana, Slovenia*
[5]*Ministry of Health, Independent Sector for IT and e-Health, Ksaver 200a, HR-10000 Zagreb, Croatia*

ABSTRACT

Connected Health is a paradigm shift looking after the individual and community health that is arising all over the world, applicable in a wide spectrum of care services, including primary care and electronic prescriptions (e-prescriptions), which are a core part of this ecosystem. Electronic prescribing has great potential as it reduces legibility problems associated with handwritten prescriptions, can automatically flag potential interactions, adverse effects improving patient safety and can facilitate operational pharmaceutical management including potentially reducing stock-outs. While e-prescriptions have been widely adopted, the manner of their application and monitoring needs further attention to optimize quality and reduce healthcare costs. The aim of this study was to investigate physicians' opinions on e-prescriptions their usage. The additional goal was to identify possible problems that might occur during the integration. Using a trusted online survey website, a brief questionnaire comprising of 10 multiple choice questions have been sent to 120 physicians through e-mail in thirteen different countries. Survey was conducted from October 1st until December 18th 2018. As of December 2018, ninety-three questionnaires were complete and analysed. The survey showed that implementation of e-prescriptions is feasible. The time for writing e-prescriptions is not shortened comparing to handwritten prescriptions and the support system is not adequate. Nineteen (20%) of the physicians lacked computer and e-prescribing skills. The use of e-prescriptions was more difficult for older physicians, the correlation between lower age and less needed time for e-prescribing was statistically significant ($P < 0.03$). To increase its use, e-prescriptions users need to be informed how electronic health systems can be beneficial for the easier use of e-prescriptions. Adequate technical support and special training must be provided for older physicians, as they need more time for e prescribing.

KEYWORDS

e-Prescriptions, e-Prescribing, Connected Health, Older Physicians, Survey, e-Health

1. INTRODUCTION

Medication prescription for acute and chronic conditions is part of the process of rational treatment, as defined by World Health Organization (WHO). The process requires a logical, common sense approach and includes other components, such as specifying the therapeutic objective, and informing patients [de Vries et al.]. Electronic prescription (e-prescription) is the term for prescription of drugs, which is issued by an authorised professional person and is securely transmitted electronically to the pharmacy for dispensing [Leavitt], as illustrated on Figure 1. E-prescription is one part of the Connected Health that is the paradigm shift underpinning individual and community health in the health journey continuum of the person, through their lifespan, leveraging information technologies to produce a safe and effective treatment [ENJECT].

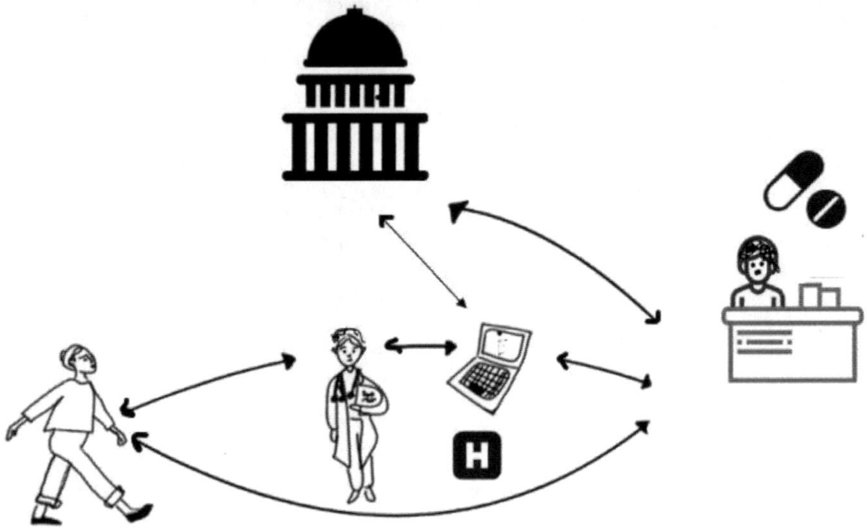

Figure 1. E-prescriptions is the term for electronically transmitted receipt for prescribed drugs prescribed by the physician in the cooperation with the patient to the pharmacist. The national institutions such as ministry of health and insurance companies regulate the process in all directions

E-prescribing is defined as a prescriber's ability to electronically send an accurate, error-free and understandable prescription directly to a pharmacy from the point-of-care. E-dispensing is defined as the act of electronically retrieving the prescription, and dispensing medications as indicated in the corresponding e-prescription. Once the medication has been dispensed, the system should also generate a report enabling monitoring of therapeutic adherence [Leavitt].

The potential benefits are multiple, ranging from reducing illegible prescriptions to facilitating real-time operational pharmaceutical management [Leavitt]. Depending on the level of clinical decision support available, potential interactions may be flagged, enhancing patient safety yet they can also potentiate error due to new process architecture [Gagnon et al.].

According to The Pharmaceutical Group of the European Union (PGEU), more than 160.000 pharmacies are computerised and have an internet connection for the daily usage of e-Health tools. These include dispensing e-prescribed medications, checking medication interactions, accessing electronic medication records, providing support for adherence via a mobile app or telephone call, or acting as the patient's entry point into the health system [Lyell et al.]. Not all medications can be e-prescribed, for example, narcotics and amphetamines, due to the potential for abuse, as well as safety aspects [Dhavle et al.]. As a result, despite appropriately regulatory framework for e-prescriptions, simultaneous manual prescriptions writing is required.

The main concern is the time required for the process of e-prescribing, which can greatly influence e-prescribing practice among different physicians. Another possible factor that may affect adoption of e-prescriptions is lower digital literacy in some health professionals [Dhavle et al.]. The e-prescribing can be time consuming especially for triplicates, as for narcotics and amphetamines prescriptions and potentially because of the repetition of the process, the problem may not only be the length of the process but also prescribing error can occur [PGEU].

To date, few studies mainly from North America and Europe have investigated the process of implementation of e-prescriptions [Gagnon et al.] [Lyell et al.] [Coleman et al.].

When a decision support system was introduced to aid the process of e-prescriptions, human factors were identified, like automation bias and reduced verification of medication alerts, which also contributed to prescribing errors [Kupperman et al.] [Kalantari and Fayaz-Bakhsh].

The aim of the study was to investigate European medical practitioners' perception of e-prescriptions. In moving from written to electronic prescriptions our research aimed to investigate physicians' use of e-prescriptions, the problems that might arise, to gain insight into the strengths and limitations, and to provide feedback, address error and facilitate streamlining the process. Our study addressed possible solutions for better e-prescribing.

2. METHODS

The authors who are physicians identified many problems during the period of transition to e- prescriptions at work. The problematics was discussed among authors and with their co-workers. Based on literature and findings of these discussions a questionnaire about e-prescribing was composed to obtain wider and non-biased view of the topic. The authors tested the validity of the questionnaire with establishment of face validity, the pilot test and clean the dataset. The questionnaire was given to the different non-medical professionals to review it. The questionnaire was then given to ten physicians to answer and evaluate the questions. Those were not the same physicians that were included in the survey and were not the authors of the questionnaire. After the testing of the survey questionnaire we did not remove any questions. The physicians included in the survey were from 13 EU countries and every participant agreed to be included in the survey.

Using a trusted online survey website (SurveyMonkey, UK), a brief anonymous survey was devised and a link to it was e-mailed to 120 medical practitioners working in various specialities in hospitals from different European countries. The only criterion was to choose the physicians from the low- and high-income countries all over Europe, they were randomly chosen from available e-mail contacts of hospitals where authors work or have connections. From the e-mail lists of physicians, the authors have access to; a cluster sample was made as much as possibly representative of the European physicians working with prescriptions. We did not include the physicians that do not prescribe drugs like radiologists. We studied the possible differences between physicians before forming a questionnaire so general questions were used. The survey consisted of 10-multiple choice questions, which, at the discretion of the authors and our collaborators who also use e-prescriptions, summarize the opinion on the usefulness of e-prescriptions and their limitations. The survey was conducted from 1st October until 18th December 2018.

The registration of the study and ethical committee application were not applicable because it was anonymous and did not include patients. There were also no incentives offered.

The personal demographics (age, gender, and country of work), personal computer skills and e-prescriptions usage was addressed. Additional questions addressed experience with e-prescriptions, as well as an assessment of self-efficacy pertaining to personal computer skills, ability to prepare the computer for e-prescriptions, time required to use the system, whether they liked the system and their assessment of time consumption with e-prescriptions. We also asked about the possibility of error during e-prescribing and how well they think e-prescriptions could help to monitor prescribed drugs. A possible cause for reverting to the handwritten prescriptions was addressed. The survey structure with questions and answer options, as well as scale used for each of the ten questions, is given in Table 1.

Data analysis included descriptive statistics for each survey item and $\chi 2$ tests to compare the distribution of responses on categorical variables and their correlations. The results of descriptive statistics are given in absolute numbers and percentages in brackets. Statistical significance was set at $P < 0.05$.

Table 1. Survey structure with questions and answer options

Question	Answer option	Scale
1. What is your age?	18 to 24 25 to 34 35 to 44 45 to 54 55 to 64 65 to 74 75 or older	Seven-point for belonging
2. In what country do you work?	<all countries>	One selection (drop-down)
3. What is your gender?	Female Male	Dichotomous (two-point for belonging)
4. I think my computer skills are sufficient for using a computer at work:	I strongly agree I agree I don't know I disagree I strongly disagree	Likert (five-point for agreement)
5. I am using e-prescriptions:	Yes, 5-10 years Yes, 1-5 years Yes, less than 1 year No	Likert (four-point for frequency)
6. I like e-prescriptions:	Yes Neutral No	Three-point (yes-no-neutral)
7. I can prepare the computer for e-prescription easily:	Yes, I can (I am the only one who is using the computer) Yes, I can (I am not the only one who is using the computer) No, I can't (I am the only one who is using the computer) No, I can't (I am not the only one who is using the computer)	Likert (four-point for likelihood)
8. I think e- prescriptions are time-saving:	I agree I use the same amount of time I disagree	Three-point (yes-no-neutral)
9. I think I make less mistakes with prescriptions than with classical prescriptions:	Yes Neutral No	Three-point (yes-no-neutral)
10. I think e-prescribed drugs are more accurately monitored (e.g. two different specialists cannot prescribe the same drug at the same time):	Yes Neutral No	Three-point (yes-no-neutral)

3. RESULTS

Hundred questionnaires were answered (83%), of which 93 (93%) were included in the study, as seven (7%) were incomplete. Participants included 56 (60%) female and 37 (40%) male physicians. Participants were stratified by age range, with 25-34-year group comprising 46% (n= 43), followed by the 35 to 44 years comprising 32 (34%), 45-54 years 10 (11%), six (6%) from the 55-64-year range and just two (2%) were from the '24 years and lower' age group. Age groups of the survey participants are given in Table 2. Participants were from 13 different European countries. Most participants came from Slovenia, Croatia and Malta. Distribution of the survey participants by country is given in Table 3.

Table 2. Age groups of survey participants

Age	Count (N)	Percentage (%)
24 years and less	2	2%
25-34 years	43	46%
35-44 years	32	34%
45-54 years	10	11%
55-64 years	6	7%
Total	**93**	**100%**

Table 3. Distribution of survey participants by country

Country	Count (N)	Percentage (%)
Austria	2	2%
Bosnia and Herzegovina	2	2%
Bulgaria	1	1%
Croatia	23	25%
Great Britain	2	2%
Ireland	1	1%
Italy	5	5%
Malta	10	11%
Norway	1	1%
Portugal	2	2%
Slovenia	43	46%
Switzerland	1	1%
Grand Total	**93**	**100%**

Most participants, 46 of them (49%), considered the process of e-prescribing to be time saving, while 29 participants (31%) disagree and eight participants (9%) think the duration is like paper prescriptions. Ninety participants (96%) consider themselves sufficiently computer-literate to provide e-prescriptions. According to this result, physicians' self-efficacy does not correlate with their perception for time-efficiency of e-prescriptions.

Participants duration of e-prescriptions usage were 5-10 years in 4%, 1-5 years in 22%, with 56 % using it for under a year, and 18 % stating that they are not using it. Survey participants' years of using e-prescriptions in given in Table 4.

Table 4. Survey participants' years of using e-prescriptions

Years	Count (N)	Percentage (%)
6-10 years	4	4%
1-5 years	20	22%
Less than 1 year	52	56%
Not using e-prescriptions	17	18%
Total	**93**	**100%**

Fifty-five (59%) responders like e-prescriptions, 12 (13%) do not like them, 26 (28%) are neutral. There is no significant difference in liking between physicians who are using e-prescriptions and those who do not.

Thirteen participants (14%) have no subjective problems in computer usage and are not sharing their computer with co-workers – other medical professionals who also use e-prescriptions, 56 participants (60%) can use computer without problems, and they share their computer with co-workers. Five (5 %) of them lack the appropriate computer skills for smoothly issuing e-prescriptions even though they have their own computers. Nineteen participants (20%), who are sharing computer with other users, lack computer skills.

The opinions of physicians about the ability of the system to support error prevention are split: forty participants (43%) think they make less mistakes with e-prescriptions, 10 participants (11%) think that with e-prescriptions more mistakes occur and 43 participants (46%) believe there is no difference regarding error prevention between handwritten prescriptions and e-prescriptions.

The majority, 69 participants (74%) think the monitoring of drugs is better with e-prescriptions, while only seven participants (8%) find it worse and 17 participants (18%) think it is the same.

We did not notice any statistical difference in computer skills ($P = 0.72$) and physicians liking of e-prescriptions ($P = 0.39$) among different age groups. Only a correlation between with a lower age and less needed time for e-prescribing ($P = 0.03$) was statistically significant. The data was presented in the table (see Multimedia Appendix 2).

4. DISCUSSION

The survey showed that implementation of e-prescriptions is feasible. The time for writing e-prescriptions is not shortened comparing to handwritten prescriptions and the support system is not adequate. The correlation between lower age of physicians and less needed time for e-prescribing was statistically significant.

Although various studies have investigated national perspectives, this study includes participants from various European regions enabling a more varied representation and overview of medical practitioners' view of e-prescriptions.

In our study, predominately women were involved. In 2014, there were 13 EU Member States where male physicians predominated, while 15 countries reported a higher share of female physicians. Because of the higher number of female physicians in 12 countries from which responses were obtained, the number of female responders was higher than male; alternatively, slightly biased [Eurostat]. The problem of the electronic literacy was described in 20%, comparing to the study in Kuwait where the problems were in more than 50% [Almutairi et al.]. Results of this study revealed that 81.31% responders of the survey currently prescribe electronically. The number of users is different in different countries. The computer skills of the physicians are according to their perspective in 96% sufficient for use of e-prescriptions. The results show that most of them are using e-prescriptions less than a year, which indicates that the implementation is still in the process. One of the biggest problems of e-prescriptions is the support system (computer, other programs, user-unfriendly application) and the fact that only 74 % of interviewed physicians believe that they can use the computer in order to write e-prescriptions, which is indicative of low ease of use. The physicians who are working in the wards are usually using multiple – not personal computers and that fore nonstop inscription of the usernames and passwords is necessary. In addition, end-users indicated that they were experiencing problems with authorisation procedures in a multi-user computer scenario but not if the computer was for solo usage. Despite many positive overall perceptions, they found that many e-prescribers reverted to handwritten prescriptions at least occasionally due to various reasons. For the minority of physicians (17%), these difficulties resulted in abandoning the use of e-prescriptions altogether [Wang et al. 2009].

The study results demonstrate that 43% consider that the system reduces prescriptions error as compared to traditional prescribing, 74% consider that prescriptions duplication could be reduced by e-prescriptions of the drugs. The literature shows that patients had the same opinion [Cochran et al.]. Prevention strategies for diminishing errors should involve physicians as well as community pharmacists [Ryan et al.] [Fischer and Rose]. E-Prescriptions are not going to be abandoned. The improvements that fore need to be done [García Alfaro and Carballeira Rodríguez].

The older physicians need longer time to prescribe drugs electronically. In aging the skills and implementation of novelties takes more time, so the older physicians should have more support and special training [Meng et al.] [Wang et al. 2019]. In addition, the importance of learning as much computer skills as possible is very important for physicians in the future [Morton et al.].

Although various studies have investigated national perspectives [Kupperman et al.] [Ryan et al.], this study includes participants from various European regions enabling a more varied representation and overview of medical practitioners' view of e-prescriptions. Even though this study is one of the first such studies trying to establish an overview of the usefulness of e-prescriptions in some European countries, it also has its limitations.

Firstly, we did not include every European country in the study and number of participants from some countries is low; the e-prescriptions could have also some minor differences between countries as well as in the support system. Furthermore, group of physicians that participated in the study were not completely randomly chosen, instead we sent the surveys to the random chosen colleagues from the email lists of hospitals where authors work or have connections. That may have introduced a bias, and therefore the results of this study cannot be conclusive as population opinion about e-prescriptions, but as opinion of a fragment of the people who answered the questionnaire. However, participation in the study was voluntary and for this reason, our sample may be biased toward physicians who are interested in this topic; hence, they may have been strong opponents or proponents of e-prescriptions. Clearly, these figures are not generalizable since they only represent the participant group. Apart from personal experience, differences in years of usage are also due to variation in year of introduction of e-prescriptions into the country, which varies from country to country.

5. CONCLUSION

The findings support the belief that many physicians are using e-prescriptions, as seen in Table 4. The problem of preparing the computer and IT infrastructure, as well as time consumption needed for e-prescriptions, should be further addressed and improvements implemented. The results are strengthened by the fact that most e-prescriptions users in our sample had used their system for less than one year, which suggests that e-prescriptions are in the beginning of successful implementation once physicians learn how to use them. To increase its use, prescribers need to be informed how electronic health systems can be beneficial for the easier use of e-prescriptions. Special support and training should be provided for older physicians. We believe that, by addressing these issues will leverage the adoption and correct usage of e-prescriptions, they can better reach their potential to reduce medication error and prescriptions fraud, save time, and improve health care quality.

ACKNOWLEDGEMENT

This work has been supported by COST Action TD1405 – "European Network for the Joint Evaluation of Connected Health Technologies" (ENJECT) and COST Action CA16226 – "Indoor living space improvement: Smart Habitat for the Elderly" (SHELD-ON).

REFERENCES

Almutairi BA, Potts HWW, Al-Azmi SF, 2018. Physicians' Perceptions of Electronic Prescribing with Electronic Medical Records in Kuwaiti Primary Healthcare Centres. *Sultan Qaboos Univ Med J.* 2018 Nov;18(4):e476-e482. Mar 28. PMID: 30988966; PMCID: PMC6443264. [Medline]

Cochran GL, Lander L, Morien M, Lomelin DE, Brittin J, Reker C, Klepser DG, 2015. Consumer Opinions of Health Information Exchange, e-Prescribing, and Personal Health Records. *Perspect Health Inf Manag.* 2015 Nov 1; 12 (Fall):1e. PMID: 26604874; PMCID: PMC4632874. [Medline]

Coleman JJ, Hodson J, Thomas SK, Brooks HL, Ferner RE, 2015. Temporal and other factors that influence the time doctors take to prescribe using an electronic prescribing system. *J Am Med Inform Assoc.* 2015 Jan;22(1):206-12: 25074989; PMCID: PMC4433371. [Medline]

Dhavle AA, Ward-Charlerie S, Rupp MT, Kilbourne J, Amin VP, Ruiz J, 2016. Evaluating the implementation of RxNorm in ambulatory electronic prescriptions. *J Am Med Inform Assoc.* 2016 Apr;23(e1):e99-e107. Epub 2015 Oct 28. PMID: 26510879; PMCID: PMC4954631. [Medline]

ENJECT, 2016. http://enject.eu/connected-health-defined-at-enject-porto-workshop/ [accessed 2019-09-09]

Eurostat, 2017. Healthcare personnel statistics – physicians. Retrieved from http://ec.europa.eu/eurostat/statistics-explained/index.php/Healthcare_personnel_statistics_-_physicians [accessed 2019-04-04]

Fischer S, Rose A, 2017. Responsible e-Prescribing Needs e-Discontinuation. *JAMA*, 317(5), 469. doi:10.1001/jama.2016.19908 PMID: 28170486 [Medline]

Gagnon MP, Nsangou ÉR, Payne-Gagnon J, Grenier S, Sicotte C., 2014. Barriers and facilitators to implementing electronic prescription: a systematic review of user groups' perceptions. *J Am Med Inform Assoc.* 2014 May-Jun;21(3):535-41. PMID: 24130232; PMCID: PMC3994867. [Medline]

García Alfaro I, Carballeira Rodríguez JD, 2019. Electronic prescription: limitations and possible improvements to ensure greater adherence to treatments. *Rev Esp Salud Publica.* 2019 Sep 25;93. [Medline]

Kalantari E, Fayaz-Bakhsh A, 2012. Can an electronic prescribing system detect doctors who are more likely to make a serious prescribing error? *J R Soc Med.* 2012 May;105(5):191. PMID: 22637768; PMCID: PMC3360530. [Medline]

Kupperman E, Vigil D, Yazdani A, Baldwin K, 2016. Model of Current Practice Regarding Prescriptions of Controlled Substances and the Perceived Benefits of E-Prescribing in an Academic Medical Center. *J Med Syst.* 2016 Dec;40(12):278. [Medline]

Leavitt MO, 2014. *The centers for medicare and medicaid service (CMS)*. US Congress Report, USA. https://www.cms.gov/Medicare/E-Health/Eprescribing/Downloads/E-RxReporttoCongress.pdf [accessed 2019-09-09]

Lyell D, Magrabi F, Raban MZ, Pont LG, Baysari MT, Day RO, Coiera E, 2017. Automation bias in electronic prescribing. *BMC Med Inform Decis Mak.* 2017 Mar 16;17(1):28. PMID: 28302112; PMCID: PMC5356416. [Medline]

Meng X, Li G, Jia Y, Liu Y, Shang B, Liu P, Bao X, Chen, 2019. Effects of dance intervention on global cognition, executive function and memory of older adults: a meta-analysis and systematic review. *Aging Clinical and Experimental Research.* 2019 Apr 13 [Medline]

Morton CE, Smith SF, Lwin T, George M, Williams M, 2019. Computer Programming: Should Medical Students Be Learning It? *JMIR Med Educ.* 2019 Mar 22;5(1):e11940. doi: 10.2196/11940. PMID: 30901000; PMCID: PMC6450476. [Medline]

PGEU – Pharmaceutical Group of European Union, 2016. *PGEU Statement: eHealth Solutions in European Community Pharmacies.*

Ryan C, O'Mahony D, Kennedy J, Weedle P, Byrne S, 2009. Potentially inappropriate prescribing in an Irish elderly population in primary care. *Br J Clin Pharmacol.* 2009 Dec;68(6):936-47. PMID: 20002089; PMCID: PMC2810806. [Medline]

de Vries TPGM, Henning RH, Hogerzeil HV, Fresle DA, 1994. *Action Programme on Essential Drugs: Guide to Good Prescribing – A Practical Manual.* World Health Organization (WHO), Geneva, pp. 115. URL: http://apps.who.int/medicinedocs/pdf/whozip23e/whozip23e.pdf [accessed 2019-09-09]

Wang S, Bolling K, Mao W, Reichstadt J, Jeste D, Kim HC, Nebeker C, 2019. Technology to Support Aging in Place: Older Adults' Perspectives. *Healthcare (Basel).* 2019 Apr 10;7(2):60. doi: 10.3390/healthcare7020060. PMID: 30974780; PMCID: PMC6627975. [Medline]

Wang CJ, Patel MH, Schueth AJ, Bradley M, Wu S, Crosson JC, Glassman PA, Bell DS, 2009. Perceptions of standards-based electronic prescribing systems as implemented in outpatient primary care: a physician survey. *J Am Med Inform Assoc.* 2009 Jul-Aug;16(4):493-502. PMID: 19390106; PMCID: PMC2705252. [Medline]

WHAT BRINGS WOMEN INTO EHEALTH? WOMEN'S CAREER TRAJECTORIES IN DIGITAL TRANSFORMATIONS IN HEALTHCARE

Åsa Cajander[1], Hilde G. Corneliussen[2], Gunilla Myreteg[3] and Kari Dyb[4]
[1]*Uppsala University, Department of Information Technology, Lägerhyddsvägen 2, Uppsala, Sweden*
[2]*Western Norway Research Institute, Box 163, 6851 Sogndal, Norway*
[3]*Uppsala University, Department of Business Studies, Kyrkogårdsgatan 10, Uppsala, Sweden*
[4]*Norwegian Centre for E-health Research, Sykehusveien 23, Tromsø, Norway*

ABSTRACT

Digital transformation of health care services is addressed world-wide in order to more efficiently meet the patients' information and health care needs. However, little is known about the people working with this transformation, where two traditionally gendered fields meet; health care and IT. While work with digitalization generally is dominated by men, digitalization of health care services involves a large number of women. This case study explores the career trajectories of women working with the digital transformation of eHealth services. Who are the women in this eHealth project, and how did they come to working with this digital transformation? The analysis shows different types of trajectories that brought the women into eHealth transformations: The first illustrating women who were pushed into working with eHealth by their job descriptions, the second showing women using eHealth as an escape route from something else, and the last trajectory showing how women stumbled across eHealth and decided to stay on. This has implications for the educational system, and points to the need for being able to study computer science later in life. It also calls for a better understanding of what drives women in transformation processes.

KEYWORDS

Career Trajectories, Digital Transformation, Ehealth, Gender

1. INTRODUCTION

A digital transformation introducing eHealth into health care services has changed health care, and is also crucial to meet the challenges of the future with an ageing population and increased health care needs. Health care and ICT, however, are two highly gendered fields, in particular in the Nordic countries including Sweden. While only one in five ICT specialists were women in 2018 (Eurostat, 2019), women completely dominate health care professions numerically, for instance in nursing where about 90% are women (Socialstyrelsen, 2019). Ehealth requires competences from both health care and ICT, but since eHealth is still new to the educational field, few employees in health care have a special eHealth education. More research is needed about people working in eHealth generally, and we know very little about the women working in eHealth or how they are brought into this field. Hence this paper explores the career trajectories of women in an eHealth case study related to a digital transformation. Who are these women, how and why are they involved in this digital transformation?

The eHealth context of this study concerns the implementation of Patient Accessible Electronic Health Records (PAEHR) in Sweden. This was quite a controversial change process that started in the 1990's. The implementation included changes in national law and has been referred to as a paradigm shift in health care services that met strong opposition (Erlingsdottir et al., 2014; Erlingsdóttir and Lindholm, 2015, 2015; Hägglund and Scandurra, 2017).

Ehealth services to patients are implemented globally to empower patients and make care more efficient (see for example Okun and Wicks, 2018; Walker et al., 2019). Sweden is an example of a country with a very ambitious agenda for the future. This involves the aim at being "best at" digitalisation (Regeringskansliet, 2017), and even "world leading" in eHealth services (Regeringskansliet, 2016). However, not all eHealth projects succeed and there are numerous studies describing problems with IT implementation (Hennemann et al., 2017). Often eHealth technology is perceived to be difficult to use by health care staff (Scandurra et al., 2014; Wass and Vimarlunda, 2018), and eHealth has had problems with user adoption (Dixon, 2007; Hoque et al., 2017), while studies also show a high satisfaction from patients. One should note that the problems related to eHealth implementation are not always technical in nature, but are also related to social challenges (Nilsson et al., 2016).

The Swedish national eHealth strategy states that "all residents from 16 years should by 2020 have access to all information documented in county-funded health and dental care" by 2025 (Regeringskansliet, 2016). However, when the eHealth service PAEHR was implemented in Sweden, it was quite controversial, and there were diverse reactions among the stakeholders (Cajander et al., 2018; Grünloh et al., 2018, 2016). With this service, patients could read their medical records notes from visits, test results, diagnosis, referrals, and a log listing health care staff who had read the electronic patient record (Moll et al. 2018). Many patients were in favour of PAEHR, and read their medical notes to keep informed and to be a part of the care process (Moll et al., 2018)), but clashes were experienced between the professional autonomy and patient empowerment (Erlingsdottir et al., 2014; Erlingsdóttir and Lindholm, 2015). Also, many health care professionals were worried about their work environment and the effects on patients when reading, for example, about cancer diagnosis online (Cajander et al., 2018; Cajander and Grünloh, 2019). Due to the political system in Sweden with autonomous regions there are local varieties to how the eHealth services are designed and implemented, both technically and organizationally. Hägglund and Scandurra (2017) describes this as that the "development and implementation of PAEHR in Sweden is based on local initiatives".

Previous research has established that the work environment in health care is problematic (Hsu, 2016). Alarming research shows that among young nurses, every fifth nurse strongly intended to leave the profession five years after graduation, and the intention was associated with levels of burnout (Rudman et al., 2014). Research has indicated that there is a lack of interest in care in the computing discipline, and that care historically has been relegated as the concern of women, working-class people, and ethnic minorities (Tronto, 2010). The literature also demonstrates that while male nurses tend to sign up for technology intensive care practices, like anaesthesia or intensive care (Mrayyan and Al-Faouri, 2008), female nurses and midwives tend to downplay or under communicate the technology dimension of their care practices (Dyb, 2009). In addition, women's careers are often more winding and interrupted than men's (Holth, 2012), and reduced gender inequalities require management control to support the agency of women (Myreteg, 2019). In order to prepare and recruit women into eHealth, it is important to generate new knowledge and search for insights about the career trajectories of women working with digital transformation in health care today.

This study investigates women who worked with the above-mentioned major transformation, aiming to understand who they are and where they came from. Our analysis identifies different types of trajectories that brought the women into eHealth transformations: some were "pushed into" working with eHealth, some used eHealth as an "escape route" from something else, while others "stumbled across" eHealth and decided to stay on. Below, we start by describing the background to the study before we present the three trajectories to eHealth, and finally we discuss the implications of our findings.

2. METHOD

Eight women, working as managers with the implementation of PAEHR process were interviewed. The in-depth interviews were about their career choices and background, and their experiences from working in eHealth. The researchers developed the interview template collaboratively and it included questions about working life, perceptions of gender at work and experiences from working in the PAEHR project. A formal document of consent was signed by the respondents. They were reminded that they could refrain from answering any specific question or choose to end the interview at any point in time, if so wished. To take part was voluntary. Participants were recruited with the help of a national organisation for eHealth services, and we used purposive sampling (Brown and Clarke, 2013; Bryman, 2008).

The eight women interviewed came from six different regions of PAEHR implementation with our intention to provide a broader understanding of this process. All interviews were recorded, transcribed and analysed using thematic analysis (Braun and Clarke, 2013). We interpreted the statements made by the interviewees as constituting their realities and subjectivities (Brown & Clarke, 2013). An informed consent was sent out and signed by all respondents, including information about anonymity, voluntariness and GDPR. The analysis was carried out collaboratively within the research group. Each respondent's answers were interpreted and coded, and the research team wrote the paper collaboratively. Citations in this paper have been made more readable.

3. RESULTS

The eight women interviewed in this study all worked with regional implementation of PAEHR. In this work they were guided by a national regulatory framework, a recommendation, on what information should be available online to patients. However, there is still great variety in what information a patient is provided within different regions, as the regions have implemented the regulatory framework differently, or as Hägglund and Scandurra (2017) concludes: "development and implementation of PAEHR in Sweden is based on local initiatives". One such difference is patients' access to psychiatric medical records online (Petersson and Erlingsdóttir, 2018a, 2018b). Furthermore, the teams in the different regions were set up differently, and the process of implementation was different in the different regions. The women interviewed came from six different regions in Sweden. They all worked in a management team of people with different roles in the implementation. The teams varied in size, but typically they consisted of 3-8 people with different roles and responsibilities. Parts of the work with implementation of PAEHR is technical, and includes standards for information sharing, interoperability issues between many different information systems where data is stored, security issues, and usability issues. All regions have their own mix of different IT systems that store information that needs to be extracted to the PAEHR online. These systems can be off-the-shelf systems or systems that the region has procured, and depending on the system the process of creating an interface to PAEHR is different. Other parts of the work with implementation are social, and are related to creating processes for maintenance of the system, external communication to users, discussions of new ways of working with different health care units, and internal communication related to the change process.

The eight women all had a background from health care, and two of them had also studied IT (Table 1). All of them were a part of the management team for the implementation project in the respective region, and they presented themselves using different job titles. Their actual tasks in the implementation project varied depending on the phase of the projects, and also depended on their expertise in relation to the expertise of their co-workers in their local team. All interviewed women chose to enter the health care field as their primary interest. Examples of their backgrounds were physician, medical secretary or nurse. None of them entered the field of eHealth services through a first degree in computer science or informatics, however two of them had developed their interest for IT through their encounter with eHealth that made them enter higher education programs in computer science or informatics later on in their career. Some describe themselves as interested in technology per se, but a large majority are more interested in the use of technology and its practical possibilities.

The analysis shows different types of trajectories that brought the women into eHealth transformations: The first illustrates women who were pushed into working with eHealth through their job description, the second shows women using eHealth as an escape route from health care work, and the last trajectory shows how women stumbled across eHealth and decided to stay on. These career trajectories are further presented below.

Table 1. Overview of the tasks of the different managers, and their background and expertise

Theme	No	Examples of task	Background and expertise	IT degree
Pushed into eHealth	INF1	Work environment perspective, patient security, development of national regulatory framework	Physician, union representative	No
	INF2	Health care strategies, business developer	Assistant nurse, nurse in emergency care, extensive management expertise.	No
	INF5	Product owner, investigator, team leader	Medical secretary, information officer, health care administration manager	No
eHealth as an escape route	INF3	Technical decisions, implementatior from a sociotechnical perspective, risk analysis	Medical secretary, information officer	No
	INF4	Education, support, communication, risk analysis	Assistant nurse, nurse, midwife, administrator and manager,	No
Stumbled across eHealth	INF6	Investigations, information structure informatics generally, process management,	Medical secretary, computer engineer, leadership and change management	Yes
	INF7	Change management, technical decisions, risk analysis, communication	Physiotherapist and IT developer	No
	INF8	Business development, informatics generally	Medical secretary, strategy officer	Yes

3.1 Pushed into eHealth Through Taking on a Role

Three of the interviewed women described that their trajectory to working with digitalisation came through taking on a role that included working with eHealth. These roles did not traditionally, or initially, include eHealth and it was not the women's active initiative to become involved in IT, but with the rise of digitalisation, eHealth has become a part various roles and positions. The first example of this is a physician who took on the assignment as a union representative (INF1). A union representative in Sweden traditionally works with social and physical work environment issues. She had no previous experience with IT, and there were no technical parts in her education. As a union representative, however, she started to work with digital transformations from a work environment perspective. "*I just got into it [eHealth] as a part of being a union representative".* During the interview she describes that her first impression of working with eHealth and PAEHR system was negative, that she was sceptical to the idea, and that she felt a strong resistance towards some of the services, such as the replacement of the physical meeting with written information about diagnoses in an ICT system: *"Well, the first feeling was negative. Hmmm, it was quite a strong negative feeling".* In her work she has since been involved in the implementation of the system, while carefully paying attention to the work environment of physicians.

The second example of a person being pushed into eHealth by taking on a role is INF2. This person describes that she is motivated by helping and meeting people, but also by changing health care when it comes to increased accessibility from the patient perspective. She does not see herself as someone that is interested in technology, but wants technology to be a vehicle for change. She describes her career as consisting of two phases, care and technology: *"it was care for quite long, and now it has been IT quite a long time".* However, she had not planned to move into eHealth and IT work. The main reason to the shift was due to her being offered a role that included IT implementation work, which is described by this quote: *"it has not been planned, but it was interesting assignments that have made me change my career in that way".*

Informant INF5 got a new job as a manager in a health care clinic that included traditional management duties. After some time they were going to implement a new system at the clinic, and she became involved in this work. Since then she has continued to work with implementation of eHealth.

All three women illustrate how they were "pushed" into working with digitalization in their respective organizations without a particular intention of working with IT, however, they took on the challenge and through their active engagement with eHealth transformations they have established a work profile that included IT.

3.2 eHealth as an Escape Route from Health Care Work

Informant INF3 describes that she was headhunted by her former manager into working at the management office, which included working with eHealth. She chose to accept the position due to it being so boring to stay on working at the clinic when her former manager had changed jobs: *"when he left his position it became very boring to stay at that place. Later he started working at the local medical management's office, and then he headhunted me so I followed him there. And then I was recruited to this IT center"*. After this job shift she has continued to work with eHealth for almost 15 years.

Informant INF4 started working in the nursing profession as an assistant nurse. She first studied to become a nurse, and later, to become a midwife. After this she moved on to working with eHealth. She describes this trajectory as a result of being displeased with the work conditions as a midwife – they did not fit with her life situation. The combination of having toddlers while working odd hours did not work out, so she moved on when she was offered a project employment at IT: *"It was an escape from ... having to work nights and weekends and evenings."*

For these two women, working with IT appeared as a solution for an otherwise dissatisfacting work situation.

3.3 Stumbled Across eHealth and Decided to Stay On

Three of the women who were interviewed described that they stumbled across eHealth and related work tasks already when they worked with caring tasks. Through this work they caught an interest and eventually found eHealth to be fascinating. Thus, once they had started working with eHealth transformations, they stayed on. Two of them (INF6 and INF8) became so interested that they later enrolled in IT related subjects (computer science and informatics) at university level. This also created a foundation for them to move on to the IT unit of the organisation, where they became managers.

One of these two women (INF6), really liked her job in health care, so she describes her moving on to something else not as an escape route, but as a trajectory to something that seemed even more interesting. She had always been very interested in technology, and liked her job as medical secretary but came to a point where she felt that she needed to do something different. *"Well, [...] I am very interested in technology and I really liked my job as a medical secretary. But then I came to the point: What do I do next? What do I want to do in my life? [...] And then I stumbled across an information about engineering."* This motivated her for enrolling in IT education herself, thus securing a platform for her continuous engagement in eHealth.

The third woman in this category (INF7) has no formal IT degree, but just as the other two, she stumbled across eHealth work and became very interested in this area. In this category, all women were satisfied with the previous work situation with caring tasks, which is a difference from the previous trajectory, but they were attracted to the IT field and wanted to carry on working there instead of in their former roles in health care.

4. DISCUSSION AND CONCLUSION

None of the women in our study came from a traditional computer science or informatics background, despite that the job they do is quite technical and requires technical competence. Instead, they all entered the field of eHealth coming from a background in health care – as physiotherapists, nurses, midwives or physicians. This study has identified three main routes to eHealth were the women (1) took on a specific role, which included the technological aspect, that they (2) escaped from what they experienced to be a poor work situation in health care (cf. Rudman et al, 2014), or that they (3) stumbled across the area by chance but was captured and stayed on.

Previous research has established that the work environment for nurses in health care is problematic (Hsu, 2016). Alarming research shows that among young nurses, every fifth nurse strongly intended to leave the profession five years after graduation, and the intention was associated with levels of burnout (Rudman et al., 2014). Thus, it is not surprising that nurses leave their profession to do something else, but it is interesting to find that some of them leave their caring tasks for IT work, by getting involved in tasks related to eHealth services.

Notably, none of the respondents in this study described the choice of working with eHealth as intentional and an active choice at the beginning of their career. Of the three identified trajectories, the last is the most positive one where they describe that they move to something better rather than the trajectory based on escaping something that is not working well. The third trajectory also lies close to an active choice. At least the women here describe how they were very interested in technology and IT services. A reason why the interviewed women did not actively choose technology as primary work might be that computer science is stereotyped as a masculine line of work (Flanagan, 2015), and might therefore be connected to a stereotype threat (cf. Bergeron et al., 2006). This mechanism could lead to these women constructing other types of explanations to why they got involved in the eHealth area. In this article we, however, only analyse the trajectories of women. This is a limitation. We do not know whether men would describe a more active choice, than these women did, to why they are involved in development of eHealth services.

Women are a key part of the workforce and research has highlighted the possibilities and values of maintaining a diverse workforce (Flanagan, 2015). The findings call for a better understanding of how transformation processes involve a combination of competences, and the need to be able to collaborate with people coming from different backgrounds. We also need to open up the discussion regarding stereotypes of men's and women's capabilities regarding technology and IT work (Schorr, 2019). Therefore, the findings also have implications for the educational system, and in particular suggests an effort is needed to avoid that pupils already at an early age in life identify women as less "suited" for technology work as future health care work also requires knowledge about technology. The findings further point to the need for adults to be able to study computer science or informatics later in life.

ACKNOWLEDGEMENT

This work was partly funded by NordForsk through the NordWit Centre of Excellence.

REFERENCES

Bergeron, D.M., Block, C.J., Echtenkamp, A., 2006. Disabling the able: Stereotype threat and women's work performance. Human performance 19, 133–158.

Braun, V., Clarke, V., 2013. Successful qualitative research: A practical guide for beginners. sage.

Bryman, A., 2008. Social Research Methods, 3rd ed. Oxford University Press, New York.

Cajander, Å., Grünloh, C., 2019. Electronic Health Records Are More Than a Work Tool: Conflicting Needs of Direct and Indirect Stakeholders, in: CHI 2019, May 4-9, 2019, Glasgow, Scotland. ACM, New York. https://doi.org/10.1145/3290605.3300865

Cajander, Å., Moll, J., Englund, S., Hansman, A., 2018. Medical Records Online for Patients and Effects on the Work Environment of Nurses., in: MIE. pp. 271–275.

Dixon, B.E., 2007. A Roadmap for the Adoption of e-Health. E-service Journal 5, 3–13.

Dyb, K., 2009. Jordmorskap og teknologi. Tidsskrift for kjønnsforskning 33, 176–193.

Erlingsdóttir, G., Lindholm, C., 2015. When patient empowerment encounters professional autonomy: The conflict and negotiation process of inscribing an eHealth service. Scandinavian journal of public administration 19, 27–48.

Erlingsdottir, G., Lindholm, C., Ålander, T., 2014. eHealth services, patient empowerment and professional accountability-An empirical study on the changing patient-doctor relationship in the digital world, in: International EIASM Public Sector Conference. pp. 1–21.

Eurostat, 2019. ICT specialists in employment [WWW Document]. URL https://ec.europa.eu/eurostat/statistics-explained/index.php/ICT_specialists_in_employment (accessed 4.27.20).

Flanagan, J., 2015. Gender and the workplace: The impact of stereotype threat on self-assessment of management skills of female business students. Advancing Women in Leadership Journal 35, 166–171.

Grünloh, C., Cajander, Å, Myreteg, G., 2016. "The Record is Our Work Tool!"—physicians' framing of a patient portal in Sweden. Journal of medical Internet research 18, e167.

Grünloh, C., Myreteg, G., Cajander, Å, Rexhepi, H., 2018. "Why do they need to check me?" Patient participation through eHealth and the doctor-patient relationship: qualitative study. Journal of medical Internet research 20, e11.

Hägglund, M., Scandurra, I., 2009. Patients' Online Access to Electronic Health Records: Current Status and Experiences from the Implementation in Sweden., in: MedInfo. pp. 723–727.

Hennemann, S., Beutel, M.E., Zwerenz, R., 2017. Ready for eHealth? Health professionals' acceptance and adoption of eHealth interventions in inpatient routine care. Journal of health communication 22, 274–284.

Holth, L., 2012. Den raka och den krokiga vägen till ingenjörsyrket: Om rationella kvinnor och passionerade män. Segregationens seghet och dess föränderliga former: En vänbok till Lena Gonäs 25, 141–159.

Hoque, M.R., Bao, Y., Sorwar, G., 2017. Investigating factors influencing the adoption of e-Health in developing countries: A patient's perspective. Informatics for Health and Social Care 42, 1–17.

Hsu, M.-Y., 2016. A quality of working life survey instrument for hospital nurses. Journal of Nursing Research 24, 87–99.

Moll, J., Rexhepi, H., Cajander, Å Grünloh, C., Huvila, I., Hägglund, M., Myreteg, G., Scandurra, I., \AAhlfeldt, R.-M., 2018. Patients' Experiences of Accessing Their Electronic Health Records: National Patient Survey in Sweden. JMIR 20, e278. https://doi.org/10.2196/jmir.9492

Mrayyan, M.T., Al-Faouri, I., 2008. Nurses' career commitment and job performance: Differences between intensive care units and wards. Journal of Research in Nursing 13, 38–51.

Nilsson, L., Eriksén, S., Borg, C., 2016. The influence of social challenges when implementing information systems in a Swedish health-care organisation. Journal of nursing management 24, 789–797.

Okun, S., Wicks, P., 2018. DigitalMe: a journey towards personalized health and thriving. Biomedical engineering online 17, 119.

Regeringskansliet, R. och, 2017. För ett hållbart digitaliserat Sverige - en digitaliseringsstrategi [WWW Document]. Regeringskansliet. URL https://www.regeringen.se/informationsmaterial/2017/05/for-ett-hallbart-digitaliserat-sverige---en-digitaliseringsstrategi/ (accessed 4.14.20).

Regeringskansliet, R. och, 2016. Vision e-hälsa 2025 [WWW Document]. Regeringskansliet. URL https://www.regeringen.se/informationsmaterial/2016/04/vision-e-halsa-2025/ (accessed 4.14.20).

Rudman, A., Gustavsson, P., Hultell, D., 2014. A prospective study of nurses' intentions to leave the profession during their first five years of practice in Sweden. International journal of nursing studies 51, 612–624.

Scandurra, I., Hägglund, M., Persson, A., \AAhlfeldt, R.-M., 2014. Disturbing or Facilitating? -On the Usability of Swedish eHealth Systems 2013., in: MIE. pp. 221–225.

Schorr, A., 2019. Pipped at the post: Knowledge gaps and expected low parental IT competence ratings affect young women's awakening interest in professional careers in information science. Frontiers in psychology 10, 968.

Socialstyrelsen, 2019. Statistik om legitimerad hälso- och sjukvårdspersonal 2018 samt arbetsmarknadsstatus 2017 [WWW Document]. Socialstyrelsen. URL https://www.socialstyrelsen.se/globalassets/sharepoint-dokument/artikelkatalog/statistik/2019-9-6311.pdf. (accessed 4.27.20).

Tronto, J.C., 2010. Creating caring institutions: Politics, plurality, and purpose. Ethics and social welfare 4, 158–171.

Walker, J., Leveille, S., Bell, S., Chimowitz, H., Dong, Z., Elmore, J.G., Fernandez, L., Fossa, A., Gerard, M., Fitzgerald, P., 2019. OpenNotes after 7 years: patient experiences with ongoing access to their clinicians' outpatient visit notes. Journal of medical Internet research 21, e13876.

Wass, S., Vimarlunda, V., 2018. I got 99 problems, and eHealth is one, in: MEDINFO 2017: Precision Healthcare Through Informatics: Proceedings of the 16th World Congress on Medical and Health Informatics. IOS Press, p. 258.

AIRDOC: SMART MOBILE APPLICATION FOR INDIVIDUALIZED SUPPORT AND MONITORING OF RESPIRATORY FUNCTION AND SOUNDS OF PATIENTS WITH CHRONIC OBSTRUCTIVE DISEASE

Rute Almeida[1], Cristina Jácome[1], Diogo Martinho[2], Pedro Vieira-Marques[1], Tiago Jacinto[1,3,4], Ana Ferreira[1], Ana Almeida[2], Constantino Martins[2], Mariana Pereira[1,3], Ana Pereira[1], José Valente[1,3], Rafael Almeida[1,3], Ana Vieira[2], Rita Amaral[1,4], Ana Sá-Sousa[1], Ivânia Gonçalves[1,4], Pedro Rodrigues[1], Magna Alves-Correia[1,3], Alberto Freitas[1], Goreti Marreiros[2], Susana Caldas Fonseca[3], Altamiro Costa Pereira[1] and João Almeida Fonseca[1,3]

[1] *CINTESIS - Center for Health Technology and Services Research, Faculty of Medicine, University of Porto, Porto, Portugal; & MEDCIDS - Department of Community Medicine, Health Information and Decision, Faculty of Medicine, University of Porto, Porto, Portugal*

[2] *GECAD - Research Group on Intelligent Engineering and Computing for Advanced Innovation and Development, Institute of Engineering - Polytechnic of Porto, Portugal*

[3] *MEDIDA - Serviços em Medicina, EDucação, Investigação, Desenvolvimento e Avaliação, LDA, Porto, Portugal*

[4] *Porto Health School, Polytechnic Institute of Porto, Porto, Portugal*

ABSTRACT

Current tools for self-management of chronic obstructive respiratory diseases (CORD) are difficult to use, not individualized and requiring laborious analysis by health professionals, discouraging their use in healthcare. There is an opportunity for cost-effective and easy-to-disseminate advanced technological solutions directed to patients and attractive to different stakeholders. The strategy of AIRDOC is to develop and integrate self-monitoring and self-managing tools, making use of the smartphone's presence in everyday life. AIRDOC intends to innovate on: i) technologies for remote monitoring of respiratory function and computerized lung auscultation; ii) coaching solutions, integrating psychoeducation, gamification and disease management support systems; and iii) management of personal health data, focusing on security, privacy and interoperability. It is expected that AIRDOC results will contribute for the innovation in CORD healthcare, with increased patient involvement and empowerment while providing quality prospective information for better clinical decisions, allowing more efficient and sustainable healthcare delivery.

KEYWORDS

mHealth, Self-Monitoring, Self-Management, Individualized Coaching, Remote Monitoring

1. INTRODUCTION

Worldwide, the burden of respiratory diseases is extremely high: originates 6 million (M)/year of premature deaths and over 10% of all disability-adjusted life-years (Roth *et al.*, 2018). About 400M individuals suffer from chronic obstructive respiratory diseases (CORD) throughout the life cycle, asthma commonly starting at early ages and chronic obstructive pulmonary disease (COPD) from the middle-age onwards. CORD are the third worldwide leading cause of death (non-communicable diseases) and have a considerable impact on patients' quality of life. Patients are continuously at risk of health deterioration, becoming more susceptible to exacerbations, which can cause frequent hospital admissions. Thus, patients with CORD require regular medical check-ups and monitoring of their health status. In Europe, the total annual cost of respiratory diseases amounts to more than €380 billion (European Respiratory Society, 2013). Traditionally, healthcare is delivered through face-to-face interactions with health professionals. With the growing prevalence of CORD and continuous pressure from healthcare authorities and insurance companies, an increasing number of patients are being managed at home in their own environment and most of the time being left alone with traditional self-management materials (books, leaflets, videos, and web-based technology). Now more than ever, technological developments empower patients to self-monitoring and management their disease. Smartphones, in addition to the advantages of being portable, personal, connected, affordable and with increasing computational power, may now be found in almost every home and person - including older persons, who progressively have more digital literacy. This creates a clear opportunity for the development of customized mobile health (mHealth) solutions. Tools taking advantage of the smartphone embedded sensors, with no external devices, would be easier to integrate into everyday lives and to disseminate without additional costs (Majumder and Deen, 2019).

For successful management of chronic diseases, sustained adoption of recommended behavior, including adherence to prescribed medication, regular monitoring of symptoms and healthy lifestyles is decisive. Automatic systems, designed to change and/or reinforce, attitudes or behaviors, namely based on the persuasive system design model, have shown to be useful in helping people to change and improve their behavior (Oinas-kukkonen, 2009; Gorini, Caiani and Pravettoni, 2020). Coaching solutions delivered through smartphones (mCoaching) appear to be ideal to deliver both simple and effective self-management interventions and to support recommended behavior, while maintaining or improving quality of care and reducing costs. These solutions allow the combination of data gathering and processing, gamification elements for user engagement and support to behaviors change. Moreover, there is evidence that healthcare delivery through health technologies has beneficial effects on chronic disease management, including CORD (Deterding *et al.*, 2011; Bashshur *et al.*, 2014; Tinschert *et al.*, 2017; Watson, Tribe and Shennan, 2019). However, solutions currently available do not take full advantage of the available technology, are not supported by clinical evidence and/or lack clinical validation (Majumder and Deen, 2019). Different studies have shown patients with CORD have interest in and willingness to use information technologies to help in the management of their disease (Fonseca *et al.*, 2006; Khurana *et al.*, 2016; Dumais *et al.*, 2019). In addition, patients' interest in real-time tracking of health status and progress, reminders, positive reinforcement and support have been reported (Simpson *et al.*, 2017). Moreover, regarding a mHealth intervention to manage COPD healthcare professionals (pharmacists, nurses, and physicians) agreed on its potential health benefits for patients (Alwashmi *et al.*, 2019). Both patients and healthcare professionals, reported "overwhelming support for mHealth to assist asthma self-management". Patients referred monitoring over time and data collection to share with healthcare teams as the more relevant features, while healthcare professionals preferred the deterioration alerts and the advises for seeking medical attention. Nevertheless, barriers such as the credibility of the systems, technical issues related to interoperability problems and the requirement of high-speed connections, as well as privacy and confidentiality concerns, need to be addressed. mHealth technologies for CORD should involve monitoring and managing signs and symptoms of the disease, while coaching solutions are urgently needed to empower patients to recognize the early signs of exacerbations and develop skills to better manage their disease.

AIRDOC project (2018-2021) is developing technologies to cope with these needs, which will be delivered in an integrated mHealth system. It is also part of the ITEA 3 - 16040 PHE - Personal Health Empowerment project, through the use case on CORD. PHE international consortium includes a total of eight partners from Portugal, Spain, Belgium and Turkey. The key consortium members were identified by the necessary competencies. For the success of AIRDOC project, it is important to take advantage of the previous experience in the development of the clinical decision support system FENO Interpretation Aid (CORD diagnostic) (Jacinto, 2010), the mobile application (app) m.CARAT for asthma and rhinitis self-management (Burnay *et al.*, 2013) and the Inspirers brand products – InspirerMundi (Jácome *et al.*, 2017), an app for asthma symptoms and medication adherence monitoring; and InspirersArcade (Almeida *et al.*, 2020), a microphone spirometry serious game. The AIRDOC system is integrable into the Inspirers brand.

2. CONCEPT AND PROPOSED SOLUTION

Traditional healthcare has improved the quality of care through evidence-based diagnostic and treatment procedures. However, the current system is reactive and process-oriented, treating patients with predefined clinical pathways, often ignoring patients' individual needs and capabilities. Additionally, healthcare tends to be more often directed for patients in acute (reactive) state, rather than timely assessing risks and preventing adverse (proactive) health outcomes. Technological developments combined with the ubiquity of smartphones, the potential of their sensors (e.g. microphone, camera) and computational power, allowing advanced processing implementations and communication capabilities, create the opportunity to develop smartphone-based monitoring solutions, without additional devices, integrated into the everyday lives of people with CORD. The current healthcare for CORD creates an opportunity for secondary prevention and lifestyle mCoaching applications to effortlessly reach a large number of people. New perspectives on psychology and gamification can be used for engaging people for more time, as required to achieve and maintain health goals.

In this regard, the AIRDOC system is developing and validating a health self-monitoring and coaching suite for people with CORD, based on mobile devices and their integrated sensors. By using these low-cost technologies in order to implement patient-centered healthcare, the proposed system has the potential to contribute to a more personalized, efficient and innovative CORD management, and be useful for patients, healthcare professionals and healthcare services alike. The proposed solution is outlined in Figure 1.

With the patient at the center of this paradigm, AIRDOC aims to empower patients with CORD for self-monitoring and management through a mobile app, which is the most visible face of a more complex integrated underlying system. The app allows the acquisition of several self-reported parameters of the patient's characteristics, their behaviors and circumstances. The sensors embedded on the smartphone will be central for data collection. The coaching component combines processed data with a clinical knowledge-base to provide information and suggest recommended behaviors / actions to the patient via the app. An interface for accessing relevant clinical information, always authorized by the patient, provides a privileged tool for the healthcare professional to access patient data and interact with the patient if needed. AIRDOC adopts good practices regarding health information related standards, privacy and confidentiality to promote interoperability and to guarantee security throughout (Ferreira, Almeida and Muchagata, 2020).

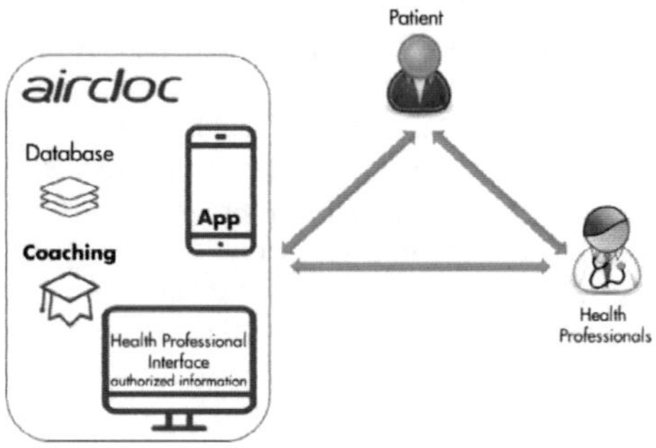

Figure 1. Solution proposed by AIRDOC

2.1 Goals and Innovation

AIRDOC is developing a set of information technologies with innovative and efficient approaches to patient's remote monitoring and chronic disease management support. It is not intended to support the CORD diagnosis or initial treatment, but to complement the delivery of high-quality healthcare.

Current respiratory function monitoring tools provide low-sensitivity parameters limiting its widespread use for monitoring patients with CORD. In this sense, AIRDOC aims to 1) deploy innovative, effective and reliable technologies 1.1) for self-monitoring by patients with CORD, based on existing sensors in the smartphone and 1.2) to extract parameters for characterization of respiratory function and computerized lung auscultation (Technology readiness levels - TRL 2-4). These technologies coupled with smartphone abilities can assist in managing data and promote its flow between patients and healthcare professionals and services, while promoting more informed and shared disease management.

Intelligent individualized support solutions, involving patients with chronic illnesses in self-management of disease signs and symptoms, are needed and will facilitate the early recognition of signs of acute illness and the development of skills to better manage their disease. Therefore, AIRDOC is proposing 2) an intelligent individualized coaching component (TRL 2-4), supported and driven by clinical evidence. These strategies are expected to motivate sustained behavioral changes in patients with CORD, with recognizable health benefits, and to provide an innovative solution for the efficient management and modernization of healthcare delivery, with improved clinical outcomes. In order to effectively deliver the monitoring technologies and coaching solutions to the patient with CORD and healthcare services, AIRDOC 3) produces an integrated smartphone app-based system (TRL 2-4) that implements the innovative technologies developed in 1) and 2). This digital health app will gather patients' information, through self-monitoring features, and the system will allow a more informed physician-patient interaction, that creates the opportunity for the delivery of additional, high quality, innovative healthcare, with great potential for improving health systems.

In this context, the privacy and security of health information are essential. Thus, AIRDOC 4) applies a new access control model that enables more personalized, secure, and context-sensitive access. This access control accounts for the risk and visualization requirements of each situation (TRL 2-4) and meets the needs of the security and privacy for the user and business. To ensure that the innovative technologies developed are suited to the needs and expectations of individuals with CORD, AIRDOC will 5) validate and test these technologies with end-users in a real environment. An initial validation phase in a clinical laboratory environment more focused on the feasibility, usefulness and usability of the app, and a second testing phase in the community, focusing on the validation and assessment of the effect magnitude (TRL 5-6).

3. CURRENT SOLUTIONS

The currently available solutions for CORD self-monitoring based on mobile applications rely mostly data collection through devices connected to smartphones and not on the use of its embedded sensors (Tables 1 and 2). Some portable spirometers with connection to the smartphone are available (MIR, no date; Larson *et al.*, 2013; Tran, Ngo and Pham, 2015; Thap *et al.*, 2016; AioCare, 2018; Hernández *et al.*, 2018; Inofabheath, 2020). These spirometers were considered promising technologies due to the high degree of innovation, potential impact and because they were considered acceptable by patients. A number of competing solutions for automatic respiratory function analysis using smartphones have also emerged, such as SpiroSmart (USA) (Hernández *et al.*, 2018) and the BKSpiro (Vietnam) (Tran, Ngo and Pham, 2015). However, these solutions are only being developed in an academic context and are not commercially available. SpiroSmart is in a more mature phase of development, being currently waiting for approval by the Food and Drug Administration. However, it lacks the ability to analyze data in real-time and its performance depends on customization to the patients' characteristics (Hernández *et al.*, 2018). Regarding computerized lung auscultation, we did not find solutions that use only the built-in smartphone sensors. Solutions integrating the connection of a stethoscope to a smartphone, such as CliniCloud Stethoscope (CliniCloud, 2019) (USA and Australia) and the methods proposed by Chamberlain et al (Chamberlain *et al.*, 2015) (USA and India) have been described but are not currently commercially available. The same stands for ResAppDx, which is already certified as medical device (CE mark) (States and Resapp, 2017). ResAppDx intends to support the diagnosis of respiratory diseases using respiratory and cough sounds collected without the need for external devices.

To the best of our knowledge, solutions similar to those proposed by AIRDOC are not currently available. AIRDOC can be one of the first solutions based solely on software and on the use of smartphone embedded sensors, avoiding the use of additional equipment connected equipment to mobile devices. The already developed lung function acquisition system includes feedback on the quality of the maneuver and the Portuguese version is currently freely available in Google Play store (Almeida *et al.*, 2020). The technologies integrated by AIRDOC as software-only solutions, with no additional costs associated, have a promising potential of innovation, allowing easy and wide spread. This is supported by the review of the National Institute for Health Research Horizon Scanning Research and Intelligence Center Work Program, where they evaluated and selected the latest technologies of respiratory function monitoring as some of the most promising tools for the diagnosis and management of patients with COPD (Dixon *et al.*, 2016).

4. AIRDOC SYSTEM

The AIRDOC system architecture is outlined in Figure 2. The mobile app allows the patient to input data that is then transmitted to a server that accommodates data storage and processing, coaching engine and system analytics. A healthcare professional interface allows access to data whenever authorized by the patient, the insertion of additional health information (e.g. diagnostic tests). An internet connection is required for communication with the server, however the local database on the smartphone device allows the basic offline use, with the ability to gather data, create and receive reminders. Once online, the data will be sent to the server and coaching recommendations will be given, if necessary.

Table 1. Existing CORD monitoring solutions for evaluation lung function using smartphones. FVC - Forced vital capacity, FEV1 – forced expiratory volume in the first second. PEF – peak expiratory flow

Product	Main Features	Validation	Strengths	Weaknesses
AioCare - MySpiroo, HealthUp, Poland (AioCare, 2018)	Real time Spirometry Decision support for COPD diagnosis Exacerbation predictive models	Several studies including one with 215 screened patients by 21 general practitioners	Air quality monitoring Digital assistant Medication reminders Doctor app	Bluetooth connected external devices required
Air-Smart Spirometer Nuvoair AB, Spain and Sweden (Hernández *et al.*, 2018)	Real time Spirometry	A two-centers study with 200 patients (Spain)	Automated treatment plan. Feedback and instructions for maneuver correction.	External devices required
BKSpiro (Tran, Ngo and Pham, 2015) Hanoi University of Science and Technology - Vietnam	Lung function based in Short-time Fourier transform from smartphone microphone Extraction of the FVC, FEV1, PEF, and FEV1/FVC.	No validation with patients, data from 17 university students was used	No need for external devices	Limited accuracy reported
Spirohome, inofabhealth, Turkey (Inofabheath, 2020)	Real time Spirometry	Not found	Progress track. Bluetooth connected. Share results with physician	External devices required. Limited information on performance
Smart One, Medical International Research, Italy (MIR, no date)	Real time FEV1 and PEF	Performance tests according to the American Thoracic Society standards	Bluetooth connected. Symptoms registration	External devices required Limited information on performance
SpiroSmart (Larson *et al.*, 2013) University of Washington - USA	Lung function based in Short-time Fourier transform from smartphone microphone Extraction of the FVC, FEV1, PEF, and FEV1/FVC	Validation with 52 subjects reporting having none or only mild lung conditions	No external devices. Consistent measures of FEV1, PEF, and FEV1/FVC Results shareable with physician	No verification, incentives or motivations for correct maneuver execution No real time Performance depends on personalization. Decreased accuracy in FVC even after personalization No commercial or publicly available version found (academic only)
Wonkwang University School of Medicine, Korea (Thap *et al.*, 2016)	Lung function from smartphone microphone based in frequency complex demodulation Extraction of the FVC, FEV1, PEF, and FEV1/FVC	Validation with 13 healthy subjects and 13 COPD patients	Able to estimate accurately FEV1/FVC	External devices required. Limited results for FVC, FEV1, and PEF No commercial or publicly available version found (academic only)

Table 2. Existing CORD monitoring solutions for lung auscultation assessment

Product	Main Features	Validation	Strengths	Weaknesses
CliniCloud Stethoscope USA and Australia	Computerized stethoscope	Not found	Sound recording at medical diagnostic quality Results shareable with physician	External devices required No automatic analysis Turned off at 30/09/2019
Mobile Stethoscope MIT, USA and India (Chamberlain *et al.*, 2015)	Computerized stethoscope	Validation with 38 patients, 11 auscultation locations	Sound recording at medical diagnostic quality Automatic detection of wheeze sounds	Connected external device No commercial or publicly available version found (academic only)
ResAppDx (States and Resapp, 2017)	Smartphone-based machine learning algorithms from cough and respiratory sounds	Validation with 979 patients with acute or chronic respiratory diseases	No external devices required. Accurately identified acute exacerbations identifications in patients with COPD or asthma, and screening for COPD in a broad general population CE mark, as medical device	Limited information on performance and methodology No commercial or publicly available version found (academic only)

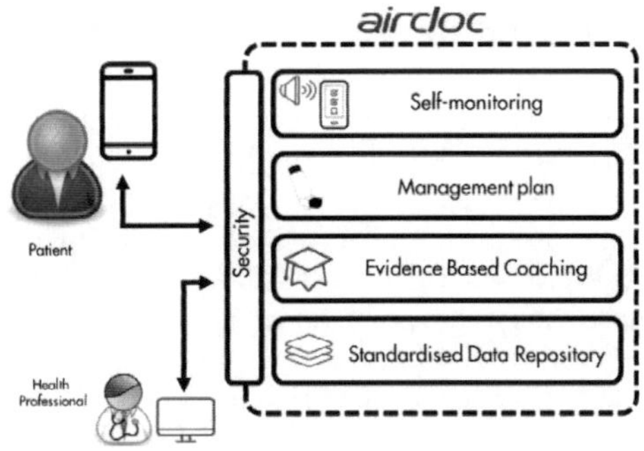

Figure 2. AIRDOC System Architecture

A security layer is in charge of authentication and authorization, managing all data exchanges in-between modules, by using OAuth 2.0 and a self-hosted identity server with openID framework, to promote security and trust. A secure and flexible delegation model for sharing health data between patients and healthcare professionals is included within that security layer (Almeida, Vieira-Marques and Ferreira, 2020). The delegation feature has been devised and prototyped within AIRDOC, to help empower the patient to perform a fine-grained control to what data can be shared, to whom and for how long. On the whole, the security layer was devised and developed with the main aim to be modular and independent, in order to be easily adapted to similar scenarios. Moreover, HL7 FHIR standard for health care data modeling and exchange is used at all levels, from app backend to health data storage, and used for data exchange between system modules. Using this approach, the system promotes interoperability allowing for semantic rich data storage, sharing and reuse, while guaranteeing data quality and security. The system takes advantage of the embedded sensors of the smartphones, with no external devices required for full use and mainly relying in self-reporting. The integration of user authorized data from external monitoring devices, other electronic health recordings and mobile apps data is possible, especially for systems also using HL7 FHIR standard. Information from other sources, such as air quality and meteorological data, may also be retrieved when subscribed. Individualized support based on the data provided by the patient and healthcare professional, clinical evidence and guidelines will be delivered, following the persuasive system design model principles of personalization, self-monitoring, strategies of reminders and suggestions, with trustworthiness, credibility and verifiability (Oinas-kukkonen, 2009).

Taking advantage of previous experience of the consortium in clinical decision support system and mHealth, the app integrates some of the previous developed features, namely from InspirerMundi app (Jácome *et al.*, 2017). The main modules for self-monitoring, management plan and evidence based coaching, which together with the data repository constitute the AIRDOC system, are detailed in the next subsections.

4.1 Self-Monitoring Module

The system mainly relies on data inserted in the AIRDOC app. To achieve personalization, the patient's relevant characteristics, previous diagnoses and circumstances should be retrieved. The system will suggest reporting of symptoms, exacerbations, health services use, habits and burden factors, in order to follow-up the CORD evolution. Validated questionnaires, such as the Control of Allergic Rhinitis and Asthma Test (CARAT) (Thap *et al.*, 2016), screening tests (Sá-Sousa *et al.*, 2019), and other patient-reported outcomes may be used.

The embedded microphone will be used to acquire the sounds of forced expiratory maneuvers (microphone spirometry) and computerized lung auscultation. The processing of the sound includes automatic quality check in particular regarding the correct performance of the maneuver, as well as extraction of features with potential clinical value (Fonseca *et al.*, 2015; Pinho *et al.*, 2018; Almeida *et al.*, 2019, 2020). These sounds can also be made available to the healthcare professional for further analysis. Automatic metrics can be immediately reported to the patient whenever available, and both summaries and graphical reports will be produced, allowing to monitor the health status and changes. Healthcare professionals, upon patient authorization, will receive relevant information for remote monitoring actions and decision making.

Health professionals indicated by the user will be able to complete the data reported by the patient adding details on diagnostic exams and health services use among others, using their specific interface.

4.2 Management Plan Module

The setup of the management plan may be introduced manually by the patient on the app, or defined in the healthcare professionals' interface. Both patient and healthcare professional can approve or suggest editions over the plan introduced by the other. Based on this plan, personalized notifications (reminders) will be generated for medicine intake, healthy behaviors and other health interventions.

The long-term goal of CORD management is to reduce symptoms and the severity and frequency of exacerbations, to which those patients are prone to. Adequate medication adherence is critical for that purpose, in particular to inhaled medication, that is the cornerstone of CORD effective treatment. AIRDOC is taking advantage of previously developed tools for objective monitoring of medication adherence. An enhanced version of the image-based dose recognition developed for InspirerMundi app will be integrated into the AIRDOC app (Vieira-Marques *et al.*, 2020). Reports and statistics on the medication adherence will be available to the patient, being shareable with the health professionals.

In spite of being designed for medication related to CORD, the tools of self-management can also be used to register and monitor adherence to any medication included in the treatment plan for any comorbidity. Different information would be shareable with different healthcare professionals.

4.3 Coaching Module

As a patient-centered system, AIRDOC will produce individualized support for self-management of CORD, giving recommendations and positive reinforcement of adequate behaviors. In this context, the coaching module has been proposed (Vieira et al., 2019) and further extended to process data from the self-monitoring module on the app and from healthcare professional interface. A rule-based system using JBoss Drools (JBoss, 2020) framework has been developed with the main goal to combine clinical knowledge-base with patient data to detect events and produce adequate recommendations. The alignment of those recommendations with the clinical evidence and updated guidelines is of utmost importance for the credibility of the system and the usefulness of the coaching. Furthermore, the system will monitor patients' behaviors and clinical data to match the suggestions to the user's profile. Additionally, the coaching component is expected to respond dynamically to measurements, interactions and environmental factors by selecting the coaching activities that best suit the patient's condition. In the specification of these algorithms, based on patient behavior and data collected, different activities can be proposed to promote wellbeing and ensure adherence to the treatment plan. The expected outcome is the definition of personalized coaching plans designed with healthcare professional support and proposed to the patient. These plans include a set of goals associated to a specific topic of CORD management, with different difficulty levels. The patient's actions will be tracked and successful/unsuccessful interaction (achieved/ non-achieved goals) will result in an automatic update of established goals for upcoming days with increased or decreased difficulty.

The patient will be able to express satisfaction with the recommendations and register events representing the compliance with them. Relevant interactions between the patient and the AIRDOC system will be analyzed in the coaching module to understand and automatically adapt future interactions based on user preferences and progress for any established coaching plan. Reports and statistics on the recommendations produced, patient response, completed plans, rate of compliance, achieved goals, and user progress will be also available.

5. PILOT STUDIES DESIGN

The conduction of studies with real patients with CORD (adults, including older people ≥65 years) to validate the prototype of the integrated system and its modules, and evaluate the final version is crucial.

Ethics and administrative approvals in the institutions involved in data collection and patient recruitment will be obtained. Patients will sign a written informed consent prior to any data collection. In addition to following the directions outlined in the Portuguese Data Protection legislation on the processing of personal and identifiable patient data (Law No. 67/98 of 26 October 1998, DR - I Series - A, No. 247), AIRDOC will also, under the European Data Protection Regulation (Regulation (EU) 2016/679 of 27 April 2016, Official Journal of the European Union, L119, 59), present in a clear and objective manner who is going to collect, store and able to access the data. Moreover, whenever there are any changes to those procedures, the patient will be fully and promptly informed and will have the chance to opt-out at any time.

Pilot studies will be performed in three phases. The initial two validation phases will have a smaller scale (minimum 12 adults) and be more focused on the feasibility, usefulness, usability and acceptance of the system. This was submitted to an ethical commission of a hospital and waits for authorization. The third phase is a medium term prospective real-life environment, focusing on validation and estimation of the magnitude of the effect of the app use and personalized coaching on symptoms and exacerbations with at least 70 adults. In spite of the detailed characteristics of the pilot studies are still to be finalized, the main design will be as follows.

5.1 Cross-Sectional Validation

In the first pilot study, the prototype of the app, particularly the self-monitoring technologies, will be validated in a clinical environment. This study will be conducted with a minimum of 12 adults with CORD. In a single visit, recruited patients will be informed on how to use the prototype app and will be asked to use it to register data, namely, acquisition of respiratory function parameters and sounds, symptoms and treatment plans. A healthcare professional will then evaluate the respiratory parameters by standard procedures (e.g., spirometry and lung auscultation) to allow further validation of measurements. Results will be treated statistically and will be used to introduce adjustments in subsequent studies. Patient feedback will be considered to refine and improve the developed mobile functionalities.

5.2 Short Term Prospective Validation

This pilot intends to evaluate the integrated system, namely the data acquisition and processing and the coaching components, by patients with CORD in a real-world setting. A minimum of 12 adults with CORD are expected to be recruited and will have the same initial visit described in the cross-sectional validation. Additionally, they will be invited to use the prototype of the integrated system to self-monitor their disease in the community for one week. Patients will be instructed by a healthcare professional to routinely input data to the app, including the recording of respiratory sounds in their usual context (e.g., home, community) and also to adhere to the coaching component. After one week of use, patients will be invited to participate in a focus group and answer usability questionnaires. Results will be processed and can be used to make adjustments to the final study pilot. Users' feedback will be used to refine and improve the system.

5.3 Medium Term Prospective Assessment

The last pilot study intends to estimate the magnitude of the effect of the developed product on disease management, including the reduction of symptoms and exacerbations. A minimum sample of 70 adults with CORD will be recruited for a 4-month observational study conducted in the community. At the initial visit, similarly to the first pilots, all patients will be instructed to routinely use the integrated system in their daily life. A subset of patients will receive instructions to adhere to the coaching component, while the remaining will not receive this individualized support. This design will not only allow to compare patients throughout the use of the integrated system but also to have a rough estimate of the specific effect of mCoaching, apart from the effect of self-monitoring. The data generated by this study will also create new opportunities for research and development of innovative healthcare. In particular, respiratory sound recordings will be organized and made available to the scientific community in the form of a fully anonymized signal database.

6. EXPECTED RESULTS

The AIRDOC project is expected to produce several tools and products, as an integrated mHealth system. The clinical knowledge-base that encompasses current guidelines, the clinical evidence rule editor to facilitate knowledge-base updates, the computational model for individualized coaching of patients with CORD, based on clinical evidence gathered in the knowledge-base and the delegation model to provide secure health data sharing were completed in 2019. The system architecture was defined and the security layer with the delegation model was implemented. The coaching engine, based on the knowledge-base constructed, is fully functional. A prototype app including forced expiratory maneuver and lung auscultation acquisition systems, and a questionnaire (as prove of concept) is available for testing monitoring data collection. Both monitoring and coaching modules are using HL7 FHIR standards for data exchange, interacting with data repository. Secure sharing of patients' clinical data with the healthcare professionals, allowing better, informed clinical decisions, and, ultimately, making healthcare more efficient and sustainable.

During 2020 are expected to be finalized the prototype of monitoring data acquisition implemented in the app, including the different patient reported outcomes and tolls for coaching personalization, and the computational models to design dynamic coaching plans and associated goals that can track and evaluate patient progress. These will allow to have full prototypes of the mobile self-monitoring system and of individualized mCoaching to be tested. The health professional interface will also be completed.

By the end of project (2021) the validated self-monitoring and self-managing system for patients with CORD, including individualized coaching should be ready to further testing and clinical care.

7. CONCLUSION

AIRDOC aims to develop and validate a set of information and communication technologies with innovative and efficient approaches for the remote monitoring and self-management of patients with CORD. These mHealth technologies will:

- empower the patient with CORD to self-monitor the disease, with effective ways, supporting early identification of risks and prevention of exacerbations;
- produce and deliver clinical-based recommendations, through individualized coaching;
- support the patient in the self-management of CORD, promoting treatment adherence and the adoption of healthy behaviors;
- provide updated information to the healthcare team, promoting efficient and safe communication with the patient and facilitating the implementation of timely interventions.

AIRDOC is expected to contribute to the delivery of digital health services and innovation in remote patient monitoring technologies. The use of the AIRDOC system should provide greater involvement and empowerment of patients in their treatment and collects quality prospective data. Therefore, is expected to have a significant impact and innovation in CORD health services, providing support for shared clinical decision making, and making healthcare more effective, efficient and sustainable. Moreover, in spite of being designed for CORD treatments, some developed tools could also be used to register and monitor the treatments of any comorbidity, or adapted for other chronic diseases.

ACKNOWLEDGEMENT

This work is supported by project NORTE-01-0247-FEDER-033275, financed by the North Portugal Regional Operational Programme (NORTE 2020), under the PORTUGAL 2020 Partnership Agreement, and through the European Regional Development Fund (ERDF). AIRDOC is with ITEA 3 16040 PHE - Personal Health Empowerment project consortium. R. Almeida is supported through the operation POCI-01-0145-FEDER-029130 (mINSPIRERS- mHealth to measure and improve adherence to medication in chronic obstructive respiratory diseases - generalisation and evaluation of gamification, peer support and advanced image processing technologies) funded by the COMPETE2020 and by National Funds through FCT (Fundação para a Ciência e a Tecnologia). A. Ferreira is also with TagUBig - Taming Your Big Data (IF/00693/2015) from Researcher FCT Program funded by National Funds through FCT.

REFERENCES

AioCare (2018) *MySpiroo*. Available at: https://attendris.com/wp-content/uploads/2018/05/RIS2018_DigitalTech_HealthUp.pdf?utm_source=RIS_Agenda&utm_medium=Presentation&utm_campaign=RIS_2018&utm_content=RespiratoryInnovation (Accessed: 3 December 2019).

Almeida, R. *et al.* (2019) 'Quality assessment and feedback of Smart Device Microphone Spirometry executed by children', in *Proceedings of IEEE 6th Portuguese Meeting on Bioengineering (ENBENG)*. IEEE, p. 8692452. doi: 10.1109/ENBENG.2019.8692452.

Almeida, R. *et al.* (2020) 'Automatic Quality Assessment of a Forced Expiratory Manoeuvre Acquired with the Tablet Microphone', in *IFMBE Proceedings MEDICON 2019*, pp. 1394–1398. doi: 10.1007/978-3-030-31635-8_170.

Almeida, R., Vieira-Marques, P. and Ferreira, A. (2020) 'Patients to mobilize their data: Secure and flexible mHealth delegation', in *ICISSP 2020 - Proceedings of the 6th International Conference on Information Systems Security and Privacy*, pp. 552–560. doi: 10.5220/0008972905520560.

Alwashmi, M. F. *et al.* (2019) 'Perceptions of Health Care Providers Regarding a Mobile Health Intervention to Manage Chronic Obstructive Pulmonary Disease: Qualitative Study', *JMIR mHealth and uHealth*, 7(6), p. e13950. doi: 10.2196/13950.

Bashshur, R. L. *et al.* (2014) 'The Empirical Foundations of Telemedicine Interventions for Chronic Disease Management', *Telemedicine and e-Health*, 20(9), pp. 769–800. doi: 10.1089/tmj.2014.9981.

Burnay, E. *et al.* (2013) 'Challenges of a mobile application for asthma and allergic rhinitis patient enablement-interface and synchronization.', *Telemedicine journal and e-health*, 19(1), pp. 13–8. doi: 10.1089/tmj.2012.0020.

Chamberlain, D. *et al.* (2015) 'Mobile stethoscope and signal processing algorithms for pulmonary screening and diagnostics', *Proceedings of the 5th IEEE Global Humanitarian Technology Conference, GHTC 2015*. IEEE, pp. 385–392. doi: 10.1109/GHTC.2015.7344001.

CliniCloud (2019) *CliniCloud Stethoscope*. Available at: https://clinicloud.com/store/stethoscope/ (Accessed: 2 September 2019).

Deterding, S. *et al.* (2011) 'Gamification: Using game design elements in non-gaming contexts', in *Conference on Human Factors in Computing Systems - Proceedings*. New York, New York, USA: ACM Press, pp. 2425–2428. doi: 10.1145/1979742.1979575.

Dixon, L. C. *et al.* (2016) 'New and emerging technologies for the diagnosis and monitoring of chronic obstructive pulmonary disease', *Chronic Respiratory Disease*, 13(4), pp. 321–336. doi: 10.1177/1479972316636994.

Dumais, K. M. *et al.* (2019) 'Preferences for Use and Design of Electronic Patient-Reported Outcomes in Patients with Chronic Obstructive Pulmonary Disease', *The Patient - Patient-Centered Outcomes Research*. doi: 10.1007/s40271-019-00376-9.

European Respiratory Society, S. (2013) 'The economic burden of lung disease', in Gibson, G. et al. (eds). European Respiratory Society, pp. 16–27.

Ferreira, A., Almeida, R. and Muchagata, J. (2020) 'How Secure Is Your Mobile Health?', in *IFMBE Proceedings - MEDICON 2019*, pp. 1377–1384. doi: 10.1007/978-3-030-31635-8_168.

Fonseca, J. A. *et al.* (2006) 'Asthma patients are willing to use mobile and web technologies to support self-management', *Allergy: European Journal of Allergy and Clinical Immunology*. Munksgaard International Publishers, 61(3), pp. 389–390. doi: 10.1111/j.1398-9995.2006.01016.x.

Fonseca, J. F. *et al.* (2015) 'Automatic Analysis of Lung Function Based on Smartphone Recordings', *Biomedical engineering systems and technologies BIOSTEC 2015*, 574, pp. 390–402. doi: 10.1007/978-3-319-27707-3.

Gorini, A., Caiani, E. G. and Pravettoni, G. (2020) 'Psycho-cognitive Factors Orienting eHealth Development and Evaluation', in *P5 eHealth: An Agenda for the Health Technologies of the Future*. Cham: Springer International Publishing, pp. 109–121. doi: 10.1007/978-3-030-27994-3_7.

Hernández, C. R. *et al.* (2018) 'Validation of the portable Air-Smart Spirometer1. Hernández CR, Fernández MN, Sanmartín AP, Roibas CM, Domínguez LC, Rial MIB, et al. Validation of the portable Air-Smart Spirometer. PLoS One. 2018;13(2):1–11.', *PLoS ONE*, 13(2), pp. 1–11. doi: 10.1371/journal.pone.0192789.

Inofabheath (2020) *spirohome*. Available at: https://spirohome.io (Accessed: 3 December 2019).

Jacinto, T. (2010) *A Decision Support System for Exhaled Nitric Oxide Values Model and evaluation.* Faculty of Medicine of University of Porto.

Jácome, C. *et al.* (2017) 'Inspirers: An app to measure and improve adherence to inhaled treatment', in *Proceedings of the International Conference on E-Health, EH 2017 - Part of the Multi Conference on Computer Science and Information Systems 2017*.

JBoss (2020) *Drools - Business Rules Management System (Java^{TM}, Open Source)*. Available at: https://www.drools.org/ (Accessed: 11 December 2019).

Khurana, L. *et al.* (2016) 'Patient preference for using computers, smartphones, and internet to participate in COPD clinical trials', *European Respiratory Journal*, 48, p. PA2899.

Larson, E. C. *et al.* (2013) 'SpiroSmart: Using a Microphone to Measure Lung Function on a Mobile Phone', *Proceedings of the 2012 ACM Conference on Ubiquitous Computing - UbiComp '12*, 4(2012). doi: 10.1145/2370216.2370261.

Majumder, S. and Deen, M. J. (2019) 'Smartphone sensors for health monitoring and diagnosis', *Sensors (Switzerland)*. MDPI AG, pp. 1–45. doi: 10.3390/s19092164.

MIR (no date) *Smartone*. Available at: https://www.spirometry.com/ENG/Products/smartone.asp.

Oinas-kukkonen, H. (2009) 'Persuasive Systems Design : Key Issues , Process Model , and System Features Persuasive Systems Design : Key Issues , Process Model , and System Features I . INTRODUCTION', *Persuasive Systems Design: Key Issues, Process Model, and System Features*, 24(March 2009), pp. 485–500.

Pinho, B. *et al.* (2018) 'Automatic Quality Assessment of Smart Device Microphone Spirometry', in *Proceedings of the 8th International Joint Conference on Pervasive and Embedded Computing and Communication Systems*. SCITEPRESS - Science and Technology Publications, pp. 115–122. doi: 10.5220/0006866701150122.

Roth, G. A. *et al.* (2018) 'Global, regional, and national age-sex-specific mortality for 282 causes of death in 195 countries and territories, 1980–2017: a systematic analysis for the Global Burden of Disease Study 2017', *The Lancet*. Lancet Publishing Group, 392(10159), pp. 1736–1788. doi: 10.1016/S0140-6736(18)32203-7.

Sá-Sousa, A. *et al.* (2019) 'Adult Asthma Scores—Development and Validation of Multivariable Scores to Identify Asthma in Surveys', *Journal of Allergy and Clinical Immunology: In Practice*. American Academy of Allergy, Asthma and Immunology, 7(1), pp. 183-190.e6. doi: 10.1016/j.jaip.2018.06.024.

Simpson, A. J. *et al.* (2017) 'Perspectives of patients and healthcare professionals on mHealth for asthma self-management', *European Respiratory Journal*. European Respiratory Society. doi: 10.1183/13993003.01966-2016.

States, U. and Resapp (2017) *ResAppDX*. Available at: https://www.resapphealth.com.au/positive-results-from-australian-prospective-clinical-study-confirm-resapps-algorithms-deliver-high-accuracy-when-diagnosing-respiratory-disease-in-adults (Accessed: 3 December 2019).

Thap, T. *et al.* (2016) 'High-resolution time-frequency spectrum-based lung function test from a smartphone microphone', *Sensors (Switzerland)*, 16(8). doi: 10.3390/s16081305.

Tinschert, P. *et al.* (2017) 'The Potential of Mobile Apps for Improving Asthma Self-Management: A Review of Publicly Available and Well-Adopted Asthma Apps', *JMIR mHealth and uHealth*. JMIR Publications Inc., 5(8), p. e113. doi: 10.2196/mhealth.7177.

Tran, H. A., Ngo, Q. T. and Pham, H. H. (2015) 'An application for diagnosing lung diseases on Android phone', *ACM International Conference Proceeding Series*, 03-04-Dece, pp. 328–334. doi: 10.1145/2833258.2833289.

Vieira-Marques, P. *et al.* (2020) 'Combined Image-Based Approach for Monitoring the Adherence to Inhaled Medications', in *IFMBE Proceedings - MEDICON 2019*, pp. 1399–1404. doi: 10.1007/978-3-030-31635-8_171.

Vieira, A. *et al.* (2019) 'Defining an Architecture for a Coaching Module to Support Self-Monitoring of Chronic Obstructive Respiratory Diseases', in *Studies in Health Technology and Informatics*. IOS Press, pp. 130–133. doi: 10.3233/SHTI190034.

Watson, H. A., Tribe, R. M. and Shennan, A. H. (2019) 'The Role of Medical Smartphone Apps in Clinical Decision-Support: A Literature Review', *Artificial Intelligence in Medicine*. Elsevier B.V., p. 101707. doi: 10.1016/j.artmed.2019.101707.

ANALYSIS OF DATA QUALITY ON ADVERSE DRUG REACTIONS AND OF METHODS FOR ITS IMPROVEMENT

Mark Tkachenko, Natalia Dobrenko, Daria Butorina, Elvira Bulatova,
Aleksandra Vatian and Anton Arazashvili
ITMO University, Sankt-Petersburg 197101, Russia

ABSTRACT

This article presents the results of quality analysis of data on adverse drug reactions (ADRs) for groups of drugs, which are most frequently discussed by users of Internet resources.
Several Russian-language resources were used as data sources. An open database MedDRA, translated into Russian, was used as a database of ADRs. To collect information about ADRs from Internet resources, a rule-based algorithm has been developed and implemented. Venn diagrams, multigraphs of pairwise as well as interaction scaling analysis were used to compare data on ADRs obtained from different sources. To assess the quality of the data, conditionally objective indicators were used, such as relevance, integrity, compatibility, measurability, and source characteristic.
As a result, significant violations of the integrity and measurability of data were identified, and a comparative assessment of the characteristics of data sources was carried out. The ways of improving the quality of data on ADRs have been formulated.

KEYWORDS

Adverse Drug Reactions, Pharmacovigilance, Data Quality

1. INTRODUCTION

Adverse drug reactions (ADR) are a significant source of morbidity and mortality, and their constant monitoring and accounting in real medical practice is required in accordance with the recommendations of the World Health Organization (WHO).

Information on ADRs used to be collected by health professionals during clinical trials of drugs. However, research and clinical practice in recent years expanded the list of ADR sources and the time limit for accepting reports. Pharmacovigilance now must not only record ADR data provided by doctors, but also collect patient opinions. As data sources, not only official protocols can now be used, but also social networks, forums, blogs and other means of Internet communication.

ADRs need to be monitored after official approval, especially due to new drugs appearing on the market and the spreading threat of self-medication. Such heterogeneity and diversity of collected data can objectively reduce its quality, which is not always taken into account to a sufficient degree when processing this data and using it for practical applications.

The quality of medical data in general [Romero, Na Hong, Van den Berghe, Mavrogiorgou] and ADR data in particular [Soyer, Jeong, Oh] is a subject or worldwide research. However, it is only beginning to gain traction in Russia [Antipov]. This article presents the results of quality analysis of data on adverse drug reactions (ADRs) for two groups of drugs: lipid-lowering drugs, antiarrhythmic drugs, which are most frequently discussed by users of Internet resources.

2. BACKGROUND AND RELATED WORKS

2.1 Means for Collecting ADR Information

Data collection systems for ADRs vary widely across countries. One of the most advanced countries in this regard is the United States of America. Russia was chosen for comparison for the following reasons. In the USSR, there was a system of state control of the quality of medicines, but with the collapse of the USSR it was almost entirely destroyed and in Russia is at a new stage of development [Lepakhin]. In addition, Russian legislation on the sale of medicine can be considered rather liberal, and many drugs that are available in other countries only with a prescription are sold in Russia without one. This allows for self-treatment and treatment based on data found on the Internet, where patients self-diagnose themselves and choose which drugs to take based on the opinion of other social network users.

The results of the analysis of available information sources on ADR and means of their processing in English and in Russian are briefly described in detail below. More complete information can be found in [Vatian].

There are several channels for collecting information about ADRs, which can be divided into two categories: information from official sources and information from individuals. In 2016, Russia entered the Good Pharmacovigilance Practice rules [GVP]. According to its norms, the participants of the circulation of medicines should report to the specialized center [Roszdravnadzor] about all cases of adverse effects not listed in the drug leaflet, serious and unexpected adverse reactions, special interactions with other drugs encountered in clinical trials and during clinical use of drugs. However, this system is aimed not so much at fixing individual cases of ADRa as on withdrawing poor-quality medicine batches from sale. The individual doctors' and patients' experience is not included in this database. In this regard, the role of practitioners and patients as information providers for the pharmacovigilance system increases sharply [Foster].

People express their opinion about medications, including ADR, primarily in the form of texts. It can be specialized medical texts, including doctors' conclusions and health records, as well as entries in social media like forums, blogs, etc. Natural lan-guage processing (NLP) methods are used worldwide due to their effective automated processing. NLP methods have seen success when used for the task of processing medical texts, English texts regarding ADRs in particular. At the same time, with respect to other languages, the progress is slower, which has negative consequences in the context of globalization and population migration growth.

Steps are being taken to improve this situation, primarily to unify worldwide approaches to ADRs. For example, American Medical Informatics Association launched a special didactic panel called "Clinical Natural Language Processing in Languages Other Than English" [AMIA]. In the manifesto of this panel, Germanic, Romance, Sino-Tibetan and Afro-Asiatic languages are mentioned, but not Russian.

In the Russian-language segment, studies on the use of NLP in medicine are underway, but only a few are focused on ADR detection. For example, in [Baranov], clinical records are analyzed in order to solve problems such as indicating the absence or the presence of disease in the patient and the severity of the course of the disease. The work [Lushnov] is aimed at monitoring the patient's state of health in different contexts: the treatment and diagnostic process, clinical trials of the new treatment regimen, monitoring the work of the clinic by its management.

2.2 Means for Evaluating ADR Degree

Problems associated with evaluating the degree of ADRs are widely discussed in various works [32-35]. The main problem researchers face is determining the causal relationship between drugs and undesirable clinical events, and in this regard, clinical trials are ongoing [36, 37].

There are official guidelines for evaluating drug-drug interactions (DDI) at the drug development level. DDI can result in alterations of either drug pharmacokinetics (PK), pharmacodynamics (PD) or both, and evaluations can be done in vitro and in vivo studies [Talbot, Marante, Khan, Prueksaritanont]. However, at the stage of using drugs in clinical practice, the main means of estimation is expert judgment [Khan]. Methods are being developed to improve the quality of group examination be increasing the intra rater agreement of a group of experts, but the opinion of each individual expert is evaluated in a binary scale [Khan].

Based on expert opinions or the results of the analysis of the chemical structure of the active substances, matrices of pairwise interactions between drugs are constructed, which allow algorithmically estimating ADR levels. For example, in the article [Zhang], several matrices of interactions between medicinal substances are built on the principle of "each with each". The authors used only official sources of information: the FAERS database and the PubChem chemical structure database. Thus, each drug was represented by a binary adverse effect profile (4,192-dimensional for labeled effect, 10,093-dimensional for off-label effect, 881 for chemical structures). The authors proposed label propagation algorithms addressing the following problem: given an undirected weighted network with n nodes where a small portion of them are labeled, estimate the labels of the rest of the unlabeled nodes. Thus, it becomes possible to evaluate the toxicity of any pair of unlabeled drugs. However, to achieve this, the parameters that determine the influence of a node's neighbors relative to its provided label need to be known, as well as the importance of each drug similarity matrix.

Based on this and similar methodologies, software has been developed and applied to evaluate DDIs, such as [Micromedex]. However, all of these tools are proprietary, are based on US medical standards and are not localized for non-English speaking countries, which makes them less accessible for medical professionals in Russia.

Various scales are proposed for assessing the danger level of ADR manifestation. For example, [Micromedex] uses two scales: a four-level scale of severity levels and a three-level scale of documentation levels (scientific evidence), by which the scientific validity of the claimed ADR degree is understood. A wider range of scales is offered by WHO [Marante]. WHO introduces four general categories of causality: Related, Probable (Likely), Possible, Unlikely, Unrelated. In addition, according to WHO, there are two more types of causality categories, which are not applicable for clinical research: Conditional/Unclassified (if more data is needed for proper assessment or when additional data is being analyzed) and Non-accessible/Unclassifiable (if the information is insufficient or contradictory and cannot be completed or verified). WHO argues that causality can be classified only as “Related” or “Not Related” for clinical research in some countries and spontaneous reports.

Finally, the initial information for scaling is context-sensitive [Marante]. All of the above complicates the problem of constructing an ADR rating scale, localized for a particular country and accessible to the end user, whether it is a doctor or a patient.

Thus, the authors of the article set themselves the following tasks:

- conduct a comparative analysis of data on ADR obtained from different sources available in Russia;
- at a substantive level, evaluate the quality of ADR data available in Russia;
- outline ways to improve Russian ADR data quality.

3. METHODS AND MATERIALS

The following resources were used as data sources: official sources – Russian Medicines Register [RMS] and manufacturers' instructions for specific medicines, and Internet sources – large Internet forums containing reviews about medicines [Doctors], articles about specific medicines on the Wikipedia.org website, as well as the APIs of VK, Russia's largest social network. We used an open database MedDRA, translated into Russian, as a database of side effects.

To collect information about side effects from Internet resources, a rule-based algorithm has been developed and implemented. The algorithm considers not only the mentions of side effects, but also the presence of words that may deny the side effect. The experimentally estimated accuracy of the constructed parser was 83%.

In order to compare data on ADR obtained from different sources we used (1) Venn diagrams showing the number of matches in the identified ADR from different sources; (2) graphs of pairwise interactions and analysis of their connectivity; (3) scaling interactions analysis. We used two types of graphs. The first type was based on the entities “medicine – side effects” and shows the presence of a connection between them. To build the graph, we used the networkx (graph construction) and plotly (graph visualization) Python libraries. An example of a graph is presented in Figure 1. The violet-colored nodes correspond to drugs, and pink-colored nodes correspond to their side effects.

The second type of graph is based on the entities "drug - drug", and the link between a pair of vertices reflects the presence of contraindications to their joint use. The graph is based on a complete database of drug interactions, whose full structure is not given in the article due to cumbersome. The database is built on data taken from the Russian Medicines Register site and contains 2187 drug interactions. To build the graph, the tables drug_substance, drug_substance2, type and the python pandas library were used. The graph in a simplified form is shown in Figure 2. Based on the full graph, one can construct partial subgraphs, which related to a specific drug or a specific degree of interaction (Figure 3, a, b). The edges between the vertices were colored to mark the degree of mutual influence of drugs.

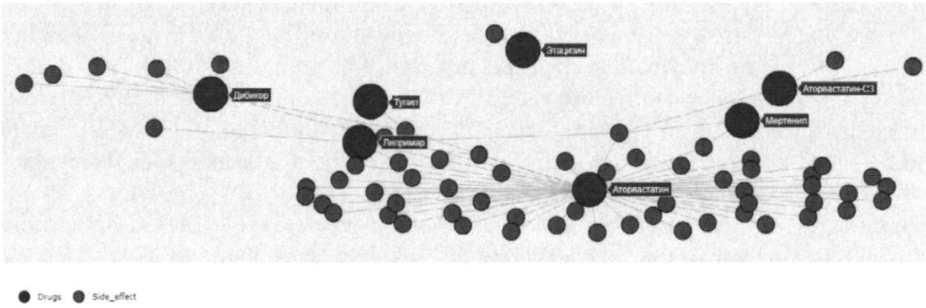

Figure 1. An example of a "drug – side effects" relationship graph

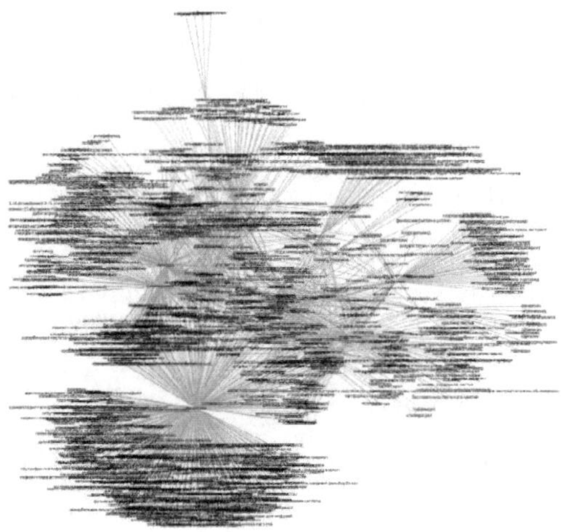

Figure 2. Graph of interactions of medicial substantions under consideration

Figure 3. Examples of graphs of interactions of medicinal substances: a- disopyramids, major interactions; b - lappaconitine hydrobromide, moderate interactions

To assess the quality of the data [Batini, Mosley, Romero, Na Hong, Van den Berghe], we used the following indicators:

- Actuality of data – a designated or indirectly determined point in time in which the data reflects the real state of the target object.
- Relevance of data – how the data on the real state of the target object corresponds to the problem being solved.
- Data integrity – the completeness of reflection of the real state of the target object in terms of semantic consistency of data and compliance with a given format.
- Measurability of data – the presence of qualitative or quantitative characteristics in the data of the real state of the target object.
- Source characteristics.

4. RESULTS AND DISCUSSION

4.1 Analysis of ADR Data Integrity

As shown by the analysis, data completeness is one of the key problems that reduce the quality of ADR data, and it can take place with varying degrees of severity [Soyer]. Note that the problem affects not only spontaneous sources of information about ADRs [Oh], but also professional ones. For example, [Soyer] states discrepancies between the ADRs identified by medical records technicians and the ADRs identified by the pharmacovigilance team.

Data integrity is not fully ensured even in an official source of medical information as the Russian Medicines Register. For example, among the groups of drugs considered in this article, adenosine and lappaconitine hydrobromide are presented, the simultaneous administration of which can cause side effects, however, mention of ADR is contained only in the description of lappaconitine hydrobromide and is not in the description of adenosine. In addition, a number of drugs are presented in the Russian Medicines Register, the interactions of the constituent substances of which are unexplored. It should be noted that the Russian Medicines Register serves as a source for drawing up instructions for medicines that are sold in pharmacies and are easily accessible to end consumers without medical supervision.

To study the integrity of ADR data most accessible to end consumers, we com-pared the data related to Atorvastatin from all the sources mentioned in Section 3. The eliciting of ADR data from the Wikipedia article and from the official instruction for the drug was carried out by experts. To retrieve ADR data from online forums and from the social network VK a rule-based algorithm was implemented. After text preprocessing, the algorithm implements the following rule: a side effect is present if the word denoting it is present in the text and there are no words negating the side effect. The algorithm was preliminarily tested on a Russian-language database compiled by an expert method, with the final F-score of 0.82. The results are presented in the Table 1. The Algorithm line corresponds to the application of the written algorithm to Internet forums, the Algorithm+VK line corresponds to the application of the algorithm to the VK network.

Table 1. Information on ADR for Atorvastatin obtained from different sources

Source	**Adverse effects**
Wikipedia	insomnia, headache, asthenic syndrome; nausea, diarrhea, abdominal pain, dyspepsia, flatulence, constipation; myalgia
Instruction	constipation, flatulence, dyspepsia and abdominal pain. headache, asthenic syndrome, insomnia, dizziness
SIDER	nasopharyngitis, pain in the extremities, urinary tract infection, musculoskeletal pain, muscular spasm, insomnia, pharyngolaryngeal pain
Algorithm	necrosis, myocardium, atherosclerosis, pain, surgery
Algorithm+VK	pain, hemiplegia, myocardium, irritability, inflammation, gout, bedridden, arthritis, separation, atrophy, osteoporosis, stress, disability, drop, aneurysm, hypertension, paralysis, cough, ulcer, tachyarrhythmia, depression, thyroiditis, hepatitis, hypertonic, atherosclerosis, trauma, alcoholism, obesity, numbness, dehydration, pancreatitis, toothache, myasthenia gravis, gas, diuresis, acne, drowsiness, infectious disease, eczema, overdose, psoriasis, constipation, headache, fatigue, suicide, irritability, death, migraine, lethargy, asthma,, neuropathy, hallucination

The comparison results are graphically presented in Figure 4 in the form of Venn diagram. The diagram shows the coincidence in the lists of side effects of the same medication (Atorvastatin), mentioned in different sources. To determine the similarity measure of side effects detected from various sources, we used the Jacquard coefficient $K_{ij} = ca + bc$, where a is the number of side effects in group i, b is the number of side effects in group j, c is the number of side effects values common for groups i and j. On the Venn diagram, for each two, three, four, and five sources c values are shown.

Table 1 shows that the coincidence in the lists of adverse effects of the same drug, mentioned in different sources, is rather weak. This is also confirmed by the Jacquard coefficients (see Figure 4). It is noteworthy that the list of side effects highlighted by users of the VK network significantly exceeds the lists provided by official sources. This fact can serve, as an indirect confirmation that the information on ADRs received from end consumers, at least, cannot be ignored.

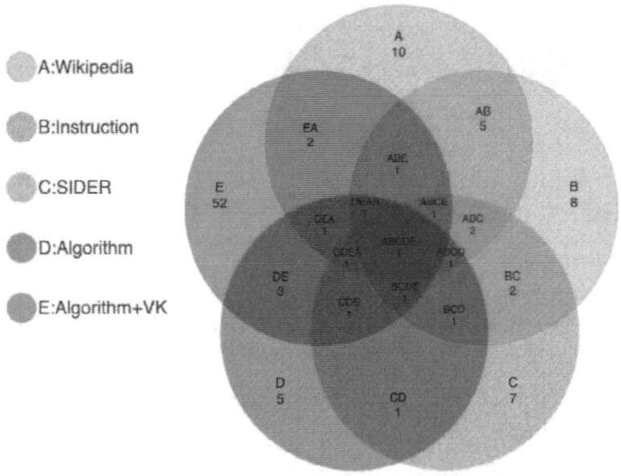

Figure 4. Venn diagram for comparing information on side effects of astrovastatin obtained from various sources

4.2 Analysis of ADR Data Measurability

Measurability and source characteristics as indicators of ADR data quality are interconnected and context-dependent. Therefore, we consider them together.

In official sources, such as instructions for medicines, different scales are used. The measured parameter can be the frequency of occurrence of a side effect, but it can be measured on different scales. For example, for Atorvastatin, one gradation is used – either the frequency of ADR occurrence is more or less than 2%. For Nifedipine, a linguistic scale "often – rarely – very rarely" is used.

There is also scaling according to ADR severity. For example, for Aspirin Cardio the gradations used are "contraindicated" and "with caution". However, we did not find simultaneous scaling of ADRs in terms of frequency of occurrence and severity of consequences, as required by the recommendations of the WHO (see section 2.2).

The characterization of official sources of ADR information is the responsibility of the relevant authorities and is outside the scope of this article. However, our analysis shows that now it is no longer possible to ignore network sources of information about ADRs where end-users of medicines leave messages, whether they are doctors or patients.

Our study showed a decline in user activity on online forums related to ADRs. So, the most popular drug on the forums (Atorvastatin) had a little more than 30 reviews, which is critically small for a full analysis of text data. At the same time, there is a growing popularity of discussing issues related to ADR in social networks. For example, from the social network VKontakte it was possible to collect 1.97 times more reviews than from all medicine forums combined.

In addition, we noticed the peculiarity of the comments about medicines in the Russian-speaking audience compared with the English-speaking one: in the first case, opinions are more often expressed in polar ("perfectly - terrible"), in the second case (for example, on DailyStrength Internet-resource) a wider range of opinions is presented. This must be taken into account when forming a scale for assessing the reliability of information about ADRs.

4.3 Summary Characteristic of ADR Data Quality from Various Sources

The analysis performed allows us to characterize the quality of ADR data available in Russia at a substantive level.

The actuality of the data is not fully respected; even in official sources, there are examples of late updates of data on specific medicines, leading to ADR information out of sync.

The relevance of ADR data in the realities of modern Russia cannot be provided only by official sources. As analysis shows, data on the use of drugs after clinical trials, i.e. in real practice, published by end consumers, are practically not involved in official sources. At the same time, extracting data on ADRs from network sources today is quite easily implemented in software: modern Russian-language NLP systems oriented to extracting ADRs provide fairly good accuracy (F-score 0,8-0,9), and new algorithms are constantly being offered to improve it.

Data integrity is significantly violated (see section 4.1). The examples we demonstrate raise the question of the degree of trust in publicly available data on side effects. It should be noted that the problem of the integrity of ADR data is acute not only for Russia (see, for example, [Soyer, Mavrogiorgou, Jeong, Oh]), but, as far as we know, it is not considered in the Russian scientific literature even at the level of problem statement.

The measurability of data in official sources is supported, but different scales are used, which makes it difficult to share data in digital processing of medical information.

As can be judged by the results of the analysis, the data on ADRs obtained from official sources weakly correlate with similar data from spontaneous sources, at the same time, the popularity of messages on ADRs in the VK network is quite high, and this source of information on ADRs can no longer be ignored. However, for its introduction into clinical practice, it is necessary to develop appropriate methodological support.

5. WAYS TO IMPROVE ADR DATA QUALITY

The analysis confirms the opinion [Lepakhin] that the system for collecting and recording ADR data in Russia needs to be improved, which requires significant resource investments and organizational changes. The following are measures that can be seen as steps towards improving the quality of ADR data from spontaneous sources.

(1) In the database on the interaction of medicinal substances, it is advisable to enter the field "The degree of reliability of the primary source". A fragment of such a base is shown in the figure 5.

The drug_substance1 and drug_substance2 tables contain information about medicinal substances. Each drug substance may be contained in several drugs. The type field - a scale for marking the degree of ADR - and the credibility field - the reliability of information on the interaction of drugs, are added to the Connection table. It is a lightweight and scalable solution that can be easily integrated into different types of databases. At the same time, it is consistent with WHO recommendations.

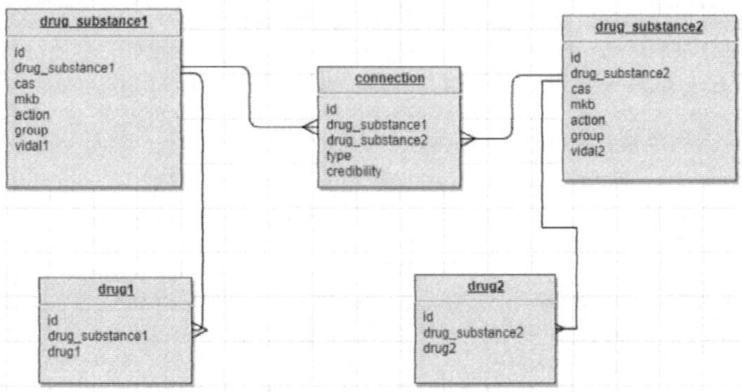

Figure 5. Database of drug interactions with fields describing the degree of interaction and the degree of reliability of primary information

(2) For assessing the danger level of ADR manifestation, it is advisable to introduce a four-level scale similar to [Micromedex], and for the convenience of consumers, to label it with colors. An example of such a scale is shown in Figure 6 (note that this marking is used in Figure 3-4).

The use of such a scale simplifies the visualization of interaction matrices between drug substances using graphs such as Figure 3-5, allows the doctor and patient to integrally assess the situation and facilitates the decision-making on a specific drug. At the same time, to assess the degree of reliability of the primary source, it is sufficient to use a two-point scale, which corresponds to the WHO recommendations (see section 2.2) and at the same time is easily visualized on the drugs interaction graph.

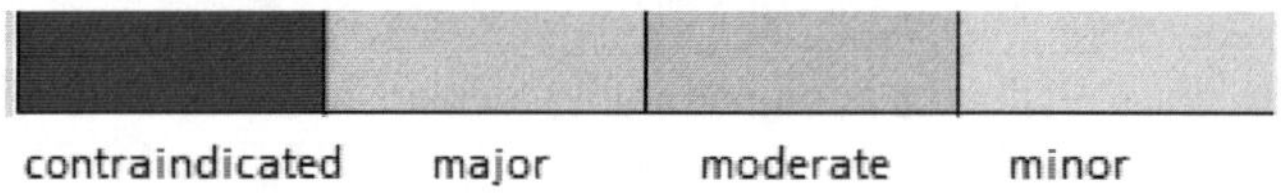

Figure 6. Proposed ADR grade scale

6. CONCLUSION

The results of quality analysis of data on ADRs for groups of drugs which are most frequently discussed by users of Internet resources were presented. The tasks set in the article have been fully carried out. Namely:

- A comparative analysis of data on ADR obtained from different sources available in Russia was conducted;
- The quality of ADR data available in Russia was evaluated. Problematic quality parameters of ADR data were identified - data integrity and measurability, as well as insufficient methodological support for retrieving and using the information from spontaneous data sources
- Ways are proposed to improve the quality of ADR data from spontaneous sources of information, which, on the one hand, will facilitate their use by end users, whether they are doctors or patients, and on the other hand, it will be easier to combine with ADR data from official sources.

ACKNOWLEDGEMENT

This work was financially supported by Russian Science Foundation, Grant 19-19-00696.

REFERENCES

[Antipov] Antipov E.A., Pokryshevskaya E.B. The effects of adverse drug reactions on patients' satisfaction: Evidence from publicly available data on Tamiflu (oseltamivir). International Journal of Medical Informatics 125 (2019) 30–36

[Baranov] Baranov A.A. et al. Technologies for Complex Intelligent Clinical Data Analysis. Annals of the Russian Academy of Medical Science. 2016; 71 (2). 160–171. doi: 10.15690/vramn663

Batini C. et al. Methodologies for data quality assessment and improvement. ACM Comput. Surv. 2009; 41, 3: Article 16, 52 p.

[Clinical] Clinical NLP in Languages Other Than English. https://amia2017.zerista.com/event/member/388973

[Doctors] Doctors rewiev about drugs. https://protabletky.ru/; Drug rating, http://topmeds.ru/, last accessed 2020/02/19.

Federal Service for Surveillance in Healthcare (Roszdravnadzor) Homepage, https://www.roszdravnadzor.ru/en, last accessed 2020/02/19.

[Foster] Foster JM, van der Molen T, de Jong van den Berg L (2007) Patient Reporting of Side Effects may provide an Important Source of Information in Clinical Practice. European Journal of Clinical Pharmacology 63: 979–980.

Good pharmacovigilance practices. European medicines agency. https://www.ema.europa.eu/en/human-regulatory/post-authorisation/pharmacovigilance/good-pharmacovigilance-practices, last accessed 2020/02/19.

[Jeong] Jeong Yun Park. Analysis of data on capecitabine-related adverse drug reactions from the Korean adverse event reporting system database. European Journal of Oncology Nursing 34 (2018) 55–60

[Khan] Khan L.M et al. Dilemmas of the causality assessment tools in the diagnosis of adverse drug reactions. Saudi Pharmaceutical Journal. Volume 24, Issue 4, July 2016, Pages 485-493

[Lepakhin] Lepakhin V.K. et al. The landmarks of control and licensing system develioment for medicinal products un Russia (25 years since the establishment of the first state institution for expert evaluation of vedicinal products). Scientific Centre for Expert Evaluation of Medicinal Products Bulletin. 2016. №1. Pp. 3-10/

Lushnov [] Lushnov M. et al. Medical Knowledge Representation for Evaluation of Patient's State Using Complex Indicators // International Conference on Knowledge Engineering and the Semantic Web. – Springer International Publishing, 2016. – C. 344–359.

[Maheshwari] Maheshwari Paramasivam R. Sivabalakrishnan, N. Pooja. Evaluation of drug - drug interactions in urban area community pharmacies. Research Journal of Pharmaceutical, Biological and Chemical Sciences 7(4):1420-1424. January 2016

[Marante] Marante K.B. The Challenges of Adverse Drug Reaction Evaluation. J Pharmacovigil 2018, 6:3.

[Mavrogiorgou] Mavrogiorgou A. et al. Analyzing data and data sources towards a unified approach for ensuring end-to-end data and data sources quality in healthcare 4.0. Computer Methods and Programs in Biomedicine 181 (2019) 104967

[MetaMap] MetaMap - A Tool for Recognizing UMLS Concepts in Text. https://metamap.nlm.nih.gov/, last accessed 2020/02/19.

[Micromedex] Micromedex Drug-Reax (2017). IBM Watson Health, Greenwood Village, Colorado, USA. Available at: https://www.micromedexsolutions.com/, last accessed 2020/02/19.

[Mosley] Mosley M., & Brackett, M. (2010). The DAMA guide to the data management body of knowledge (DAMA-DMBOK guide), first edition. Bradley Beach, N.J.: Technics Publications.

[Na Hong] Na Hong et al. A computational framework for converting textual clinical diagnostic criteria into the quality data model. Journal of Biomedical Informatics 63 (2016) 11–21

[Oh] Oh I-S. et al. (2019) Differential completeness of spontaneous adverse event reports among hospitals/clinics, pharmacies, consumers, and pharmaceutical companies in South Korea. PLoS ONE 14(2): e0212336. https://doi.org/10.1371/journal. pone.0212336

[Prueksaritanont] Prueksaritanont T. et al. Drug-drug interaction studies: regulatory guidance and an industry perspective. AAPS J. 2013 Jul;15(3):629-45. doi: 10.1208/s12248-013-9470-x. Epub 2013 Mar 30.

[Romero] Romero F.P. et al. An approach to web-based Personal Health Records filtering using fuzzy prototypes and data quality criteria. Information Processing and Management 48 (2012) 451–466

[RMR] Russian Medicines Register. https://www.rlsnet.ru/rlsnetshop_show_details_82.htm, last accessed 2020/02/19.

[Soyer] Soyer J. et al. Identification of discrepanies between adverse drug reactions coded by medical records technicians and those reported by the pharmacovigilance team in pediatrics: An intervention to improve identification, reporting, and coding. Archives de Pediatrie 26 (2019) 400–406

[Talbot] Talbot J., Aronson J.K. (Eds.) Stephens' Detection and Evaluation of Adverse Drug Reactions: Principles and Practice, Sixth Edition (2011)

[Van den Berghe] Van den Berghe S., Van Gaeveren K. Data quality assessment and improvement: a Vrije Universiteit Brussel case study. Procedia Computer Science 106 (2017) 32 – 38

[Vatian] Vatian A. et al. (2018) Adaptation of Algorithms for Medical Information Retrieval for Working on Russian-Language Text Content. In: Sojka P., Horák A., Kopeček I., Pala K. (eds) Text, Speech, and Dialogue. TSD 2018. Lecture Notes in Computer Science, vol 11107. Springer, Cham

[Zhang] Zhang P. et al. Label Propagation Prediction of Drug-Drug Interactions Based on Clinical Side Effects. Nature. Scientific Reports 5, 12339 (2015). https://doi.org/10.1038/srep12339

ENGAGEMENT AND USAGE PATTERNS OF A MOBILE APPLICATION TO MONITOR DISEASE AND TREATMENT ADHERENCE IN PATIENTS WITH ASTHMA

Rute Almeida[1], Cristina Jácome[1], Pedro Vieira-Marques[1], Ana Pereira[1], Ana Sá-Sousa[1], Rita Amaral[1, 2], José Valente[1, 3] and João Almeida Fonseca[1, 3]

[1]*CINTESIS - Center for Health Technology and Services Research & MEDCIDS - Department of Community Medicine, Health Information and Decision, Faculty of Medicine, University of Porto, Porto, Portugal*

[2]*Porto Health School, Polytechnic Institute of Porto, Porto, Portugal*

[3]*MEDIDA - Serviços em Medicina, EDucação, Investigação, Desenvolvimento e Avaliação, LDA, Porto, Portugal*

ABSTRACT

Inhaled therapies are the cornerstone of effective asthma treatment and adequate inhaled medication adherence (IMA) is critical. mHealth technologies have shown to be promise for asthma self-management, including IMA improvement. InspirerMundi is a mobile application designed to support self-management of patients with asthma. It aims to transform the adherence to treatment into a positive experience through gamification and social interaction while allowing for ubiquitous verified IMA monitoring. The app includes an image-based inhaler usage detection tool, tools for reporting symptoms and burden of the asthma (disease monitoring) and Game & Peer support features. Still, the effectiveness of these tools depends on a regular app use and a real-life assessment of patient engagement and pattern of use is needed.

This work evaluates the patient engagement, usage, and acceptance of InspirerMundi app during a real-world multicentre feasibility study. The app use was recommended for a 4-months period to 77 participants with persistent asthma. From those, 72 installed the app, with 67% of them beginning to use both the inhaler usage detection tool and the disease monitoring components within the first week. Over 95% used it more than once, with the period of usage (from first use to last registered monitoring in 2018) being over 30 days for almost 70% of the users, and over 90 days for around 35%. Nevertheless, the usage rate (ratio of the number of days with app usage and the period of use) had a median value of 0.6 and was above 75% for only 35% of the users, revealing room for improvement.

In general, the users started to use the app features right after installation and the usage patterns and retention rates indicate that InspirerMundi is well accepted among patients with asthma.

KEYWORDS

mHealth, User Engagement, Medication Adherence, Asthma

1. INTRODUCTION

Recent technology developments have been promoting patient empowerment for self-care more than ever before. The ubiquitous presence of smartphones and tablets, at increasingly low cost and computational power, create a clear opportunity for the development of customized mobile Health (mHealth) solutions for self-monitoring and management of chronic diseases. Solutions taking advantage of the embedded sensors of smart devices, with no external accessories, are even more attractive as they are easy to integrate into everyday lives and to disseminate without additional costs for patients.

Asthma is an increasing incidence, high-burden, chronic obstructive respiratory disease, starting most often in early life and affecting 434M individuals worldwide (*Forum of International Respiratory Societies*, 2017). The long-term goals of asthma management are to reduce symptoms and the severity and frequency of exacerbations, to which those patients are prone to (GINA, 2020). Inhaled therapies are the cornerstone of effective asthma treatment and adequate inhaled medication adherence (IMA) is critical. Insufficient IMA is associated with poor symptom control, exacerbations, emergency department visits, and hospitalizations, leading to disease progression, additional social burden and health costs (Antoniu, 2010; Doz *et al.*, 2013).

Adherence to medication for chronic diseases has reported rates from 15% to 70% (Engelkes *et al.*, 2015; Bender, 2016). Considering the impact and the strong evidence showing that poor IMA is a major obstacle to reduce that burden, the development of effective strategies to improve IMA is critical. Recognizing this problem, key stakeholders stated the urge of putting IMA on the EU policy agenda (van Boven *et al.*, 2017).

To improve IMA, we need to measure it objectively. Research shows that clinicians cannot accurately estimate patient adherence and that self-reported adherence to treatments tends to be fictionally high (by more than 50%) (Ose *et al.*, 2012; Bender, 2016; Ferreira *et al.*, 2019; Jácome *et al.*, 2019). The objective measurement of IMA is difficult, and several attempts have been made to make it more effective. Electronic monitoring devices that register IMA by logging the date and time an inhaler is actuated gained some acceptance, providing detailed information about patterns of medication use, gained some acceptance (Chan *et al.*, 2015). Randomized trials in asthma demonstrated that these devices have benefits on IMA (up to 84%), but they were not sustained over long periods of time (Morton *et al.*, 2017). Also, their wide dissemination is difficult and costly. Implementation of low-cost, ubiquitous solutions for measuring IMA, easy to disseminate and accept in healthcare services, would impact large numbers of patients and are therefore warranted.

mHealth technologies proved promising for asthma self-management, including IMA improvement, allowing the combination of data gathering and computation, gamification elements for user engagement and support to behavior change (Deterding *et al.*, 2011; Marcano Belisario *et al.*, 2013; Haase, Farris and Dorsch, 2017; Sardi, Idri and Fernández-Alemán, 2017), although, they generally do not take full advantage of the available technology (Tinschert *et al.*, 2017). Gaining and maintaining a person's awareness and attention is the first step in promoting behavior change, through the addition of elements of fun together with peer support to promote behavior change (DeSmet *et al.*, 2014), namely on IMA (Bousquet *et al.*, 2017; Kosse *et al.*, 2017; Miller *et al.*, 2017). With that in mind, the InspirerMundi app (Figure 1) was conceived as an attempt to transform the adherence to treatment into a positive experience through gamification and social interaction, while allowing for ubiquitous verified IMA monitoring. This work evaluates the patient engagement, usage and acceptance of the app during a real-world multicentre feasibility study.

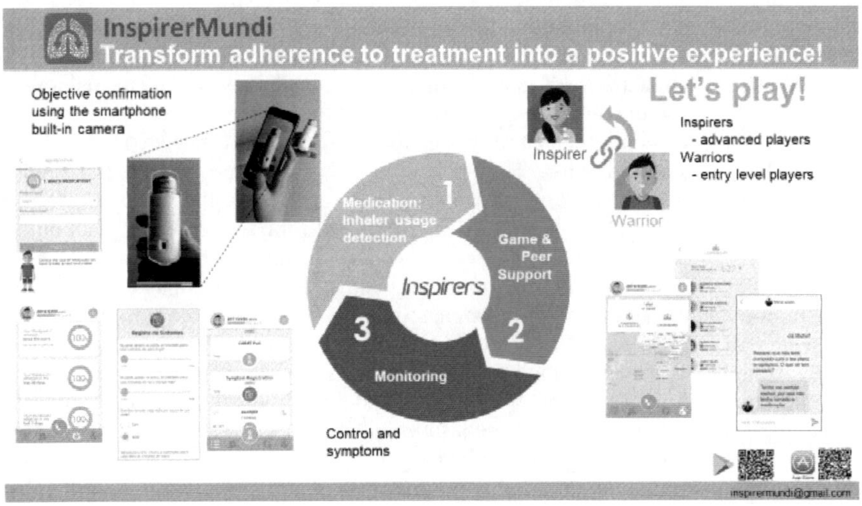

Figure 1. InspireMundi components

2. METHODS AND DATA

InspirerMundi aims to promote beneficial behavior changes in asthma patients' daily routine, with immediate and enjoyable feedback, and the delivery of easy-to-disseminate measures of IMA. The development phases have been previously described (Jácome et al., 2017). The app currently integrates different components regarding: 1. Medication, allowing to schedule alarms and including an image-based inhaler usage detection tool, 2. Game & Peer support features and 3. Monitoring, to report asthma symptoms and burden (Figure 1). The functioning of the app is illustrated in videos available at the YouTube Chanel (InspirerMundi, 2020).

To address the need for objective IMA measures, the Inhaler detection tool uses the embedded mobile camera to capture a short video stream of the inhaler presented by the patients. Each frame is processed to confirm the type of inhaler (out of 6 available) using a double template, one with the inhaler outline and other with the counter outline. For inhalers without a correspondent template on the App, as well as for other types of medication, the usage detection module just collected a photo of the medication. Along with the confirmation of the type of inhaler, the app has a character recognition tool that can be used to assess the number of remaining doses for any inhaler providing a number-based dose counter (Vieira-Marques *et al.*, 2020)

As is common in other gamified products, the app makes use of several game mechanics like points, badges and a leaderboard to engage the user. These are part of the game & peer support component dynamics. Points can be earned using app features, except for unplanned use of reliever medication. Reaching certain goals is rewarded with virtual badges. By getting to higher game levels the user earns the possibility of evolving from Warrior, a beginner player, to Inspirer, a higher-grade player that has the possibility to socially influence others. Each Warrior can associate himself to an Inspirer, seeking peer support; allowing to exchange messages. These features enable Inspirers to stimulate Warriors to be compliant with the therapeutic plan. Users can see their position on the game leaderboard and visualize the network of influence on a world map.

In addition to the introduction of the therapeutic plan (medication name, dose, treatment duration, dosage, and schedule, when applicable), the app allows to register intakes, both for scheduled (including data from the Inhaler detection tool) and reliever medication. Reminders are generated to help the user to follow the plan. Asthma symptom control and burden are monitored through patient-reported outcomes measures, including- daily and weekly questionnaires, and the CARAT-Control of Allergic Rhinitis and Asthma Test, which is presented to the user on the onboarding day, and repeated on personalized intervals of 1, 2 or 4 (default) weeks.

Introducing a therapeutic plan, registering medication intakes, and answering questionnaires are crucial to the correct use of the App, while actions related to the social component are optional. The mobile application was developed in native Swift language for iOS and Java for Android. In the web component, Drupal 7 was used as the content management system, as it is known to have a powerful ability to support Web services, and data was stored in a MySQL database. The use of the InspirerMundi App was assessed during 2018 using iOS versions 1 and 1.1 and Android versions 1.0.x, available at the mobile application stores under Health and Fitness category. These versions included templates for only 6 types of inhalers (Diskus/Accuhaler, Ellipta, Flutiform, Novolizer/Genuair, Spiromax, Turbohaler), which were the most used in Portugal.

Data from the real-world multicentre prospective observational feasibility study mInspirers 2018 study were analyzed (Jácome *et al.*, 2019). It enrolled 77 adults and adolescents (>=13 years old) with a previous medical diagnosis of persistent asthma, having an active prescription for an inhaled controller medication and followed by clinicians from the INSPIRERS group. The study did not interfere with the prescribed medications and all inhalers were admitted, even if their template was not included in the app. Patients were recruited in the first semester of 2018 on an initial face-to-face visit (T0) at one of 24 hospital care centers (Allergy, Pediatrics, Pulmonology outpatient clinics) in Portugal. The study was approved by the ethics committees of all participating centers. Eligible patients were approached by physicians during medical visits and informed and agreed to participate adults signed a consent form will adolescents signed an assent form, and a parental consent form was also obtained. They were invited to download and use the InspirerMundi app during at least a four months period. In T0 was collected data from patient and physician, including physician-reported information on prescribed control inhaled medications). Additionally, the patients received three follow-up phone interviews, at one week (T1, when additional support in App install and onboarding was given, if required), at one month (T2) and at four months (T3). Data were collected on: demographics, patients' opinions on mobile applications, satisfaction with InspirerMundi (global and regarding concerning different specific components, integration of functionalities, ease of use, and confidence; using a 5-item scale from 1, lower, to 5, highest), willingness to use and recommend the app, as well as the perceived impact of use. Data from the app database collected up to the end 2018, including a minimum of six months after inclusion in the study, were retrieved. To validate the inhaler introduced in the therapeutic plan by the user, images captured using the image-based inhaler usage detection tool were distributed by 16 reviewers, who visually checked for validity (inhaler visible in the photo and agreement with the app registered inhaler type).

The day of T0 (patient inclusion in the study) was defined as day 1 and used as a reference for engagement metrics. ***Onboarding***, defined as the day of user creation in the app (first use), as well as the dates of use of specific features were included in the analysis. A user was considered ***active*** from onboarding until the last date of registered medication use or answered questionnaire in 2018. The ***period of use*** was defined as the number of days in between. A ***day with app usage*** is a day with at least one registered medication use or answered questionnaire. The ***usage rate*** was taken as the ratio between the number of days with ***app usage*** and the period of use. The arithmetic mean and standard deviation (mean[SD]) were considered to summarize engagement metrics, with frequency counts and median values also presented when most relevant. MATLAB 2018a) was used to perform statistical analysis.

3. RESULTS

From the 77 recruited participants, 72 (94%) used the InspirerMundi app at least once. Three quarters (n=53) installed the Android version. From the app users, 57% were female and ages ranged from 13 to 70 years old (27[12.7]), with 1/3 under 18 and only 3% over 50 years old. Higher education degree was reported by 21% and high school degree by 44%. Daily use of apps was reported by 82% of the users, 75% several times a day, download and use of health/fitness Apps by 54%, but only one had previously used an asthma related App.

Onboarding occurred within the first week after inclusion for 57 (79%) of the users (5.3[10.7] days from inclusion to onboarding). Figure 2 presents active app users per study week. More than 95% of the users used the app more than once and 65% used the App for more than 10 days. The 30-day retention rate (considering all users) and 90-day retention rate (with respect to the previous 90-days) were 70% and 50%, respectively. The app usage from onboarding up to six months after inclusion (2 months after the 4-month of use requested by the study period) are displayed in Figure 3, sorted by use period. Of notice is the fact that some users did not use the app immediately after onboarding and initial weeks can also have no days with usage. Use patterns differed across users and the number of days between consecutive App uses ranged from zero to 87. Nevertheless, more than 80% of consecutive app uses were on following days, as illustrated in Figure 4. The usage rate had a median value of 0.6 (the distribution is presented in Figure 4.b). The period of use was longer than 30 days for almost 70% of the users and longer than 90 days for around 35% (67.3[60.4] days). More than 20% of the users used the app after day 120, i.e. after the requested use period within the study.

By the end of the first week of use, 74% of the users had introduced at least one medication, while 67% also answered at least one questionnaire. The proportion of users with at least a medication and a questionnaire increased to 74% by the end of the second week. Three users inserted medication, but did not answer any questionnaire, five answered at least one questionnaire, but did not insert any medication, while three did not use any of the two components. The median number of days with use of the medication or questionnaire modules was 30 days (38[39]), and the maximum was 160 days, with 47 (65%) of the users returning to the app more than 10 times to use these features.

A total of 245 medications were inserted, 53 of which were reliever medications (repeating allowed). From the 138 entries for inhaled medication, 53 correspond to medication renew (same medication). Also, 38 entries were for reliever medication, 5 entries (4 different users) had the wrong type of medication chosen and one user indicated all 3 medications in the same entry (medication registration errors). The *Add medication* tool was used to insert control medication by 64 (88.9%) of the users (Table 1). Up to 846 medication intakes were registered per user (100[140] intakes/user). Considering only users that inserted control medications, 90% registered at least one intake, almost 70% registered intakes in over 75% of days with App usage (days in which they used the App to use medication or questionnaire modules) and 31% registered intakes in all days with App usage. In the validation of the inhaler type 1251 images were reviewed from the first study participants (18 at the time of analysis). An inhaler was visible and identifiable in 84% of the images and it was in agreement with the user's choice in 96% of them.

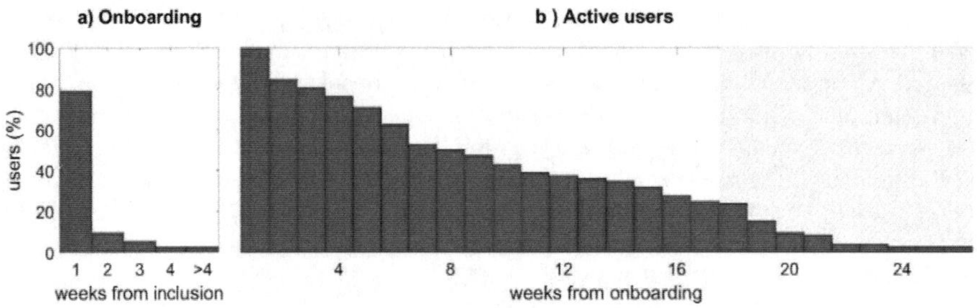

Figure 2. Panel a): Onboarding week (i.e. week of first use) from inclusion. Panel b: active users per week from onboarding (users with usage in that or posterior weeks) (the shaded area in grey identifies weeks after the four-month period of use requested by the study)

A total of 1634 questionnaires were answered by 66 users (92%), distributed as indicated in Table 2. Respectively 63% and 42% of the users answered the first daily / weekly questionnaire on the onboarding day, as suggested. Globally, the compliance with questionnaires answering, accounting for the period of use and periodicity, was 62% for daily and 21% for weekly questionnaires (58[31] and 18[18] per user, respectively). For 65% of the users, the first CARAT was answered on the onboarding day. Those with at least two answers, had a mean time interval of 4.7 weeks between CARAT responses. The number of answered CARAT was equal or above the requested minimum (one at each four weeks for the period of app use) for 23% of the users, with around 60% missing just one questionnaire.

Regarding the social support component, five Inspirers and 22 Warrior invitations were formulated by 14 (19%) study participants. Seven participants accepted to be a Warrior of other users (not necessarily participating in the study), while nine become Inspirers for someone else. Overall, only 10 (14%) were an active part of a social network of influence.

Satisfaction with InspirerMundi was reported as a five-item scale of agreement level as summarized in Table 3. Except for the social support component, most users reported to be highly satisfied with all other app features (score of 4-5), both in T2 and T3.

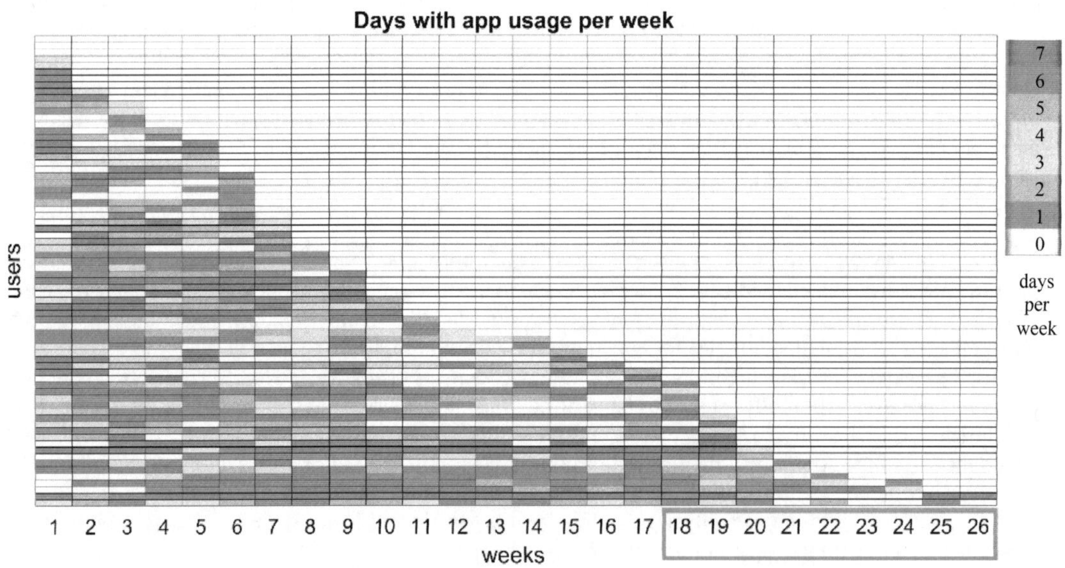

Figure 3. Weekly use, by user, from onboarding up to six months, sorted by use period (the square in grey identifies weeks after the four-month period of use requested by the study)

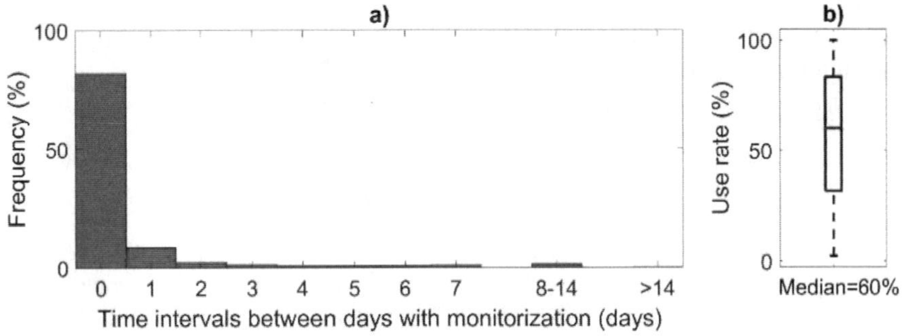

Figure 4. Panel a) Time intervals between days with monitorization (all users); Panel b) usage rate As usual in box and whiskers plot, the central box is limited by the first and third quartiles, with the median marked in red, and the vertical dashed lines reaching the minimum and maximum values of the distribution

Table 1. Description of medication inserted on the App using the "Add medication" option

Type of medication	Entries	Uses	Min-Max per user	Mean[sd] per user
inhaled	100	57	0 - 8	1.4[1.5]
pills	70	31	0 - 10	1.0[1.8]
other (e.g nasal)	22	19	0 - 3	0.3[0.6]
total	192	64	0 - 15	2.7[3.0]

Table 2. Number of questionnaires answered in the App

Type of questionnaire	Users	Answers	Min-Max per user	Mean[sd] per user
Daily	58	999	0-69	13.9 [17.1]
Weekly	65	451	0-24	6.3 [5.9]
CARAT	65	184	0-12	2.6 [2.1]1
Total	66	1634	0-91	22.7 [24.1]

Table 3. Satisfaction (4-5 rating) with InspirerMundi App

Satisfaction with respect to	T2 (n=65)	T3 (n=60)
Ease of use	86%	88%
Symptoms monitoring component	88%	87%
Willingness to recommend	85%	83%
Integration of the functionalities	86%	78%
Confidence in the App	80%	83%
Global satisfaction with the App	82%	77%
Adherence to medication component	75%	73%
Willingness to use	75%	63%
Gaming component	72%	63%
Image based component	58%	63%
Perceived impact of app use on medication adherence	58%	50%
Social support component	43%	52%

4. DISCUSSION

Most participants used the InspirerMundi App to register both medication and symptoms, and the short time between inclusion and first use, together with the reported high global satisfaction, are positive signs. As patients were enrolled in this study invited by their physicians, it was expected that most would become users. Nevertheless, effective monitoring and IMA promoting will only be possible if users adhere to the app, using it frequently and consistently. The user retention rates for Health and Fitness apps were reported in 2012 by Flurrymobile as 47% for 30-days and 30% for 90-days (Farago, 2012). In 2015, the IMS Institute for Healthcare Informatics noticed 30-day retention rates of 59% for mHealth apps tracked by the AppScript platform after informal recommendation by healthcare providers(Aitken and Lyle, 2015). They estimated an effect of the provider recommendation around 30%. More recent data, for apps across all industries, from the Adjust Mobile Benchmarks 2018 Edition (UK, France, Germany, Turkey, Japan, US, Russia, and China data) and 2018/2019 statistics from Localytics (an app analytics company in the USA), reported user 30-day retention rates below 15% (Adjust, 2018) and an average 30-days rate of 43% (29% for the average 90-days rate) (Perro, 2018), expressing the high variability of users behaviour. Also, the percentage of users who abandon an app after one session - user abandonment - was higher than 70% in the report by Adjust and higher than 80% in the one from Localytics (Adjust, 2018; Rodde, 2019). Regarding the user retention after 10 sessions, values from 31% to 39% have been reported. In this study, less than 5% of the users abandoned InspirerMundi after the first use and 65% used the App for more than 10 days, with values for the 30-day retention rate (considering all users) and a 90-day retention rates higher that the previously reported (70% and 50%, respectively). This comparatively high retention rate might be positively influenced by the app recommendation by a heath provider in the context of a clinical study, by the phone calls during the study period and the explicit recommendation to use at least for a four-month period. Although, more than 20% of the users used the App after the 120 days (approximately 4 months or 18 weeks) required by the study.

The 92% participants answering to at least one questionnaire is in line with the 85.2% (out of the 7593 enrolled users) replying to at least one survey in the Asthma Mobile Health Study (Chan *et al.*, 2017). Answering CARAT regularly is required to assess asthma control, which has been fulfilled by around 20% of the users, while 40% missed just one questionnaire. Registering the prescribed control medication is essential to take advantage of this App for treatment support and monitoring.

At least one inhaled control medication was registered by 57 (79%) of the 72 users; thirteen users did not register inhalers that were reported as prescribed by the physician. Most users registered medication intakes in the App in most days with App use. Nevertheless, for adequate IMA quantification, daily use was required. Although more than 80% of the data were introduced in consecutive days, the usage rate was above 75% in only 35% of the users. This indicates irregular monitoring, with sets of consecutive monitored days, followed by intervals of non-use or even leaving. Further analysis regarding the clinical data is required to determine if this is related to changes in the therapeutic plan or health deteriorating/improving cycles, resulting from asthma exacerbations. These results highlight the difficulty in achieving a continuous medication monitoring. As a mitigation strategy, the remaining doses counter can be used to assess all doses taken between verified intakes, comparing with the registered therapeutic plan, if the same inhaler is used, and estimate a more precise IMA rate (Vieira-Marques *et al.*, 2020).

The gamification component is intrinsic to the App use. By performing the medication and symptoms related tasks the users earned points, progressed levels and achieved related badges. Most users (81%) did not send any Inspirers or Warriors requests and, thus, they did not use the peer support resources at all. The social support component was also the one associated with lower user satisfaction. These features need to be reviewed in future app versions to become more attractive to the user and attain their value as tools that facilitates user engagement, promoting a daily, continuous and persistent app usage, which will allow a more effective IMA measure and support. The use of an app can also be evaluated through the number of actions that are performed by the users. That information is not collected by InspirerMundi, which limits the analysis to those metrics that are based on the information directly introduced by the users. Nevertheless, the main actions (related to scheduled medication and questionnaires) are initiated by the app at fixed times or according to the therapeutic plan, and, thus, the rate of fulfilment with respect to the planned might be a more meaningful approach to evaluate InspirerMundi app usage than the number of actions by itself.

Regarding the inhaler detection tool some important limitations are: 1) requiring the correct registration of the inhaled medication (done by the user), to allow the selection of the adequate template, and 2) requiring a specific position of the inhaler, during image collection, with respect to the specified template (no rotation or position adjustments are possible). Our data included entries from 5 users that were not able to insert adequately the type of medication; errors were related to the incorrect identification of the type of drug (as inhaler, pill or other) and one to the introduction of multiples medications under the same entry. These errors, although infrequent, suggested that the way the therapeutic plan was introduced by the user was not completely clear and should be improved. Moreover, when reviewing the inhaler images, it was also noticeable than even when the medication type was correctly selected as inhalation, some users did not correctly registered the specific device (thus no template was selected) and 4% of the images were associated with a wrong template. The app versions that were used in the study included templates for only 6 types of inhalers, which can also lead to errors especially. The updated version of the app has now 11 types of inhaler templates (Diskus/Accuhaler, Easyhaler, Ellipta, Flutiform, K-haller, NEXThaler, Novolizer/Genuair, Seretaide MDI, Spiromax, Turbohaler, Twisthaler), including almost all inhalers used in Europe in 2019 that have a numeric indication of the remaining dosages. Additionally, the quality of most images was very low, with many not showing the complete or not showing the inhaler at all. Previous usability tests had already shown that registering medication intake using the image-based detection tool was not straightforward and that some participants did not understand that they were supposed to use the smartphone's camera (Ferreira *et al.*, 2019). Based on these results, we modified the inclusion of the therapeutic plan in order to guide the user for the correct selection. The visual identification of the type of inhaler is now the first step when introducing the therapeutic plan; then, a full list of medications on the market that use that type of inhaler is presented. Explanatory videos (for medication registration and intake) were also made available.

5. CONCLUSIONS

The InspirerMundi app combines features of inhaler usage detection and disease monitoring allowing a reliable and ubiquitous collection of patients' health state and adherence to treatment, using only smartphone embedded sensors. To our knowledge there is no other available solution that allows an image-based objective measurement of inhaled medication adherence. In general, the users started to use the app features right after installation. Use patterns and retention rates indicate that this App can be well accepted among patients, at least if its use is suggested by a specialist. The data retrieved from the mInspirers study allowed to launch an updated version of the App, with improvements in the introduction of the therapeutic plan, currently available at the PlayStore (October 2019). Additional enhancements are required, specially regarding gamification and social components, in order to further improve user engagement.

ACKNOWLEDGEMENT

This work is partially supported by project NORTE-01-0247-FEDER-033275, financed by the North Portugal Regional Operational Programme (NORTE 2020), under the PORTUGAL 2020 Partnership Agreement, and through the European Regional Development Fund (ERDF). AIRDOC is with ITEA 3 16040 PHE - Personal Health Empowerment project consortium. R. Almeida is supported through the operation POCI-01-0145-FEDER-029130 (mINSPIRERS- mHealth to measure and improve adherence to medication in chronic obstructive respiratory diseases - generalisation and evaluation of gamification, peer support and advanced image processing technologies) funded by the COMPETE2020 and by National Funds through FCT (Fundação para a Ciência e a Tecnologia).

REFERENCES

Adjust (2018) *MOBILE BENCHMARKS 2018*.

Aitken, M. and Lyle, J. (2015) *Use, Evidence and Remaining Barriers to Mainstream Acceptance Patient Adoption of mHealth*. Available at: www.theimsinstitute.org.

Antoniu, S. A. (2010) 'Adherence to inhaled therapy in COPD: effects on survival and exacerbations', *Expert Review of Pharmacoeconomics & Outcomes Research*. Taylor & Francis, 10(2), pp. 115–117. doi: 10.1586/erp.10.10.

Bender, B. G. (2016) 'Nonadherence to Asthma Treatment: Getting Unstuck.', *The journal of allergy and clinical immunology. In practice*. Elsevier, 4(5), pp. 849–51. doi: 10.1016/j.jaip.2016.07.007.

Bousquet, J. *et al.* (2017) 'Realising the potential of mHealth to improve asthma and allergy care: How to shape the future', *European Respiratory Journal*, 49(5). doi: 10.1183/13993003.00447-2017.

van Boven, J. F. M. *et al.* (2017) 'Urging Europe to put non-adherence to inhaled respiratory medication higher on the policy agenda: a report from the First European Congress on Adherence to Therapy', *The European respiratory journal*. doi: 10.1183/13993003.00076-2017.

Chan, A. H. Y. *et al.* (2015) 'Using Electronic Monitoring Devices to Measure Inhaler Adherence: A Practical Guide for Clinicians', *Journal of Allergy and Clinical Immunology: In Practice*. Elsevier Inc, 3(3), pp. 335-349.e5. doi: 10.1016/j.jaip.2015.01.024.

Chan, Y.-F. F. Y. *et al.* (2017) 'The Asthma Mobile Health Study, a large-scale clinical observational study using ResearchKit', *Nature Biotechnology*. Nature Publishing Group, 35(4), pp. 354–362. doi: 10.1038/nbt.3826.

DeSmet, A. *et al.* (2014) 'A meta-analysis of serious digital games for healthy lifestyle promotion', *Preventive Medicine*. Academic Press, pp. 95–107. doi: 10.1016/j.ypmed.2014.08.026.

Deterding, S. *et al.* (2011) 'Gamification: Using game design elements in non-gaming contexts', in *Conference on Human Factors in Computing Systems - Proceedings*. New York, New York, USA: ACM Press, pp. 2425–2428. doi: 10.1145/1979742.1979575.

Doz, M. *et al.* (2013) 'The association between asthma control, health care costs, and quality of life in France and Spain', *BMC Pulmonary Medicine*. BioMed Central, 13(1), p. 15. doi: 10.1186/1471-2466-13-15.

Engelkes, M. *et al.* (2015) 'Medication adherence and the risk of severe asthma exacerbations: a systematic review.', *The European respiratory journal*. European Respiratory Society, 45(2), pp. 396–407. doi: 10.1183/09031936.00075614.

Farago, P. (2012) *App Engagement: The Matrix Reloaded.*

Ferreira, A. *et al.* (2019) 'How Inspiring is your App? A usability take on an App for asthma medication adherence', in *11th International Conference on e-Health (in press).*

GINA (2020) 'GLOBAL STRATEGY FOR ASTHMA MANAGEMENT AND PREVENTION'. doi: 10.15036/arerugi.45.792.

Haase, J., Farris, K. B. and Dorsch, M. P. (2017) 'Mobile Applications to Improve Medication Adherence', *Telemedicine and e-Health*. Mary Ann Liebert, Inc. 140 Huguenot Street, 3rd Floor New Rochelle, NY 10801 USA, 23(2), pp. 75–79. doi: 10.1089/tmj.2015.0227.

InspirerMundi (2020) *YouTube Channel - InspirerMundi App*. Available at: https://www.youtube.com/channel/UCCRiXqsttKoHiEGbbykLHxQ.

Jácome, C. *et al.* (2019) 'Patient-Physician Discordance In The Assessment Of Adherence To Inhaled Medication', in *EMJ Allergol Immunol.*

Kosse, R. C. *et al.* (2017) 'mHealth intervention to support asthma self- management in adolescents : the ADAPT study', pp. 571–578.

Marcano Belisario, J. S. J. S. *et al.* (2013) 'Smartphone and tablet self management apps for asthma.', in Marcano Belisario, J. S. (ed.) *The Cochrane database of systematic reviews*. Chichester, UK: John Wiley & Sons, Ltd, p. CD010013. doi: 10.1002/14651858.CD010013.pub2.

Miller, L. *et al.* (2017) 'Mobile Technology Interventions for Asthma Self-Management: Systematic Review and Meta-Analysis', *JMIR mHealth and uHealth*. JMIR Publications Inc., 5(5), p. e57. doi: 10.2196/mhealth.7168.

Morton, R. W. *et al.* (2017) 'STAAR: a randomised controlled trial of electronic adherence monitoring with reminder alarms and feedback to improve clinical outcomes for children with asthma', *Thorax*, 72(4), pp. 347–354. doi: 10.1136/thoraxjnl-2015-208171.

Ose, D. *et al.* (2012) 'Let's talk about medication: Concordance in rating medication adherence among multimorbid patients and their general practitioners', *Patient Preference and Adherence*. Dove Press, 6, pp. 839–845. doi: 10.2147/PPA.S35498.

Perro, J. (2018) *Mobile Apps: What's A Good Retention Rate?* Available at: http://info.localytics.com/blog/mobile-apps-whats-a-good-retention-rate (Accessed: 21 August 2019).

Rodde, T. (2019) *25% of Users Abandon Apps After One Use, localytics*. Available at: http://info.localytics.com/blog/25-of-users-abandon-apps-after-one-use (Accessed: 21 August 2019).

Sardi, L., Idri, A. and Fernández-Alemán, J. L. (2017) 'A systematic review of gamification in e-Health', *Journal of Biomedical Informatics*. Academic Press, pp. 31–48. doi: 10.1016/j.jbi.2017.05.011.

Sheffield, E. R. S. (2017) *Forum of International Respiratory Societies. The Global Impact of Respiratory Disease- Second Edition, Forum of International Respiratory Societies.* Sheffield: European Respiratory Society. Available at: https://www.fi (Accessed: 15 April 2020).

Tinschert, P. *et al.* (2017) 'The Potential of Mobile Apps for Improving Asthma Self-Management: A Review of Publicly Available and Well-Adopted Asthma Apps', *JMIR mHealth and uHealth*. JMIR Publications Inc., 5(8), p. e113. doi: 10.2196/mhealth.7177.

Vieira-Marques, P. *et al.* (2020) 'Combined Image-Based Approach for Monitoring the Adherence to Inhaled Medications', in, pp. 1399–1404. doi: 10.1007/978-3-030-31635-8_171.

DEVELOPING A REGIONAL SOLUTION TO DIGITAL TRANSFORMATION OF HEALTH CARE SERVICES: COLLABORATION, LEADERSHIP, AND TECHNOLOGICAL EXPERTISE

Hilde G. Corneliussen
Western Norway Research Institute
Box 163, 6851 Sogndal, Norway

ABSTRACT

This paper analyses a regional solution to digital transformation of health care services in Norway during the period 2014 to 2020. Facing the national health authorities' requirements for implementing welfare technology in public health care services, this rural region with its many small and medium-sized municipalities developed a unique framework for supporting this task across the small units. First, collaboration between organizations representing a wide set of competences, roles, and positions in the region and locally; second, a regional center responsible for developing health care services that – unlike other similar centers in Norway, took on a role to lead this work; and third, a regional emergency central originally developed for fire services taking on the role as the technical expert in the region. The analysis is based on documentation, project reports and interviews with participants from the project. While national guidelines for implementation of welfare technology have been important for motivating this work, many challenges were left for municipalities to solve, in particular the question pursued here: how should local and regional traditions be transformed to integrate technology in a sector and among professional groups that traditionally have not been associated with technology? The findings show that the regional model for collaboration supported the municipalities' health care services in this task by building on a regional model for cross-sectoral collaboration as well as regional trusted networks and channels of communication in unique ways in a national context.

KEYWORDS

Welfare Technology, Regional Development, Rural Regions, Collaborative Network

1. INTRODUCTION

Since the early 2010s, new digital technology, often referred to as care technology or welfare technology in Norway, has become increasingly important for the public health care services (NOU, 2011:11). This digital transformation has been motivated by an expected increase in the number of persons in need of care services, without a corresponding increase in health care personnel. In Norway, as in other western countries, this has created a need for delivering health care services in new ways (Moe et al., 2019). Welfare technology has been launched as one of the solutions that will enable the health care sector to maintain the quality of the services facing the increasing gap between need and personnel (Frennert & Baudin, 2019; Helsedirektoratet, 2012; Karlsen et al., 2019). The direction of this effort in Norway has been pointed out through national guidelines and recommendations, in particular emphasizing technology that can support delivery of health and care services in patients' homes, to increase safety and security (Melting, 2017; Melting & Frantzen, 2015). One factor that was not clearly defined in the early national guidelines was: how should local and regional traditions be transformed to integrate technology in a sector and among professional groups that traditionally have not been associated with technology?

This paper analyses how implementation of welfare technology in a geographically vast but sparsely populated rural region in the western part of Norway started from scratch in 2014 and has spread to all municipalities in 2020. The region is the former county Sogn & Fjordane with 26 municipalities with between 800 and 13.000 inhabitants. This region, rural and remote from the national capital, does not in all aspects reflect the larger and more densely populated regions that were among the main test areas for the early national advices; the needs in the region do not mirror a city, and many of the small and medium-sized municipalities did not have available personnel with the necessary technological expertise or the innovation and project competence to support the introduction of new technology in health care services (Corneliussen & Hove, 2017). Furthermore, the national guidelines were experienced as vague and difficult for the smaller municipalities, among other things because some of the recommended technologies were not available in small-scale versions (Hove & Corneliussen, 2017). On the other side, certain regional factors in Sogn & Fjordane contributed to the task of implementing welfare technology. This paper discusses three factors found to support the digital transformation in this region: first, collaboration between organizations representing a wide set of competences, roles, and positions in the region and locally; second, a regional center responsible for developing health care services that – unlike other similar centers in Norway, took on a role to lead this work; and third, a regional emergency central originally developed for fire services taking on the role as the technical expert in the region. This has developed a framework that has given the relevant institutions knowledge and ability not only to work with the ongoing digital transformation in health care, but also to react and move quickly when situations, like the recent Corona crisis, requires an immediate restructuring of health care services.

The analysis is based on documentation from a 6 year project for implementation of welfare technology in the region Sogn & Fjordane in Norway, including project reports, project notes and data material collected during the project period (Corneliussen, 2014b, 2015; Corneliussen et al., 2016; Hove & Corneliussen, 2017; Hove & Corneliussen, 2018). The data material includes interviews with administrative and health care personnel from municipalities' health care services, two surveys mapping the municipalities' use of and self-reporting knowledge about welfare technology, the researcher's notes from observations in the municipalities' working groups for implementation of welfare technology as well as from project meetings in the IT-forum working group described below.

Below we will explore each of the three success factors before we discuss how this has strengthened the region's ability to deal with challenging situations and crisis.

2. DEVELOPING REGIONAL KNOWLEDGE

Norway has a public health care system with the main responsibility for all nursing and health care services. Public health care represents one of the largest costs for municipalities that are responsible for providing such services, to, among others, a growing elderly population. While new digital technology includes solutions that can improve the quality of health care services, an equally important motivation for the current implementation of welfare technology is to meet the growing costs in the health care sector (Meld. St. 15 (2017 –2018); Melting, 2017; Melting & Frantzen, 2015). The goal is to use digital technology when it can benefit users as well as services. The national guidelines state that new technology can and should be a substitute for physical visits in user's home when selected on the basis of individual mapping of users' needs (Melting & Frantzen, 2015). Thus, in rural regions with long distances of travelling between users, welfare technology can have a major impact on the costs of the growing need for health care services. One such example that makes a difference for both users and services in this region, is digital medication dispensers that can remind users to take their medicine at the correct time. This can improve the quality of medication as well as reducing the number of home visits, or alternatively, providing more flexibility in the exact timing of home visits, for health care personnel (Corneliussen et al., 2016).

Profit realization and cost benefit analysis are keywords for monitoring and securing that new technology is in fact able to make savings. However, it takes time, effort, and money to figure out how to reach such goals from the first introduction of new technology, reorganizing the services, and to identify the actual benefit of new technology (Smaradottir et al., 2019). Furthermore, it requires a diversity of competences, from being familiar with technology, knowledge about innovation and restructuring processes, insight and knowledge about legal, ethical, and other practical questions and challenges that need to be at the forefront of such a digital transition. That means, this process requires competences that have not traditionally been a part of the health care personnel's education (Cajander et al., 2020). Facing these challenges, this particular region developed a solution that is different from other regions in Norway, by building a network inviting all the 26 municipalities together with a wide set of actors with different types of knowledge and competence with relevance for this work. The main building blocks that make this network special, are the way of organizing the collaborative network, the leadership of this work, and the technological expertise in the network.

2.1 Gathering Competence in a Collaborative Network

While efforts to introduce digital technology for support in health care services has a longer history, it was particularly from 2013 that the municipalities in Sogn & Fjordane started to engage in this work. In spring 2014 a collaborative initiative managed to secure funding for one year, and since then, this project has received economic support from the County Governor for the project "Welfare technology in the municipalities of Sogn & Fjordane", while the municipalities themselves contributed 50% of the funding. The group behind this was organized through a collaborative network with long roots in the region: IT-forum Sogn & Fjordane. This is a network established in the 1990s with the aim of supporting regional actors, public as well as private, to build infrastructure for and take up efficient use of digital technologies (Larsen & Corneliussen, 2016). This network has a unique way of organizing cross-sectoral collaboration that has been labelled the "Sogn & Fjordane model". The features of this model are that all relevant actors are invited, together they define the challenge and decide how to solve it, and in collaboration they work for enacting and implementing solutions (Corneliussen & Larsen, 2018). A working group with the name "welfare technology" was established under this network, and the participating partners of this group included municipalities with representatives from health care services, administration, and IT departments; Center for development of institutional and home-care services (USHT), a public organization responsible for fitting technical health care equipment for users, representatives from the county governor, the county and a high school, a regional university college and a research institute. In later years, this group has also included specialist healthcare services, a company developing exhibitions for health care technology, and an emergency central in the county.

One of the unique qualities of this network is that it is entirely voluntary to participate and the organization is quite loose. Although it is open for all relevant institutions and organizations, it is not a democratic group in the sense of representing "everybody", but rather based on "people who are willing to contribute" in the network and for the task at hand. While this can be challenging because it makes both authority and responsibility in the group vague, it also means that this collaborative network is not challenging the municipalities' or institutions' own power structures, which often seems to be one of the main challenges to overcome in such cross-sectoral and cross-institutional collaborations (Corneliussen & Larsen, 2018). At the same time, this network is recognized and respected in the region, giving it legitimacy and trust that facilitates the process of gathering the actors (Larsen & Corneliussen, 2016). Furthermore, one of the most valuable aspects of this network is the way it facilitates collaboration between organizations representing a wide set of competences, roles, and positions in the region and locally. This reminds of the bazaar in Raymond's analysis of the cathedral versus the bazaar (Raymond, 1999). While the cathedral represents a top-down structuring of a highly hierarchic organization, the bazaar represents a bottom-up organization were each representative brings her own specialty and expert knowledge. Each member of the group brought a particular background, experience, and knowledge that together made this group a creative and dynamic arena for developing competence within the digital transformation of health care services, which is a challenge that requires expert knowledge from many fields (Corneliussen & Larsen, 2018). This also meant that each member of the group could benefit from the network and bring valuable insights back to their own organization (ibid.).

The collaborative project described here has given legitimacy to the municipalities' efforts to implement welfare technology, a task that met many critical and skeptical reactions at first, from a fear for technology to replace human contact (Corneliussen et al., 2016), to a reluctancy towards "fumbling with something technical that we can't make work" instead of doing it "the old way" without technology support (Corneliussen & Hove, 2017). This group has strengthened the otherwise small actors, municipalities, and institutions, and eventually the group effort succeeded in securing funding from the National Welfare Technology Program for all municipalities in 2019 - an accomplishment that would not have been possible without the group effort. Thus, currently all municipalities in this region have united to work towards the health authorities' goal, which is to make welfare technology an integrated part of public health care services by the end of year 2020 (Helse- og omsorgsdepartementet, 2015).

2.2 A Regional Center for Development of Health Care Services

One of the success factors for this work was that the regional center, USHT, took on a role to lead this work. USHT has a mandate from the national health authorities that includes the responsibility for developing knowledge and competence among health care service institutions in the region. There are 21 similar centers across Norway with a responsibility for different regions of the country. The center in this particular region chose a different strategy than the other centers through this network, by taking a leading position in the work to encourage and help the municipalities implementing welfare technology. This gave a major advantage to the network group, as USHT was already recognized by health care units and institutions as well as the municipalities' administrative leaders across the county. USHT also has a wide network and is an arena for communication of knowledge and expertise in health care services in this region. The reason for USHT joining this group and taking on the role as a leader was in the beginning coincidental. However, recognizing that the IT-forum working group was a way of mobilizing institutions and actors that were not already within the center's own network made this collaborative group valuable also for USHT. Furthermore, it was an opportunity for USHT to see the implementing welfare technology in relation to other development processes in which the center already was guiding the municipalities. This produced a synergy in this work, strengthening the network of people who worked together. It also gave additional input to and motivation for several parallel processes of developing health care services that the municipalities were involved in by establishing recognizable patterns and methodologies for dealing with innovation and organizational change in the field (Helse- og omsorgsdepartementet, 2015).

2.3 Regional Technical Expertise

Facing the new requirements for implementing and integrating new technical solutions in health care services involves processes of fitting and developing organizational routines to a new way of providing health care. This project has illustrated that health care workers are used to working under pressure and to react quickly to solve urgent and critical situations, like one of the participants described it (Hove & Corneliussen, 2017). This gave a lot of momentum in the work once it had started. However, the same processes required a certain level of knowledge about the relevant technological solutions. As Corneliussen and Dyb have shown, the health authorities launched a discourse about welfare technology claiming that it had mostly to do with humans and organization of health care services, promoting the idea that welfare technology was something that could be implemented mainly with a focus on humans and organization (Corneliussen & Dyb, 2017). While this strategy can be seen as one way of motivating a non-technical health care profession to engage with a technology implementation process, this also undermined the need for technical competence in a way that made many health care leaders struggle with this new technology. This was particularly challenging for some of the smaller municipalities that did not have their own IT experts, but rather relied on an external organization of IT support shared between several municipalities (Corneliussen et al., 2016). One of the early participants illustrates some of the challenges created by this discourse: with a background in nursing and at the time administrative head of a municipality's health care services, this new process forced her to do what an IT person should have done, she says, and to attend the meetings that required IT competence, and to enter into negotiation with technology producers (Corneliussen et al., 2016).

This tendency of health care personnel being responsible for developing the services with new technology that they had weak knowledge about, sometimes even technology that had not been fully developed (Hove & Corneliussen, 2018), was one of the biggest challenges across the municipalities in the region. While this certainly was a challenge also in other region, it was made additionally complicated by the lack of technology experts in the municipalities' administrations. Also, this challenge found a solution that was specific to this region. An intra-municipality emergency alarm central, originally established for fire services, was already organizing the old type of analogue safety alarms aiming to support, in particular, elderly people to continue living safely in their own home. When these safety alarms gradually where exchanged for new digital alarms, this gave opportunities to connect other types of alarms in the range of more advanced welfare technology. Soon, the emergency central became a hub for knowledge about a wide set of concrete e-health solutions, and it increasingly took on a role of representing the technological expertise that the municipalities did not have themselves.

3. DISCUSSION

Implementation of welfare technology has been a long and cumbersome process for many municipalities across Norway. Health authorities optimistically claimed in 2013 that this process was to be completed by 2018. It has later become clear that it will take longer to make welfare technology a natural and integrated part of the municipalities' health care services. The project described here installed a unique regional solution to this challenge. All the municipalities have certainly put in a lot of effort in this project, and many of them would undoubtedly have managed to do this on their own as well. However, among the participating municipalities, many have claimed that they would not have been able to raise this effort as early as they did without the collaborative network and the project. The many small municipalities in this region cannot pursue this work the same way as larger units in more central and densely populated regions of the country. The IT-forum network, with its long history in the region of promoting the collaborative "Sogn & Fjordane-model" inviting a wide set of participants to an organizational loose but powerful working group, is a unique tool for achieving advanced results across the many small units in this rural region (Corneliussen, 2014a). In this project it made all the small and medium-sized municipalities able to work together as one large group, giving them strength that each one of them did not have on their own. The leadership of USHT in this process contributed by legitimating the work through recognized arenas for sharing and developing knowledge about the quality of health care services, and with a wide network of contacts among communities of practice, health care personnel and municipality administrators across the county. The emergency alarm central entered the scene in the first phase of the project with a sorely needed technological expertise that filled a gap that the health authorities' discourse of welfare technology had contributed to by undermining the need for technological competence.

While these three factors, the collaborative network, USHT's leading role, and the technical expertise of the emergency central, have been identified as crucial for supporting and helping the municipalities in this region into actively implementing welfare technology, they were not part of the national advices for this work. Instead, they grew out of the regional tradition, illustrating the innovative power in the "Sogn & Fjordane-model" of collaborating across units, institutions, and actors that represent different experiences, competences, and responsibilities, and the loose, yet widely trusted way of organizing this collaboration. The innovative work in the region has been communicated back to the national health care authorities and has contributed to a more nuanced understanding of regional and local conditions. Furthermore, the network of collaborative partners, the close contact between relevant institutions, organizations, and companies as well as the knowledge about local and regional conditions has also produced a flexible framework that can act quickly. When Corona made personal face-to-face contact difficult during the spring of 2020, this network was quick to react when a regional bank and a regional bank fund wanted to fund tablets for nursing homes in the region. With USHT in the lead, tablets were soon distributed through this network together with a quick-guide and a demo-video for employees, and user support for implementing video solutions for elderly users to speak with their relatives.[1]

4. CONCLUSION

"When I walk down the street, I feel like a superhero", one of the participants of the central hub of this network, IT-forum Sogn & Fjordane, has said, referring to how the network provides a collaborative strength to the many small units in this rural region (Corneliussen & Larsen, 2018). The framework for implementation of welfare technology in this region has indeed brought superpowers to the participants, by lifting the general level of knowledge and contributing to a recognizable network of people with knowledge and competence about welfare technology. Furthermore, the project has contributed to motivating the development of internal organization for the work in each of the municipalities. And finally, the achievements in the region have made the project and the relevant regional actors visible on the national arena. While none of the municipalities in this region were included in the initial round of funding from the National welfare technology program, all the

[1] See https://ushtsognogfjordane.no/koronavirus/velferdsteknologi/oppstart-av-videokommunikasjon-i-kommunar-i-tidligare-sogn-og-fjordane.120298.aspx and https://www.ks.no/fagomrader/helse-og-omsorg/velferdsteknologi3/nasjonalt-velferdsteknologiprogram/flere-kommuner-har-anskaffet-videolosning-sammen/

municipalities have been included in this program since 2019, as a result of the collective strength, now recognized as an active and dynamic innovative structure, rather than small units.

Furthermore, during the Corona crisis that turned the entire society upside down overnight in March 2020, the collaborative network and the municipalities' new competences about digital technology was put to the test, documenting that they were able to make quick adjustments and decisions that involve technology. This suggests that the project that initially started because of national requirements for implementing welfare technological solutions to health care services has succeeded through the unique regional framework not only in equipping the municipalities for the task, but has also established a dynamic network of expertise in the region, ready to act during a crisis.

When this project started in 2014, the participants had little knowledge about how to implement technology in health care services, and how to organize it across a rural region with many small units. Although the national advices and, in the later phases of the project, also the economic support from the national health authorities, have been vital for the positive results, so has the regional framework of collaboration under IT-forum Sogn & Fjordane, the leadership by USHT, and expertise from the emergency central. Different from the national guidelines describing the implementation process in a rather uniform way, this illustrates the necessity of taking regional factors and conditions into consideration to secure successful implementation of welfare technology across units of different sizes and shapes.

ACKNOWLEDGEMENT

This project was funded by the County Governor of Sogn & Fjordane. I want to thank all the partners in the project as well as informants participating in surveys and interviews during the six years of this project. A special thank you to Marta Strandos, leader of USHT, who has contributed to developing this analysis.

REFERENCES

Cajander, Å., Corneliussen, H. G., Myreteg, G., & Dyb, K., 2020. What Brings Women into eHealth? Women's Career Trajectories in Digital Transformations in Healthcare *Proceedings of the 12th International Conference on e-Health.*

Corneliussen, H. G., 2014a. A Success Story about a Communication Reform in the County of Sogn & Fjordane'. In C. E. Moe & M. Fossum (Eds.), *Proceedings from Scandinavian Conference on Health Informatics*, August 21-22, 2014, Grimstad, University of Agder, Norway.

Corneliussen, H. G., 2014b. Velferdsteknologi – status i kommunane i Sogn og Fjordane, Vestlandsforsking-rapport nr. 11/2014.

Corneliussen, H. G., 2015. Velferdsteknologi for kommunane i Sogn og Fjordane - Prosjektstatus, Vestlandsforsking-notat nr. 1/2015.

Corneliussen, H. G., & Dyb, K., 2017. Om teknologien som ikke fikk være teknologi – diskurser om velferdsteknologi. In J. R. Andersen, E. Bjørhusdal, J. G. Nesse, & T. Årethun (Eds.), *Immateriell kapital. Fjordantologien 2017* (165-181): Universitetsforlaget.

Corneliussen, H. G., & Hove, M. H., 2017. New Technology in Norwegian Municipalities' Health Care Services: Experiences in Small Rural Municipalities. In S. Martinez, A. Budrionis, A. Bygholm, M. Fossum, G. Hartvigsen, M. Hägglund, C. E. Moe, E. Thygesen, V. Vimarlund, & K. Y. Yigzaw (Eds.), *Proceedings from The 15th Scandinavian Conference on Health Informatics 2017 Kristiansand, Norway, August 29-30, 2017* (1-7): Linköping University Electronic Press.

Corneliussen, H. G., & Larsen, Ø. H., 2018. På jakt etter suksessformelen for samarbeidsnettverket IT-forum Sogn og Fjordane. In D. M. Berge, K. Y. Dale, & H. Hogset (Eds.), *Det regionale i det internasjonale. Fjordantologien 2018* (101-117): Universitetsforlaget.

Corneliussen, H. G., Strandos, M., & Hove, M. H., 2016. Evaluering av pilotar i prosjektet velferdsteknologi i kommunane i Sogn og Fjordane: eit prosjekt for å etablere lokal erfaring med velferdsteknologi: Vestlandsforsking-rapport 8 / 2016.

Frennert, S., & Baudin, K., 2019. The concept of welfare technology in Swedish municipal eldercare. Disability and Rehabilitation, 1-8. doi:10.1080/09638288.2019.1661035

Helse- og omsorgsdepartementet, 2015. Omsorg 2020. Regjeringens plan for omsorgsfeltet 2015-2020.

Helsedirektoratet, 2012. Velferdsteknologi. Fagrapport om implementering av velferdsteknologi i de kommunale helse- og omsorgstjenestene 2013-2030.

Hove, M. H., & Corneliussen, H. G., 2017. Velferdsteknologi i kommunar i Sogn og Fjordane – Fase 3, avsluttande rapport: Vestlandsforsking-rapport 11/2017.

Hove, M. H., & Corneliussen, H. G., 2018. Velferdsteknologi for kommunane i Sogn og Fjordane - spreiing og implementering: Vestlandsforsking-notat 3/2018.

Karlsen, C., Moe, C. E., Haraldstad, K., & Thygesen, E., 2019. Caring by telecare? A hermeneutic study of experiences among older adults and their family caregivers. Journal of clinical nursing, 28(7-8), 1300-1313.

Larsen, Ø. H., & Corneliussen, H. G., 2016. An Informal Multi-Stakeholder Network as a Collaborative Strategy Tool in e-Health. Proceedings from IADIS, eHealth conference 2016.

Meld. St. 15 (2017 –2018), Leve hele livet - En kvalitetsreform for eldre: Meld. St. 15 (2017 –2018) Melding til Stortinget.

Melting, J. B., 2017. Andre gevinstrealiseringsrapport med anbefalinger: Nasjonalt velferdsteknologiprogram, Helsedirektoratet.

Melting, J. B., & Frantzen, L., 2015. Første gevinstrealiseringsrapport med anbefalinger: Nasjonalt velferdsteknologiprogram, Helsedirektoratet.

Moe, C. E., Thygesen, E., & Aanestad, M. (2019). *Dilemmas in Scaling up Telemedical Services.* Paper presented at the SHI 2019. Proceedings of the 17th Scandinavian Conference on Health Informatics, November 12-13, 2019, Oslo, Norway.

NOU, 2011:11. Innovasjon i omsorg.

Raymond, E., 1999. The Cathedral and the Bazaar. Philosophy & Technology, 12(3), 23.

Smaradottir, B., Sjaavaag, S., Ås, I.-A. N., Bakke, B., Alfsen, L., Tangene, W., & Fensli, R. (2019). *Why is it so Hard to Integrate Telemedicine as a Part of Municipal Health Care Services? Experiences from a Norwegian Municipality.* Paper presented at the SHI 2019. Proceedings of the 17th Scandinavian Conference on Health Informatics, November 12-13, 2019, Oslo, Norway.

COMPUTER COMPETENCY AS AN INDICATOR OF HEALTHCARE INSTITUTION READINESS FOR HEALTH INFORMATION SYSTEMS: A STUDY IN BENGHAZI

Salah Awami
University of Benghazi, Benghazi - Libya

ABSTRACT

Background: Healthcare professionals' computer competency is one aspect of healthcare institutions' readiness assessment for the implementation and adoption of health information systems (HIS). Computer competent healthcare professionals normally have a positive attitude toward HIS which is considered to be an important facilitating factor for the implementation process.
Aim: To discuss computer competency in healthcare and investigate computer competency of healthcare professionals in four healthcare institutions in Benghazi.
Methods: The cross-section qualitative research administered a questionnaire to collect data from four healthcare institutions in Benghazi; aspects evaluated were related to computer use, understanding of IT concepts, and general attitudes toward computers. With a 69% response rate; 93 participants (physicians, nursing staff and public health practitioners) filled the questionnaire.
Results: The majority of the participants think that computers could be useful for their work and that they have confidence in using them. 29% of the participants use computers at work and the same percentage have attended formal training in IT. Data entry could be a challenge for using computers, particularly among physicians. There is also a knowledge gap about important underlying IT concepts among healthcare professionals in these institutions.
Conclusion: Healthcare professionals at these institutions are likely to have a positive attitude toward any future HIS implementations. Lack of use of computers at work, lack of formal IT training, and data entry challenges are some of the issues that need to be addressed before any realistic implementation can be considered.

KEYWORDS

Computer Competency, Computer Literacy, Readiness Assessment, HIS Implementation, HIS in Benghazi

1. INTRODUCTION

Healthcare professionals' positive perception of health information systems (HIS) (Dansky, Gamm et al. 1999) and their level of confidence in dealing with computers in general (Terry, Giles et al. 2009) are critical factors for the successful adoption and implementation of these systems. Thus, computer literacy of healthcare professionals is an important aspect to be considered (Van der Meijden, Tange et al. 2001; Štepankova and Engova 2006; Simon, Kaushal et al. 2007; Li, Talaei-Khoei et al. 2013; Mills, Francis et al. 2015; Correia, Lapão et al. 2017) particularly when assessing organizations' readiness for embracing information and communication technology (ICT)(Saleh, Khodor et al. 2016). Computer competent healthcare professionals with proper experience and use of computers typically have the needed computing skills, appreciate the potentials of HIS and have the necessary positive attitude (Mannan, Murphy et al. 2006; Simon, Kaushal et al. 2007; Ward, Pollard et al. 2007; Ludwick and Doucette 2009).

1.1 Computer Competency

Computer literacy is defined as the skills and competencies needed to use computer hardware and software effectively, it reflects familiarity and positive attitude toward computers (Ferrari 2012). Additionally, it enables basic computer skills such as comfortable use of the keyboard for data entry which can be a determinant factor in using HIS (Ludwick and Doucette 2009; Terry, Giles et al. 2009). Computer competency, another term used in this context, refers to using computers to accomplish job tasks, which reflects a deeper knowledge and more intellectual abilities than the mere learning of computer hardware and software (Lin 2011). Computer competency involves understanding three kinds of knowledge: contemporary skills (the ability to use available information technology applications), foundational concepts (the basic principles of IT) and intellectual capabilities (the ability to use IT for organizing, reasoning, and problem solving)(Hoffman and Blake 2003).

Formal education has been offered to qualify healthcare professionals to be computer competent (Rigby, Hulm et al. 2007), including incorporating health informatics courses within the curriculum of health science educational programs. Besides obtaining the IT knowledge and skills, such courses make students appreciate the importance, potentials, and risk of 'health ICT' and positively influence students' trust in and adoption of ICT in healthcare at their workplace in the future (Gibson and Silverberg 2000; Lam, Amon et al. 2012). ICT understanding and use can also be achieved informally which might be effective particularly when technology is related to trainees' lives. However, informal knowledge covers the simple use of the technology that is considered meaningful to the individual at that time with little information about underlying concepts (Hoffman and Blake 2003).

Information literacy is another term used in this context referring to competencies and skills required to find, evaluate, understand, and use information (Association 2000). Information literacy is supported and enabled by computer literacy and is concerned with addressing modern-day information issues such as abundance, diversity of sources, diversity of formats, and authenticity.

The term literacy is often used with other terms to refer to similar concepts such as digital literacy, ICT literacy, e-literacy, network literacy, and media literacy (Bawden 2008). Worth mentioning as well, is cyber literacy, a term emerged recently to emphasize important features of the internet that need to be considered such as speed, reach, anonymity, and interactivity as well as the social topics of gender, online rage, hoaxes, and privacy (Hoffman and Blake 2003).

1.2 Readiness Assessment

The readiness assessment is considered an essential HIS pre-implementation requirement because it highlights the importance of effective planning to evaluate the institution's preparedness for change and embracing technical innovations and awareness of and responsiveness to factors affecting HIS implementation (Terry, Giles et al. 2009). Many readiness assessment frameworks have been presented in the literature which, generally, focus on aspects related to organization infrastructure, management support, government policies, and end-users' skill level and attitudes. Factors related to end-users include time to learn, awareness of IT usefulness, trusting IT, staff exposure to computers, computer anxiety and IT education. Therefore, assessing healthcare professionals' IT skills and addressing shortages through training courses become essential (Adjorlolo and Ellingsen 2013). Assessing the IT literacy of end-users may also identify the 'engagement' construct of a readiness assessment framework which addresses issues of awareness of IT usefulness and trust (Li, Land et al. 2008).

This study explores healthcare professionals' computer competency in selected healthcare institutions in Benghazi and provides a snapshot of the extent of healthcare providers' use of computers in these institutions. Furthermore, this outcome reflects on an important aspect of these organizations' readiness for IT adoption.

2. METHODS

This cross-section qualitative research used a semi-structured self-administered questionnaire for data collection from a non-probabilistic sample from three hospitals and a medical center in Benghazi city. The participants were recruited from Al-Kwaifia Chest Hospital, Al-Hwari General Hospital, the Nephrology Center of Benghazi, and Benghazi Medical Center over the period of July 2019 to October 2019. All of these healthcare institutions are run and financed by the Libyan government. The questionnaire has close-ended, open-ended and partially structured questions to gauge understanding of healthcare professionals' ICT knowledge and skill levels and awareness and attitudes toward IT and HIS.

The study sample was diverse including participants of different genders, various healthcare specialties, and different age groups and work experiences. As a qualitative research the emphasis was on the relevance of the sample subjects rather than representativeness (Neuman 2014). Therefore, selected sample sufficiently provides us the information and understanding we seek about a social world.

The 25 item questionnaires were distributed to healthcare professionals in the designated institutions mainly via the head or staff of medical affairs departments who invited their employees to participate. The researcher collected the filled out questionnaires from the department office later on completion. Occasionally, the researcher approached the subjects directly whenever that was possible. There were two versions of the questionnaire: an English version distributed to physicians and public health specialists and an Arabic version targeted nursing staff.

Five aspects were examined to reflect on the subjects' computer competency: the use of computers, awareness of IT potentials and challenges, knowledge of computer components, aspects of information literacy, and general attitudes towards information technology application in healthcare.

The use of computers aspect was contemplated through posing direct questions of whether subjects are using computers at work, at home or even searching for work-related information on their mobile phones, in addition to posing questions about application software use and frequency of use.

Awareness of computer potentials and challenges was assessed by posing a question on choosing important features of computers that subjects appreciate most. Other questions explored their knowledge, awareness, and comments about data entry challenges, computer security threats and counteraction measures, communication support, and electronic medical records.

Knowledge of computer components (i.e. processor, RAM, HD, and screen size.) was probed through posing a question about a hypothetical situation where subjects decide what computer components they would consider if they were to buy a computer. The question also elicits subjects' knowledge of the units of measurement for every component and added the computer price as a property that might be considered. Another question in this context reflected on their awareness of their computer's connection to a LAN.

The information literacy aspect was contemplated through posing questions about whether they are using the web to search for work-related information, frequency of use, what information sources they are using, and whether they considered information validation. Similarly, their awareness and handling of web site cookies.

The general attitude of subjects about the application of IT and the use of computers was probed through direct questions on their opinion about the importance of using computers at work, ease of computer use, and trusting computers with work-related data. Participants' level of enthusiasm was also explored by eliciting their suggestions and concerns about using computers at their work.

Users also were asked whether they have attended a course to learn about computers. Another question pertained to their role with any system development process to evaluate their knowledge of the concept of user participation with the system development team.

3. RESULTS

134 questionnaires were distributed and 93 were collected, 2 of the returned questionnaires were discarded due to insufficiency. Table 1 summarizes the demographic data of the participants and Table 2 summarizes the study results.

Table 1. Demographic data of the participants

Variable	Category	Number	Percentage
Occupation	Physicians	31	33%
	Public health practitioners	12	13%
	Nursing staff	50	54%
Years of experience	< 1	6	6.5%
	1 – 5	27	30%
	6 – 10	28	31%
	11 – 15	16	17.5%
	16 – 20	5	5.5%
	> 20	9	10%
Gender	Male	26	28.5%
	Female	65	71.5%
Age group	20 – 30	23	25%
	31 – 40	57	63%
	41 – 50	9	10%
	51 – 60	2	2%

Table 2. Results of the study

Participants who	Number	Percentage (n=91)	Comments
Use computers at work.	27	29%	
Use computers at home but not at work for work-related tasks.	36	39%	
Use mobile phone for work-related tasks AND no use of computers at all.	20	21%	
Do not use computers nor mobile phone for work-related tasks at all.	7	8%	All from nursing staff.
Use MS Word for work.	59	64%	20% of them (n=12) stated daily use.
Use MS Excel for work.	30	32%	10% of them (n=3) use the software daily.
Use MS Access for work.	21	23%	Only 3 participants stated daily use.
Think that computers are either very important or important.	67	73%	
Think that computers are not important for work.	11	12%	
Think that computers are either very easy or easy to use.	62	68%	
Think that data entry and keyboard use can be a challenge.	26	28%	18 physicians think that compared with 8 nurses.
Trust computer with work data.	59	64%	
Take necessary measures to protect data.	74	81%	Of them, more are in favour of password protection (59%, n=44) than making backups (30%, n=22). Only very few (11%, n=8) take both measures.
Think that having antivirus software running in computer is very important.	59	65%	Most of the remaining participants (35%, n=32) didn't know or were undecided.
Are aware of their computers' connection to a local network.	83	91%	Stating 'yes' or 'no' they are connected.
Demonstrated adequate knowledge about computer components.	7	8 %	Specified several components that matters to their decision to purchase a computer and their measuring units.
Indicated lack of knowledge about computer components.	84	92%	56% of all participants (n=51) chose some components without specifying units of measurement. 36% of all participants (n=33) chose 'the price', didn't know, or were undecided.
Attended a training course about computers.	27	29%	
Taught themselves about computers.	52	57%	
Use electronic communication for work.	65	71%	
Use the web to search for work-related information.	58	63%	Mostly physicians and only 27% of them (n=16) search on a daily basis.
Specified web sites they visited to search for work-related information.	49	53%	75% of them (n=37) are physicians.
Handle web sites cookies.	67	73%	

Reject web sites cookies.	23	25%	Very few of them provided reasons for rejection which included: harm to computer and lack of trust.
Validate information from web sites.	25	27%	26% of all participants (n=24) do not validate and 46% (n=42) didn't answer the question.
Know about EMR.	58	63%	24% of them (n=14) stated knowing a lot about EMR.
Think that EMR will be beneficial to their work.	63	69%	
Think that they can be part of HIS development or acquisition processes.	25	27%	
Didn't know whether there is a role for them in the development or acquisition processes of EMR.	42	46%	

With regards to the computer features that participants appreciate most, a third of them chose at least three features, 40% chose only one feature and 7% chose none. The most appreciated computer features were data storage and information search and retrieval; and then the ability to share and transfer data and electronic communication. Visual presentation of data, data analysis, and decision support were the least appreciated features of the computer for the participants.

30% of participants provided extra comments and suggestions at the final question of the questionnaire. Nurses' group comments were mainly about the lack of availability of computers at work and suggestions for provision of training courses to increase their computer knowledge and skills. Some also noted that using information systems allow better patient data management.

Physicians' notes and comments varied and included supporting the activation of HIS, requests for organizing training programs, and enumerating benefits and drawbacks of using computers at work. Benefits mentioned included better patient data management and making work easier. Drawbacks mentioned included virus attacks and network failure. One physician was skeptical and commented that minimal outcomes can be achieved from using computers at work.

4. DISCUSSION

The data analysis provides insights into the healthcare professionals' computer competency and the extent of computer use in the targeted healthcare institutions which were the main objectives of this study.

The low percentage of healthcare professionals using computers at work in addition to the low frequency of use indicate that computers are not in actual and meaningful use to support work-related tasks, despite the existence of a considerable number of healthcare professionals who use computers at home to perform work-related tasks. MS Word is the application software most used by participants, which suggests that computers are mainly used for simple tasks such as document and report creation and editing. A lesser percentage of the participants use more sophisticated software such as Excel and Access. The majority appreciate the computers' ability to store data, search for information, and support communication. There seems to be a lack of understanding and appreciation of the computers' ability to analyze data and support decisions.

The majority of participants perceives computers as useful tools for their work and state that they are easy to use. Although some of the physicians thought that using a keyboard poses a challenge to using computers. This suggests that routine data entry accompanying the use of HIS might prove to be a challenge.

An important percentage of participants trust computers with data and are aware of some of the security measures that can be taken to protect them. The majority of data security-vigilant healthcare providers chose using passwords as protection. At the same time, a low percentage of participants were inclined to make data backups.

Participants' knowledge of computers' hardware is insufficient, with only a few demonstrating an adequate understanding of computer hardware components by choosing one or more components they would consider when purchasing a computer and specifying units of measure.

Some of the participants attended training courses to learn about operating systems and some productivity software. However, the majority gained their computer skills informally which could undermine their knowledge of important underlying concepts of computers and IT. This outcome was supported by having a low percentage of participants using application software providing decision support and problem-solving capabilities such as Excel and Access. In addition a high percentage of participants did not know whether there is a role for them to play in HIS development/acquisition process.

A significant percentage of participants communicate with their colleagues electronically about work-related issues. However, they use email services and social media platforms provided on the web. Consequently, these healthcare institutions are missing the opportunity of capturing these conversations as well as risking the confidentiality and integrity of work-related data.

A considerable number of participants use and interact with the web to look up work-related information, although the low frequency of use and lack of information validation suggests the requirement for more training in information literacy.

A considerable percentage of participants stated having some knowledge about electronic medical records. However, a few stated knowing a lot about EMR and it is not clear how they gained this knowledge given that none of the participating healthcare institutions are using any form of electronic medical records systems. Nonetheless, this knowledge seems to have positively affected their opinion about the importance of EMR.

The study provides insightful findings regarding the computer competency of healthcare providers at the designated healthcare institutions. The number of participants could have been greater, particularly at large institutions such as Benghazi Medical Centre, but many healthcare professionals were reluctant to participate. Perhaps there would have been more participants had the researcher approached the subjects directly. Moreover, data quality of self-administered questionnaires can suffer due to participants' misunderstanding of some of the questions, lack of concentration, bias or language barriers. Also, triangulation of research methods could have revealed more details and confirmed some findings.

5. CONCLUSION

Proper planning for HIS implementation requires conducting readiness assessment for the targeted institutions. Many readiness assessment frameworks have been suggested which consider healthcare professionals' attitudes, knowledge and skills in computers as an important aspect. Computer competency is concerned with understanding and gaining knowledge and skills in computers and IT. This study investigated the computer competency of healthcare professionals in some healthcare institutions in Benghazi to assess one aspect of these institutions' readiness for any prospective HIS implementation.

Healthcare providers in the selected healthcare institutions have a positive attitude toward computers. Their level of confidence in dealing with computers is relatively high and was inferred from the high percentage of participants who think computers are important and easy to use, as well as trusting computers with their work data. Characteristics of such a workforce enable the recruitment of enthusiastic change agents – another determinant of successful HIS implementation.

However, the low percentage of users using computers at work and dearth of knowledge about some aspects and concepts of information technology are some of the drawbacks that need to be addressed. Setting up local area networks and providing basic web services such email communication is one solution for integrating computers into the daily work practice of healthcare professionals in these institutions. Furthermore, knowledge and skill gaps of healthcare professionals can be addressed by providing adequate education and training in ICT and health informatics.

As far as readiness assessment of these healthcare institutions is concerned this study examined only one aspect related to healthcare professionals' computer competency. Other aspects of readiness assessment that need to be considered in the future to have a comprehensive assessment, include management support, government policies, and IT infrastructure.

REFERENCES

Adjorlolo, S. and G. Ellingsen (2013). "Readiness assessment for implementation of electronic patient record in Ghana: a case of university of Ghana hospital." Journal of Health Informatics in Developing Countries 7(2).

Association, A. L. (2000). "Information literacy competency standards for higher education."

Bawden, D. (2008). "Origins and concepts of digital literacy." Digital literacies: Concepts, policies and practices 30: 17-32.

Correia, J. C., L. V. Lapão, et al. (2017). "Implementation of a Telemedicine Network in Angola: Challenges and Opportunities." Journal of Health Informatics in Developing Countries 12(1).

Dansky, K. H., L. D. Gamm, et al. (1999). "Electronic medical records: are physicians ready?" Journal of Healthcare Management 44(6): 440-454.

Ferrari, A. (2012). Digital competence in practice: An analysis of frameworks, Luxembourg: Publication office of the EU. Research Report by the Joint ….

Gibson, K. E. and M. Silverberg (2000). "A two-year experience teaching computer literacy to first-year medical students using skill-based cohorts." Bulletin of the Medical Library Association 88(2): 157-164.

Hoffman, M. and J. Blake (2003). "Computer literacy: Today and tomorrow." Journal of Computing Sciences in Colleges 18(5): 221-233.

Lam, M. K., K. L. Amon, et al. (2012). The effect of e-health contents on health science students' attitude toward the efficiency of health ICT in care provision. HIC.

Li, J., L. P. W. Land, et al. (2008). "E-Health readiness framework from Electronic Health Records perspective." GlobDev 2008: 4.

Li, J., A. Talaei-Khoei, et al. (2013). "Health care provider adoption of eHealth: systematic literature review." Interactive journal of medical research 2(1): e7.

Lin, T.-C. (2011). "A Computer Literacy Scale for Newly Enrolled Nursing Students: Development and Validation."

Ludwick, D. and J. Doucette (2009). "Primary care physicians' experience with electronic medical records: barriers to implementation in a fee-for-service environment." International Journal of Telemedicine and Applications 2009: 2.

Mannan, R., J. Murphy, et al. (2006). "Is primary care ready to embrace e-health? A qualitative study of staff in a London primary care trust." Journal of Innovation in Health Informatics 14(2): 121-131.

Mills, J., K. Francis, et al. (2015). "Enhancing computer literacy and information retrieval skills: A rural and remote nursing and midwifery workforce study." Collegian 22(3): 283-289.

Neuman, L. W. (2014). Social Research Methods: qualitative and quantitative approaches. England, Pearson.

Rigby, M. J., C. Hulm, et al. (2007). Enabling the safe and effective implementation of health informatics systems-validating and rolling out the ECDL/ICDL Health Supplement. Medinfo 2007: Proceedings of the 12th World Congress on Health (Medical) Informatics; Building Sustainable Health Systems, IOS Press.

Saleh, S., R. Khodor, et al. (2016). "Readiness of healthcare providers for eHealth: the case from primary healthcare centers in Lebanon." BMC health services research 16(1): 644.

Simon, S. R., R. Kaushal, et al. (2007). "Correlates of electronic health record adoption in office practices: a statewide survey." Journal of the American Medical Informatics Association : JAMIA 14(1): 110-117.

Štepankova, O. and D. Engova (2006). "Professional competence and computer literacy in e-age, focus on healthcare." Methods of Information in Medicine 45(03): 300-304.

Terry, A. L., G. Giles, et al. (2009). "Adoption of electronic medical records in family practice: the providers' perspective." Family medicine 41(7): 508.

Van der Meijden, M., H. Tange, et al. (2001). "Development and implementation of an EPR: how to encourage the user." International journal of medical informatics 64(2-3): 173-185.

Ward, R., K. Pollard, et al. (2007). "Developing Information Technology Attitude Scales for Health (ITASH)." Studies in health technology and informatics 129(1): 177.

NAMED ENTITY RECOGNITION AND INFORMATION EXTRACTION FOR ARABIC MEDICAL TEXT

Jaafar Hammoud, Natalia Dobrenko and Natalia Gusarova
ITMO University
Sankt-Petersburg 197101, Russia

ABSTRACT

The article discusses the possibilities of solving NER (Named Entity Recognition) problem for medical texts in Arabic with limited availability of labeled datasets, as well as computational and specialized linguistic resources. To overcome them, it is proposed to use recurrent neural networks. In our experiments, we used "BERT-Base, Multilingual Cased" from Google and Pooled-GRU with Multi-lingual Universal Sentence Encoder (MUSE) from Facebook. Each network was fine-tuned with our dataset.

The used dataset was obtained from three medical volumes issued by Arabic Encyclopedia. We experimentally evaluated the effectiveness of tuned models on real NLP (Natural Language Processing) task - medical entities recognition from the Arabic Medical Encyclopedia and obtained encouraging results.

KEYWORDS

Natural Language Processing, Named Entity Recognition, Arabic, e-Health

1. INTRODUCTION

E-health has been recognized as a cost-effective and reliable form of developing health systems and services, including in relatively low-income countries and in regions where health coverage is inadequate. However, language barriers often become a real obstacle to the introduction of advanced E-health technologies in real healthcare practice in such countries. An Indication in this sense is Arabic. The gap between it and other languages such as English and European Union languages is still huge because of such reasons like orthographic ambiguity and inconsistency, morphological inflections, word order freedom, dialectal variation, and lack of adequate dataset (Maloney, 1998). These circumstances complicate the direct application of NLP methods with these languages, which are mainly developed and verified on English sources.

The NER task is one of the primary keys that used in texts processing. Many NLP applications such as question answering, information retrieval, and machine translation require a NER system to enhance their results. In 1995, at the Sixth Message Understanding Conference, the concept of the NER was first introduced, which aims to develop a new method to improve the process of information extracting. NER works to extract predefined entities from an unstructured text that can include person names, location names, organizations, percentages, temporal expressions, money expressions, and numerical expression (Alanazi, 2017).

There are two common groups of methods for dealing with entities extraction and recognition. The first uses linguistics grammar-based techniques. As a rule, these methods obtain better accuracy, but they are language dependent and require months of work by experienced computational linguists (Nothman, 2013).

The second group is based on statistical models and uses traditional machine learning in combination with the latest techniques. The biggest benefit to be gained from these methods is that they are context sensitive. Accordingly, they have advantages when processing texts in languages with complex context-sensitive grammar, such as Arabic. But they need a large amount of manually annotated training data, and the more advanced algorithms require significant computational resources.

The article discusses the possibilities of solving NER problem for medical texts in Arabic with limited availability of labeled datasets, as well as computational and specialized linguistic resources.

Consequently, In the article, we identified the following tasks

– Create a dataset from available medical texts in Arabic.
– Choose NN models that obtaining competitive performance when solving the NER problem and fine-tune them in the conditions of the above limitations.
– Experimentally evaluate the effectiveness of tuned models on real NLP tasks.

2. RELATED WORKS

Through a literary review of past and related works, NER systems in Arabic language are limited to political and economic fields, and most of the texts used were based mainly on news agencies. Moreover, NER systems were rarely used on any Arabic medical documents.

Until 2012, researchers focused on the rule-based approach, many articles used own and small corpus or ANERcorp to recognize person names, organizations, locations, date, and numerical expression (Al-Shalabi, 2009; Elsebai, 2009; Al-Jumaily, 2012; Shihdaeh, 2012; Zaghouani, 2012). To recognize the crime type, location, and nationality researchers used a rule-based approach too on Arabic Crime News Report Corpus (ANCRC) (Alruily, 2012).

On another hand, to recognize (person names, locations, organization, and miscellaneous) researchers used maximum entropy on ANERcorp (Benajiba, 2007), used conditional random fields on ANERcorp (Benajiba, 2008; Morsi, 2013) and ACE (2005) (Abdul-Hamid, 2010), and used support vector machine on UPVCorpus and ACE (2003-2005) (Benajiba, 2008; Koulali, 2012). In (Helwe, 2019) researchers used deep co-learning, and in (Ali, 2019) they boosted Arabic named entity recognition with a multi-attention layer.

In (Albukhitan, 2013) researches introduced an Arabic semantic annotation tool with integrated information extraction for food, nutrition, and health. In (Dudchenko, 2013) researchers used the Bayesian Belief Network (BBN) to extract disease name, symptoms, treatment methods, and diagnosis methods.

In (Salah, 2017) authors used two approaches, the first was the traditional ML with CRF and the second was a DNN model which used Bi-LSTM-CRF. In (Shahina, 2019) researchers used different versions of RNN cells for NER problem (e.g. Bi-LSTM, Bi-GRU, and Bi-RNN) on ANERcorp dataset.

3. DATASET

First, we have obtained the dataset from three medical volumes issued by the Arabic Encyclopedia in the Syrian Arab Republic, available as text on the Internet through the encyclopedia site[1], using the Scrapy[2] library, a free and open-source web crawling framework written in Python. The three volumes are (Respiratory System Diseases, Cardiovascular Diseases, and Skin Diseases). In Table 1, the number of articles in each volume and the number of sentences are shown.

Table 1. Source characteristics

Volume	Number of articles	Number of sentences
Respiratory System Diseas	28	6691
Cardiovascular Diseases	33	9464
Skin Diseases	22	5921

Articles are relatively long in these volumes, and one article may contain more than one partial article within it. For example, an article that talks about electrocardiograms mentions all types of electrocardiograms within separate paragraphs, and we can consider that each paragraph of the text is a special article. But in our study and because of the use of the BERT model for training, we care more about the concept of a sentence than the article concept.

[1] http://arab-ency.com.sy
[2] https://scrapy.org/

The annotation of entities (disease name, organ name, disease symptoms, and drug name) were done manually using the IOB rule by some Arabic medical students at Arab universities. Then, the English and French terms were removed from the text and only the Arabic text was retained. Figure 1 is a screenshot of the data format.

الربو	B-DISEASE	تسبب	O
حالة	O	هذه	O
التهابية	O	الحالة	O
مزمنة	O	الالتهابية	O
في	O	فرط	B-SYMPTOM
الطرق	B-ORGAN	استثارة	I-SYMPTOM
التنفسية	I-ORGAN	قصبية	I-SYMPTOM
تعتمد	O	لعوامل	O
على	O	منبهة	O
الخلايا	B-ORGAN	ينجم	O
البدينة	I-ORGAN	عنه	O
،	O	أزيز	B-SYMPTOM
محبات	O	نوبي	I-SYMPTOM
الحامض	O	وضيق	B-SYMPTOM
واللمفيات	O	نفس	I-SYMPTOM
المساعدة	O	وسعال	B-SYMPTOM
		أكثره	O
		في	O

Figure 1. Screenshot from the dataset

The Arabic Medical Encyclopedia consists of several volumes, each article in these volumes is written by a different physician, so there is no unified structure for articles, and in a large number of articles, the symptoms are mentioned in the context of the text and not in a special paragraph.

There is no unified tag for every new paragraph in the article, titles of some paragraphs tagged as "strong", others as "h2", etc. For these reasons, most of the pre-processing was done manually.

Also, one of the problems that we encountered while working with text is sentence length, especially sentences for symptoms of disease and difficulty in determine the appropriate sentence, we illustrate that in the following example Figure 2.

(9)- التهاب ما حول الظفر الحاد أو الداحس بالمبيضات: (الشكل22)، يصيب الأشخاص الذين يعملون بتماس الماء فترات طويلة، يحدث احمرار وتوذم حول الظفر مع تراجع القشيرة الظفرية، وقد يخرج القيح من الطية الظفرية، ويجب تمييزالداحس بالمبيضات من الداحس الجرثومي. وحين يزمن الالتهاب تحدث في الظفر تبدلات حثلية، واسمرار حوافه أو اخضرارها.

Figure 2. An example that shows how difficult it is to choose a sentence that describes a symptom of an illness. Translation of figure 2: Inflammation around the sharp nail or paronychia with candidiasis affects people who work in contact with water for long periods. Redness and swelling occurs around the nail as the nail scales retract and pus may come out of the nail fold. The paronychia of Candida must be distinguished from bacterial paronychia. When the inflammation is chronic, inflammation in the nail changes in its induction and darkening of its edges or being green

We notice the interconnectedness of the sentence together, and we can also choose the symptoms in more than one way. This affects the training of the model, for example:

- Redness and edema around the nail
- Redness
- edema
- nail plate retracts

Figure 3 also illustrates the problem of overlapping more than one entity together.

العلاج: يعالج السلاق في الرضع بالنيستاتين بشكل معلق فموي وهو فعال موضعياً، ولا يمتص عن طريق الجهاز الهضمي. تستعمل في الكبار مضمضات البوفيدون أو الكلورهيكزيدين، والإيميدازولات بشكل هلام فموي أو أقراص مضغ. ويعطى الفلوكونازول في الحالات الشديدة والممتدة للمريء بجرعة 100ملغ في اليوم مدة أسبوعين. ويجب تعقيم بدائل الأسنان يومياً بمحلول الهيبوكلوريت في المسنين الدرد.

Figure 3. Overlapping more than one entity together. Translation of figure 3: Treatment: Slave is treated in infants with nystatin in the form of an oral suspension, which is effective locally and not absorbed by the digestive system. It is used in adults as povidone or chlorhexidine and imidazole in the form of an oral gel or chewable tablets. Fluconazole is given in severe and prolonged cases in a dose of 100 mg per day for two weeks and sterilization alternatives should be sterilized Teeth daily with hypochlorite solution

The sentence is composed of the names of some medicines, and at the same time the sequence of words represents a treatment.

The BERT model requires for text processing (see Section 4) a special format for the data before working with the model. The sentence must start with the [CLS] token, while the [SEP] token is added at the end of each sentence. Thus, for Arabic tokenization, we used the BERT tokenizer which uses WordPiece (Wu, 2016). After that, from the token vocabulary file that BERT is provided, every token is mapped to an index.

4. CHOSEN NNS AND THEIR FINE-TUNING

First, we fine-tuned BERT model (Devlin, 2018) published by researchers at Google AI Language[3] with our dataset. This pre-trained model has raised the interest of researchers in a field of machine learning and has provided distinctive results in a wide range of NLP applications. BERT uses a mechanism that learns contextual relations between words in a text by using the transformer. The transformer includes two separate mechanisms, the first is an encoder that reads the text as input and a decoder that produces a prediction for the input. 15% of the words in each sequence are replaced with a [MASK] token then the model attempts to predict the original value of the masked words by using the context provided by non-masked words. Using BERT, the NER model can be trained by feeding the output vector of each token into a classification layer that predicts the NER label. In NER, two popular approaches:

1. A probabilistic graphical model called CRF (Conditional Random Field), in which the token is labeled considering the context.
2. Multi-class Classification-based: each token in the sentence will get a label, so we can treat our process as a multi-class classification problem, and that what we did in our work.

We used the library deployed by hugginface[4] and called transformers, this library includes pre-trained models for NLP problems that achieve state-of-the-art results. Our dataset was separated into two groups (80% as training data) and (20% as testing data) and then with the dataset we fine-tuned a pre-trained model called "BERT-Base, Multilingual Cased" which supports 104 languages, and contains 12 layers, 768 hidden, 12 heads, and 110 million parameters. For the purposes of fine-tuning, we chose the following values: Batch size = 32; Learning rate (Adam) = 2e-5; Number of epochs = 4; The epsilon parameter eps = 1e-8.

The BERT built on pre-training contextual representations that including the Transformer (Vaswani, 2017), Semi-supervised Sequence Learning (Dai, 2015), the OpenAI transformer (Radford, 2018), ELMo (Peters, 2018), and ULMFiT (Howard, 2018).

BERT was pre-trained on a large corpus of unlabeled text which includes the entire Wikipedia and a book corpus and what makes BERT unique compared to the rest of models that it is first deeply bidirectional, unsupervised language representation, and pre-trained using only a plain text corpus, on other hand, BERT was first successfully pre-train bidirectional attempt in a deep neural network. While traditional context-free models generate a single word embedding for each word, BERT would represent based on both the left and the right context.

The two biggest positive points in using the BERT model are:

- Quicker Development: the authors recommend only 2-4 epochs of training for fine-tuning BERT on a specific NLP task.

[3] https://research.google/teams/language

[4] https://github.com/huggingface/transformers

– Less Data: Because of the pre-trained weights BERT allows us to fine-tune our task on a small dataset.

As researchers refer and because the way that BERT was trained by, we can use BERT for many NLP tasks like question answering system, text classification (and in our work, we treat the task of NER as a multi-class classification as mentioned before), part-of-speech tagging, etc.

Second, in (Al-Smadi, 2020) researchers introduced a new model for working with ANER (Arabic named entity recognition)

This model consists of 6 layers and we have made some adjustments to it to fit our dataset

1. Text embedding: the model uses "Multi-lingual Universal Sentence Encoder (MUSE)" introduced by Facebook (Cer, 2018).

2. Bi-Directional-GRU: with a relatively training time faster than Long Short-Term Memory (LSTM) an enhanced version of RNN is used in this model.

3. Pooling layer: in this layer to extract the discriminative features of the input text the model used the Global Average Pooling and Global Max Pooling.

4. Concatenation layer: to keep all the possible discriminative features in this layer and after computing both values of GAP and GMP concatenate them in a single vector.

5. Fully connected network: to learn the discriminative features from the previous layer and with ReLU activation function a Dense layer with 512 fully connected neurons.

6. Classification layer work with IOB framework.

5. RESULTS AND DISCUSSION

We measure the accuracy of the model by using f1 score:

$$F1 = 2 * \frac{\text{Precision} * \text{Recall}}{\text{Precision} + \text{Recall}}$$

The Precision computed by the following equation:

$$Precision = \frac{TP}{TP + FP}$$

and the Recall computed by the following equation:

$$Recall = \frac{TP}{TP + FN}$$

where TP denotes to True Positive, FP False Positive, and FN False Negative.

Table 2. F1 Score for every entity

	disease name	organ name	disease symptoms	drug name
BERT	87.7102	86.3021	69.8519	77.4986
Pooled-GRU	88.9133	87.5129	72.6125	76.5148

All previous research that studied this issue on Arabic, relied relatively on small datasets, in addition to that, the entities were limited to the names of people, places, and organizations. In the medical field, researchers whose we were able to review their works focused on one disease. such as cancer.

However, the results were very promising, compared to the results found for the same problem in languages such as English and French.

6. CONCLUSION

The tasks posed in the article have been successfully solved. Namely:

- We created a dataset from available medical texts in Arabic.

- We have chosen two NN models of recurrent type which obtained competitive performances when solving the NER problem.
- We experimental evaluated the effectiveness of tuned models on real NLP tasks. The models provide good results compared to the existing level.

As future work, we suggest applying the rest of the Transformers models that have achieved state-of-the-art results, such as (BERT, GPT-2, RoBERTa, XLM, DistilBert, XLNet, CTRL, etc) on the Arabic for various tasks (Text classification, Text summarization, Question-answering systems, etc.)

ACKNOWLEDGEMENT

This work was financially supported by Russian Science Foundation, Grant 19-19-00696.

REFERENCES

Abdul-Hamid, A., & Darwish, K. (2010, July). Simplified feature set for Arabic named entity recognition. *In Proceedings of the 2010 Named Entities Workshop* (pp. 110-115). Association for Computational Linguistics.

Alanazi, S. (2017). A named entity recognition system applied to Arabic text in the medical domain *(Doctoral dissertation, Staffordshire University).*

Albukhitan, S., & Helmy, T. (2013, January). Automatic Ontology-based Annotation of Food, Nutrition and Health. *Arabic Web Content. In ANT/SEIT* (pp. 461-469).

Ali, M. N. A., et al. (2019). Boosting Arabic named-entity recognition with multi-attention layer. *IEEE Access*, 7, 46575-46582.

Al-Jumaily, H., et al. (2012). A real time Named Entity Recognition system for Arabic text mining. *Language resources and evaluation*, 46(4), 543-563.

Alruily, M. (2012). Using text mining to identify crime patterns from arabic crime news report corpus.

Al-Shalabi, R., et al. (2009, March). Proper noun extracting algorithm for arabic language. *In International conference on IT*, Thailand.

Al-Smadi, M., et al. (2020). Transfer Learning for Arabic Named Entity Recognition With Deep Neural Networks. *IEEE Access*, 8, 37736-37745.

Benajiba, Y., & Rosso, P. (2007, December). ANERsys 2.0: Conquering the NER Task for the Arabic Language by Combining the Maximum Entropy with POS-tag Information. *In IICAI* (pp. 1814-1823).

Benajiba, Y., & Rosso, P. (2008, May). Arabic named entity recognition using conditional random fields. *In Proc. of Workshop on HLT & NLP within the Arabic World, LREC* (Vol. 8, pp. 143-153).

Benajiba, Y., et al. (2008). Arabic named entity recognition: An svm-based approach. *In Proceedings of 2008 Arab International Conference on Information Technology (ACIT)* (pp. 16-18). Amman, Jordan: Association of Arab Universities.

Cer, D., et al. (2018). Universal sentence encoder. arXiv preprint arXiv:1803.11175.

Dai, A. M., & Le, Q. V. (2015). Semi-supervised sequence learning. *In Advances in neural information processing systems* (pp. 3079-3087).

Devlin, J., et al. (2018). Bert: Pre-training of deep bidirectional transformers for language understanding. arXiv preprint arXiv:1810.04805.

Dudchenko, A., et al. (2019). Extraction from Medical Records. *Stud. Health Technol. Inform,* 261, 62-67.

Elsebai, A. (2009). A rules based system for named entity recognition in modern standard Arabic *(Doctoral dissertation, University of Salford).*

Habash, Nizar Y. (2010). Introduction to Arabic natural language processing. *Synthesis Lectures on Human Language Technologies* 3.1, pp 1-187.

Helwe, C., & Elbassuoni, S. (2019). Arabic named entity recognition via deep co-learning. *Artificial Intelligence Review*, 52(1), 197-215.

Howard, J., & Ruder, S. (2018). Universal language model fine-tuning for text classification. arXiv preprint arXiv:1801.06146.

Koulali, R., & Meziane, A. (2012, November). A contribution to Arabic named entity recognition. *In 2012 Tenth International Conference on ICT and Knowledge Engineering* (pp. 46-52). IEEE.

Maloney, J., & Niv, M. (1998). TAGARAB: a fast, accurate Arabic name recognizer using high-precision morphological analysis. *In Computational Approaches to Semitic Languages*.

Morsi, A., & Rafea, A. (2013, May). Studying the impact of various features on the performance of Conditional Random Field-based Arabic Named Entity Recognition. *In 2013 ACS International Conference on Computer Systems and Applications (AICCSA)* (pp. 1-5). IEEE.

Nothman, J., et al. (2013). Learning multilingual named entity recognition from Wikipedia. *Artificial Intelligence*, 194, 151-175.

Peters, M. E., et al. (2018). Deep contextualized word representations. arXiv preprint arXiv:1802.05365.

Radford, A., et al. (2018). Improving language understanding by generative pre-training.

Salah, R. E., & Qadri binti Zakaria, L. (2017). A comparative review of machine learning for Arabic named entity recognition. *International Journal on Advanced Science, Engineering and Information Technology*, 7(2), 511-518.

Shahina, K. K., et al. (2019, March). A Sequential Labelling Approach for the Named Entity Recognition in Arabic Language Using Deep Learning Algorithms. *In 2019 International Conference on Data Science and Communication (IconDSC)* (pp. 1-6). IEEE.

Shihadeh, C., & ünter Neumann, G. (2012, November). ARNE: A tool for named entity recognition from Arabic text. *In Fourth Workshop on Computational Approaches to Arabic Script-based Languages (CAASL4), located at the Tenth Biennial Conference of the Association for Machine Translation in the Americas (AMTA)* (pp. 24-31).

Vaswani, A., et al. (2017). Attention is all you need. *In Advances in neural information processing systems* (pp. 5998-6008).

Wu, Y., et al. (2016). Google's neural machine translation system: Bridging the gap between human and machine translation. arXiv preprint arXiv:1609.08144.

Zaghouani, W. (2012). RENAR: A rule-based Arabic named entity recognition system. *ACM Transactions on Asian Language Information Processing* (TALIP), 11(1), 1-13.

DEEP LEARNING-BASED MOTION ACTIVITY RECOGNITION USING SMARTPHONE SENSORS

Saedeh Abbaspour[1,2], Faranak Fotouhi[2], Hossein Fotouhi[1], Maryam Vahabi[1] and Maria Linden[1]
[1]*School of Innovation, Design, and Engineering, Ma¨lardalen University, Sweden*
[2]*Faculty of Engineering, University of Qom, Qom, Iran*

ABSTRACT

MHealth systems establish a new way to transfer the health service to remote places. These systems offer significant benefits for continuous health monitoring. Motion activity recognition is one of the challenging mHealth use cases that incorporates continuous data collection and analysis of measurements. The main goal of this research is to analyze physical activity data. We employ measurements from the WISDM lab dataset [1]. These data are collected from participants performing motion activities. This data is then used by deep learning algorithms to predict special activities. In particular, CNN and CNN-LSTM algorithms are used to compare their accuracy, which resulted in approximately 95% and 97% respectively. Thus, the CNN-LSTM has higher accuracy in this analysis.

KEYWORDS

MHealth, Activity Recognition, Accelerometer, Deep Learning

1. INTRODUCTION

EHealth encompasses a wide variety of sub-domains of digital health such as telehealth and telemedicine, health IT system, electronic health record, and mobile health(mHealth). The recent advances in manufacturing mobile devices and the rise of wearable sensors embedded in people's everyday lives have led to the concept of mHealth monitoring. MHealth is defined as access and practice of healthcare via mobile telecommunication, multimedia technology, and wireless body sensors through a smartphone healthcare application (Adibi, 2015) (Choi, 2018). Figure 1 shows a schematic view of a typical mHealth system. In this system, the data collected from the sensor/actuator devices are sent to the cloud servers, where they are interpreted and analyzed. The analysis results are then submitted to both specialists in the clinic and the mobile application users. Any decision made by the specialists is then recorded in the servers and forwarded to the actuators.

The mobile applications can be categorized into two groups: (1) fitness, diet, and nutrition apps, and (2) disease and treatment management apps. According to Gartner, 40% of digital health systems in 2021 will shift to wearable devices which 90% will include fitness trackers (Lee, 2018). These mHealth apps could calculate the amount of energy consumed during physical activity. In fact, they use accelerometers, gyroscope (Fan, 2015), and GPS (Gusenbauer, 2010) (Jain, 2017)(Lee, 2018)(Sukaphat, 2013)(Rundle, 2016) to obtain and analyze the required data. For example, Hwang et al. have provided a fitness system embedded in shoes to measure steps, calorie, and biomedical information using the pressure sensors (Hwang, 2016).Adi et al. have built a system to measure the heart rate, ECG signal, step count, and calorie consumption using the Smartphone's accelerometer (Pande, 2015). Pande et al. used Smartphone sensors such as accelerometer and barometer to obtain energy expenditure (Kwapisz, 2011). Pal and Hepsiba estimated the user's energy expenditure using the Smartphone's accelerometer and GPS (Vathsangam, 2015). Duclos et al. have used accelerometer data to measure energy expenditure (Duclos, 2016). Li et al. used the ECG and accelerometer sensors to estimate energy expenditure (Li, 2016). Altini et al. used five accelerometer sensors attached to the body to assess the energy expenditure during the daily activities (Altini, 2014). Carneiro et al. used a smartphone accelerometer to estimate energy expenditure (Carneiro, 2015). As a previous work, a wearable kit is designed to calculate calorie consumption. This system used a motion sensor and a PPG

1 The WISDM Lab dataset: http://www.cis.fordham.edu/wisdm/dataset.php

(Photoplethysmography) sensor separately. The output of these sensors was sent to an android app by Bluetooth (Fotouhi-Ghazvini, 2020).

The majority of the above contributions rely on Smartphone embedded sensors to collect the physiological data, which hugely lacks precision and accuracy. For example, in (Asimina, 2018), the Fitbit was joint with a Smartphone app for step count, but it wrongly underestimated the step counts. In (Van Camp, 2017), the Fitbit was incorporated to measure the step counts though it was imprecise. Similarly, in the SmartWatch market for instance in (Schneider, 2016) and (Hamari, 2017), the ActiGraph was used which erroneously overestimated the frequency of the activity and the step counts. The Huawei and Apple watches include an accelerometer, gyroscope, heart rate, altimeter, and microphone sensors (King, 2018). In (Boletsis, 2015), the Basic B1 smartwatch was incorporated for dementia care at home. It provided the heart rate, the ambient temperature, and the activity levels. However, it did not reveal the exact activity during daily activities. In this paper, it is demonstrated how deep learning can help us to improve the accuracy of predictions. In fact, two well-known algorithms are elaborated and compared with respect to their accuracy. Thus, the main goal of this paper is to use deep learning algorithms to classify motion activities by using accelerometer data.

The rest of the paper is organized as follows. Section 2 discusses the input data used in the experiments. The deep learning algorithms used to analyze those data are explained in section 3, and the analysis results are discussed in section 4. Finally, section 5 concludes the paper and outlines future directions.

2. DATA COLLECTION

In this paper, the data from WISDM dataset (Kwapisz, 2011) is used, which consists of cell phone accelerometer data. In particular, the values of acceleration are analyzed to detect the activity type by employing deep learning algorithms. The activities include "Walking", "Jogging", "Upstairs"," Downstairs", "Sitting" and "Standing". This dataset is extracted from twenty-nine participants. In the first step, the values of the accelerometer are used to predict activity type. These data can help us to monitor the motion activity for each participant. In this case, the participants carried the Android phone in their front pants leg pocket and were asked to walk, jog, ascend stairs, descend stairs, sit and stand for specific periods of time. The accelerometer data are collected every 50ms, so the sampling rate is 20 Hz per second.

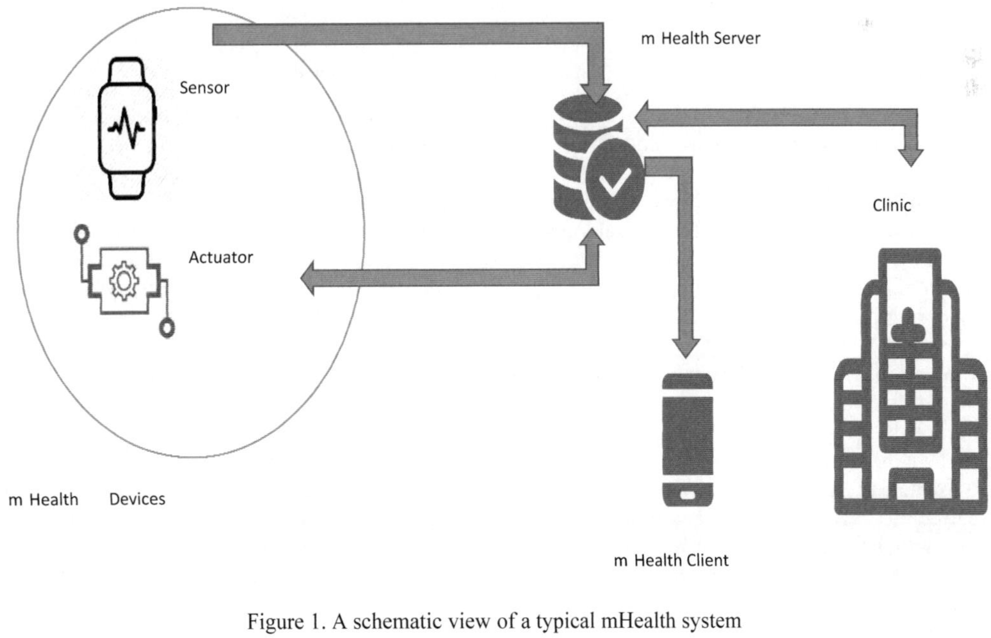

Figure 1. A schematic view of a typical mHealth system

3. DEEP LEARNING ALGORITHMS

Deep learning is a subset of machine learning that relies on several layers of artificial neural networks (ANNs) to process unstructured data and detect patterns to be used in decision making. ANNs are inspired by the structure of the human brain and consist of layers, each representing a set of features. These features help the deep learning algorithm to extract the best model characterizing the input data.

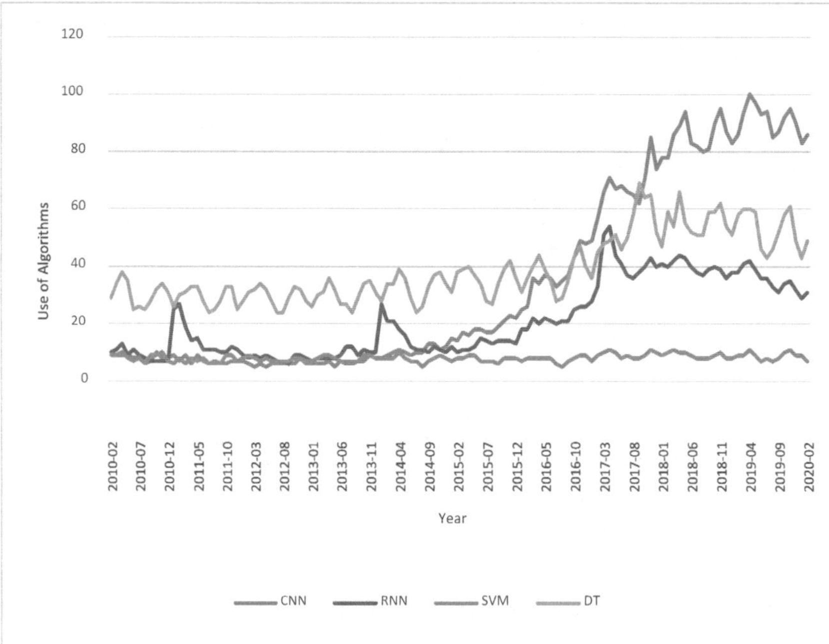

Figure 2. The use of deep learning and traditional algorithms

There are several types of deep algorithms each suitable for a different type of application. CNN (Convolutional Neural Network) and RNN (Recurrent Neural Network) are among the well-known deep algorithms. CNNs are feed forward ANNs widely used in computer vision for image recognition. However, RNNs include some feedback loops which make them more suitable for classifying time series data.

Deep learning can extract features automatically. Also, it works better than traditional machine learning algorithms. The greater the number of layers within the neural network, the better the features are extracted. And there is no need to calculate the feature manually. Traditional machine learning algorithms such as SVM (Support Vector Machine), KNN (K- Nearest Neighbors) and Decision Tree is dependent on human knowledge. However, deep learning algorithms ignore the presence of a human in the loop. Deep learning is an appropriate technique for obtaining accurate and fast classification. Figure 2 shows the use of deep learning and traditional algorithms in the last 10 years. Accordingly, the trend of using the CNN algorithm has increased enormously compared to other machine learning algorithms.

The main goal of this paper is to use deep learning algorithms to classify motion activities. These activities are extracted from WISDM dataset. In this case, acceleration values are used. These data are collected from smartphone sensor. The acceleration data are used as input of deep algorithms. Now, the feature extraction phase is performed automatically without expert supervision by using CNN and CNN-LSTM. As shown in figure 2 these two types of algorithms are more used. Figure 3 shows the structure of CNN and CNN-LSTM in our analysis.

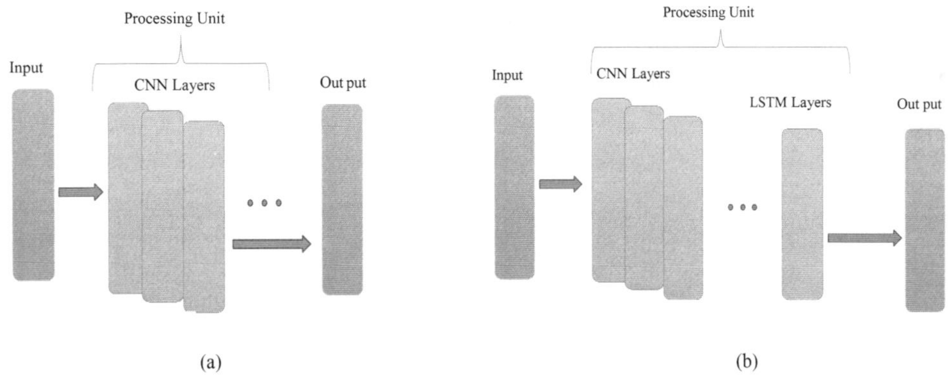

Figure 3. (a) Structure of CNN algorithm, and (b) structure of CNN-LSTM

4. RESULTS AND DISCUSSION

In this section, we discuss the results obtained from using the CNN and CNN-LSTM algorithms to recognize daily activities. As mentioned in section 2, six types of activities are analyzed by these algorithms i.e. "Walking", "Jogging", "Upstairs"," Downstairs", "Sitting" and "Standing". The analysis results are summarized in Figure 4, Figure 5 and Table 1. Figure 4 presents a comparison between the accuracy of these algorithms with respect to each activity. Accordingly, both of these algorithms are able to recognize activities with at least 90% accuracy. This is important regarding the fact that WISDM is an imbalanced dataset. However, the CNN-LSTM algorithm has a higher level of accuracy. This superiority could be related to the fact that LSTM can consider timing dependencies among the input data.

Furthermore, the results presented in Figure 4 indicate that Jogging is the easiest activity to recognize by either of the algorithms, since it involves more changes in the acceleration parameter. On the other hand, recognition of going upstairs/downstairs is more difficult for these algorithms, since there is no input data about the direction of movement.

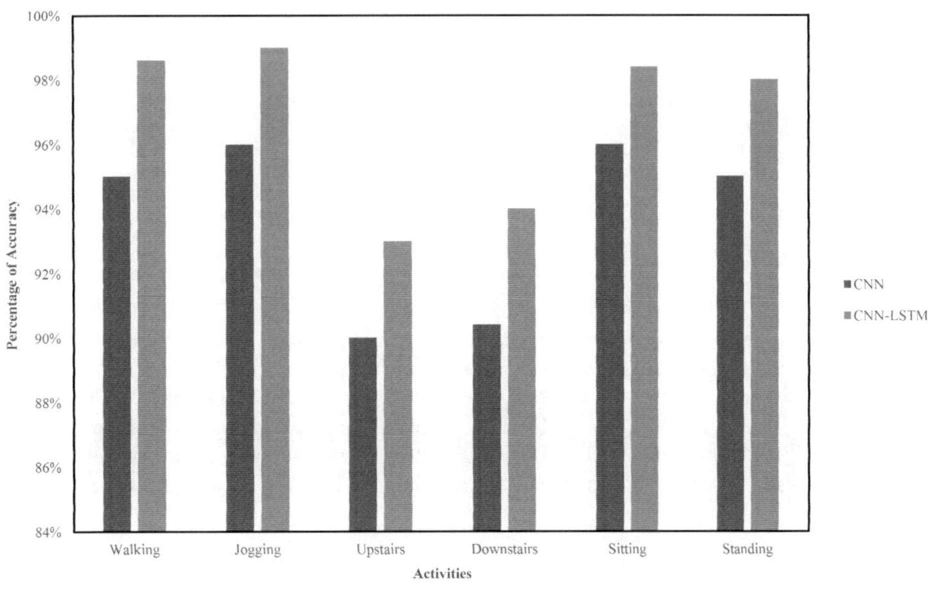

Figure 4. The accuracy of deep learning algorithms for different activities

Table 1 compares the algorithms with respect to loss value, number of layers and epochs, and average accuracy. Both algorithms have the same number of convolution layers, and as expected, the CNN-LSTM algorithm has a higher accuracy and a less error ratio in average. Figure 5 indicates the relation between the number of epochs and the loss and accuracy of each algorithm. The CNN algorithm shows the optimal performance after 200 epochs of training. However, the CNN-LSTM algorithm needs 50 more epochs to reach the highest accuracy. Form the loss value viewpoint, we witness a steady decrease for both algorithms as we increase the number of epochs.

Table 1. The number of layers, Error and Accuracy values in different deep algorithms

Type of algorithm	*Epochs*	*Error*	*Layers*	*Average Accuracy*
CNN	200	0.15	4	94%
CNN-LSTM	250	0.12	5	97%

Achieving high accuracy in learning motion activities is very important, especially in the context of mHealth systems. For example, assume that we have some patients suffering from Parkinson's disease, and they are supposed to perform some rehabilitation activities to improve their bodily functions. These activities help patients get their abilities back and regain independence. If we have a mHealth system capable of accurately recognizing these activities and giving the patients feedback in case of some mistakes, then they would be able to improve their health with less need to go to clinics.

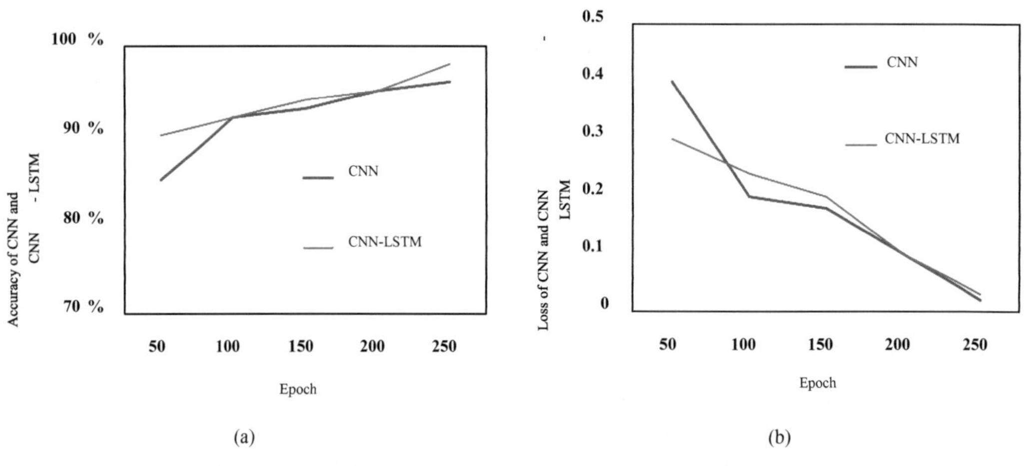

Figure 5. (a) The accuracy of deep learning algorithms, and (b) The loss of CNN and CNN-LSTM

Despite the advantages mentioned above, our proposed research has several limitations. First, the WISDM dataset is a small dataset and includes a few types of activities. Second, the data stored in the WISDM dataset is collected by a smartphone. In future work, we are going to use wearable sensors to collect more useful data and analyze the accuracy of deep learning algorithms. Third, we focused on only two deep learning algorithms i.e. CNN and CNN-LSTM in this research. Despite the strength and wide application of these models, we plan to examine other deep learning algorithms e.g. Gated Recurrent Units (GRUs) in our experiments.

5. CONCLUSION

In this paper, we presented the results of an experiment on the application of deep learning algorithm in the context of human activity recognition. In this experiment, the data recorded in the WISDM dataset were used for training two deep learning models. The first model included four layers of CNNs, but the second model was a combination of four CNN layers and one LSTM layer. The experiment results indicated that both of the models were able to recognize daily activities with a high level of accuracy i.e. more than 94%. However, the second algorithm performed more accurately.

REFERENCES

Adibi, S. (Ed.). (2015). *Mobile health: a technology road map* (Vol. 5). Springer.

Altini, M., Penders, J., Vullers, R., & Amft, O. (2014). Estimating energy expenditure using body-worn accelerometers: a comparison of methods, sensors number and positioning. *IEEE journal of biomedical and health informatics*, *19*(1), 219-226.

Asimina, S., Chapizanis, D., Karakitsios, S., Kontoroupis, P., Asimakopoulos, D. N., Maggos, T., & Sarigiannis, D. (2018). Assessing and enhancing the utility of low-cost activity and location sensors for exposure studies. *Environmental monitoring and assessment*, *190*(3), 155.

Boletsis, C., McCallum, S., & Landmark, B. F. (2015, August). The use of smartwatches for health monitoring in home-based dementia care. In *International Conference on Human Aspects of IT for the Aged Population* (pp. 15-26). Springer, Cham.

Carneiro, S., Silva, J., Aguiar, B., Rocha, T., Sousa, I., Montanha, T., & Ribeiro, J. (2015, May). Accelerometer-based methods for energy expenditure using the smartphone. In *2015 IEEE International Symposium on Medical Measurements and Applications (MeMeA) Proceedings* (pp. 151-156). IEEE.

Duclos, M., Fleury, G., Lacomme, P., Phan, R., Ren, L., & Rousset, S. (2016). An acceleration vector variance based method for energy expenditure estimation in real-life environment with a smartphone/smartwatch integration. *Expert Systems with Applications*, *63*, 435-449.

Fan, X., Huang, H., Qi, S., Luo, X., Zeng, J., Xie, Q., & Xie, C. (2015). Sensing home: a cost-effective design for smart home via heterogeneous wireless networks. *Sensors*, *15*(12), 30270-30292.

Fotouhi-Ghazvini, F., & Abbaspour, S. (2020). Wearable wireless sensors for measuring calorie consumption. *Journal of Medical Signals and Sensors*, *10*(1), 19.

Gusenbauer, D., Isert, C., & Krösche, J. (2010, September). Self-contained indoor positioning on off-the-shelf mobile devices. In *2010 International Conference on Indoor Positioning and Indoor Navigation* (pp. 1-9). IEEE.

Hamari, L., Kullberg, T., Ruohonen, J., Heinonen, O. J., Díaz-Rodríguez, N., Lilius, J., ... & Salanterä, S. (2017). Physical activity among children: Objective measurements using Fitbit One® and ActiGraph. *BMC research notes*, *10*(1), 161.

Hwang, P. Y., Chou, C. C., Fang, W. C., & Hwang, C. M. (2016, May). Smart shoes design with embedded monitoring electronics system for healthcare and fitness applications. In *2016 IEEE International Conference on Consumer Electronics-Taiwan (ICCE-TW)* (pp. 1-2). IEEE.

Jain, Y., Chowdhury, D., & Chattopadhyay, M. (2017, December). Machine learning based fitness tracker platform using mems accelerometer. In *2017 International Conference on Computer, Electrical & Communication Engineering (ICCECE)* (pp. 1-5). IEEE.

King, C. E., & Sarrafzadeh, M. (2018). A survey of smartwatches in remote health monitoring. *Journal of healthcare informatics research*, *2*(1-2), 1-24.

Kwapisz, J. R., Weiss, G. M., & Moore, S. A. (2011). Activity recognition using cell phone accelerometers. *ACM SigKDD Explorations Newsletter*, *12*(2), 74-82.

Lee, H., Kim, E., Lee, Y., Kim, H., Lee, J., Kim, M., ... & Yoo, S. (2018). Toward all-day wearable health monitoring: An ultralow-power, reflective organic pulse oximetry sensing patch. *Science advances*, *4*(11), eaas9530.

Li, M., Kwak, K. C., & Kim, Y. T. (2016). Estimation of energy expenditure using a patch-type sensor module with an incremental radial basis function neural network. *Sensors*, *16*(10), 1566.

Pande, A., Zhu, J., Das, A. K., Zeng, Y., Mohapatra, P., & Han, J. J. (2015). Using smartphone sensors for improving energy expenditure estimation. *IEEE journal of translational engineering in health and medicine*, *3*, 1-12.

Schneider, M., & Chau, L. (2016). Validation of the Fitbit Zip for monitoring physical activity among free-living adolescents. *BMC research notes*, *9*(1), 448.

Sukaphat, S. (2013). An applying of accelerometer in Android platform for controlling weight. In *Diambil dari: http://ibac-conference. org* (No. 20&, p. 20).

Van Camp, C. M., & Hayes, L. B. (2017). Identifying beneficial physical activity during school recess: Utility and feasibility of the Fitbit. *Journal of Behavioral Education*, *26*(4), 394-409.

Vathsangam, H., & Sukhatme, G. S. (2015). *U.S. Patent No. 8,965,402*. Washington, DC: U.S. Patent and Trademark Office. 8,965,402.

Rundle, A. G., Sheehan, D. M., Quinn, J. W., Bartley, K., Eisenhower, D., Bader, M. M., ... & Neckerman, K. M. (2016). Using GPS data to study neighborhood walkability and physical activity. *American journal of preventive medicine*, *50*(3), e65-e72.

Choi, E. J., & Kang, S. W. (2018). The relationship between acceptance intention toward a smartphone healthcare application and health-promoting behaviors among nursing students. *CIN: Computers, Informatics, Nursing*, *36*(10), 494-500.

CYBERSECURITY IN PANDEMIC TIMES: CHALLENGES AND OPPORTUNITIES

Ana Ferreira[1,2] and Ricardo Cruz-Correia[1,2]
[1]*CINTESIS - Center for Health Technology and Services Research*
[2]*Departamento Medicina da Comunidade, Informação e Decisão em Saúde (MEDCIDS)*
Faculty of Medicine, University of Porto, Portugal

ABSTRACT

In times of COVID-19 and isolation, many are the challenges related to a proper balance between patient's personal and sensitive data privacy and protection, and the need to use those data to better understand the disease and act in a more efficient and quick manner, to fight it. Some research is available on how to tackle a few of these issues in isolation, however, there is a need for a more comprehensive overview of what are the current security challenges, what is being done to face them, and what still needs addressing. Since the control of this pandemic depends on the actions of every individual, when there is no trust in the technology or services provided, allied to the fear and uncertainty of the current context, it is very difficult to provide an integrated and informed response to this disease. The main goal of this paper is to expose and address vulnerabilities and risks associated to patients' data protection and COVID-19. Thirty-two direct, and 25 indirect consequences of COVID-19 to data privacy and cybersecurity, were identified. There is the need for more comprehensive and resilient solutions addressed by the research community, from multidisciplinary sources of knowledge.

KEYWORDS

Cybersecurity, Privacy, COVID-19, Health Data Protection, Quarantine and Isolation

1. INTRODUCTION

When we think about COVID-19 and the current pandemic, the last thing that comes into our mind is the subject of cybersecurity. Yet, more than ever, cybersecurity is crucial for all that is connected with COVID-19, both directly, in terms of research and active treatment and containment, and indirectly, with all the side effects and consequences that it has created. One of the most direct consequences of COVID-19 in cybersecurity relates with the use and sharing of personal and health data from patients that contracted the new coronavirus. It is suggested that the more such analyses are initiated and concluded openly, in accordance with the law and with due respect for privacy and confidentiality, the greater will be the public trust and the ability to produce reliable analytic insights (Jawaid, 2020)(Ienca, 2020). Failing to do so will undermine public trust and adherence to public-health recommendations, which can result in poorer health outcomes (Ienca, 2020). Moreover, extensive gathering of diagnostic data on who is infectious may be essential to save lives, train Artificial Intelligence systems, and limit economic damages (Naudé, 2020).

Still related to this issue are patients' data shared in the context of contact tracing. Contact tracing using smartphone technology is a powerful tool that may be employed to limit disease transmission during a pandemic, enforce quarantine rules, notify risk zones or warn infected people (Abeler, 2020). However, contact tracing apps present significant privacy concerns since they collect personal data, such as location (Yasaka, 2020), which can be used to perform a high degree of surveillance and harm to the human right to privacy. There have been some works that try to find the right balance between the required data and user's privacy, by not collecting location data, but only using anonymized graph of interpersonal interactions, stored locally on the smartphone device (Yasaka, 2020).

More indirect consequences of the current pandemic to data privacy and security are associated to the current isolation of people, who are confined to their homes or a single physical location, for long periods of time. A high percentage of the humanity is isolated or in lockdown and rely on digital infrastructures to perform most of their daily activities. Working from home, home-schooling, online shopping, home-banking, contact with friends and family, and even exercising and entertaining activities, are all being performed using the Internet 24/7. There are many implications on cybersecurity and physical security that arise from this situation such as the need to deal with most pressing issues related to diseases, other than COVID-19. Healthcare professionals have to opt for other means to support their patients such as: tele-consultation, advice via email, whatsapp, or similar platforms (Rajkumar, 2020)(Kavoor, 2020)(Now, 2020). Moreover, digital home equipment, as well as their users, are most probably not prepared to securely provide this type of interaction. On the other hand, the confinement situation can also exacerbate the need for this type of support from vulnerable populations, such as chronic, oncological, and mental health patients (Jawaid, 2020)(Yao, 2020)(Rajkumar, 2020), where the usual assistance in an hospital or other healthcare institutions is not advised during this period, not only for the high risk of becoming infected with COVID-19 (more dangerous on patients with comorbidities) but also, due to the lack of healthcare resources available.

The provided examples are just a small set which clearly motivates and justifies this research work. The aims of this paper are to: 1) identify the main vulnerabilities and risks associated, both directly and indirectly, to healthcare data protection and COVID-19; 2) understand how the research community is tackling these cybersecurity challenges; 3) expose the identified gaps that still need addressing by that community; and 4) further discuss the identified challenges to press the advancement of existing knowledge, to achieve a better balance between patient's privacy and their data use, to effectively fight this pandemic.

2. METHODS

A literature review of the research performed in this area was performed. Inclusion criteria comprised works in English, published from January to May 2020, which link the pandemic with privacy/security issues (e.g., "security OR privacy OR confidentiality" AND "covid OR pandemic"). Searches were performed in 5 database engines and works were reviewed by their titles and abstracts first, with repeated works excluded (Figure 3). The selected works were analysed by: Type of work (article, letter, comment, etc); Database engine; Direct/Indirect consequence from COVID-19; Main subject; Goal(s); Problem raised; and Proposed solution, and they were divided in two categories:

- **Directly related to the combat of the disease (Figure 1)**, i.e., i) the need to use patient's personal and sensitive data for analysis, control of the disease propagation and transmission; ii) the exposure of the lack of investment in critical security infrastructures; and iii) the increase of fraud and theft via email, SMS or social networks;
- **Indirectly related to the combat of the disease (Figure 2)**, i.e., i) associated to consequences of the pandemic, mostly related to the isolation and confinement situation and their influence to other non-covid-19 related diseases, such as mental health, chronic diseases or other health conditions that need remote managing (teleconsultation, online advice; dealing with health emergencies); and ii) augmented risks to physical security, with stress on constantly using home/personal digital infrastructures and devices, for other contexts and goals, for which they are not security prepared.

Figure 1. Cybersecurity challenges/consequences DIRECTLY related to COVID-19

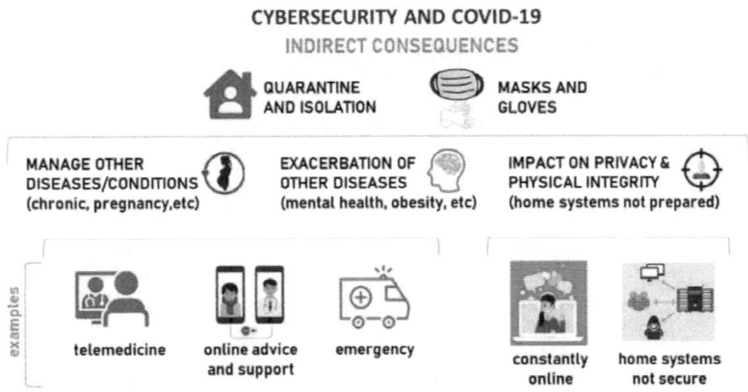

Figure 2. Cybersecurity challenges/consequences INDIRECTLY related to COVID-19

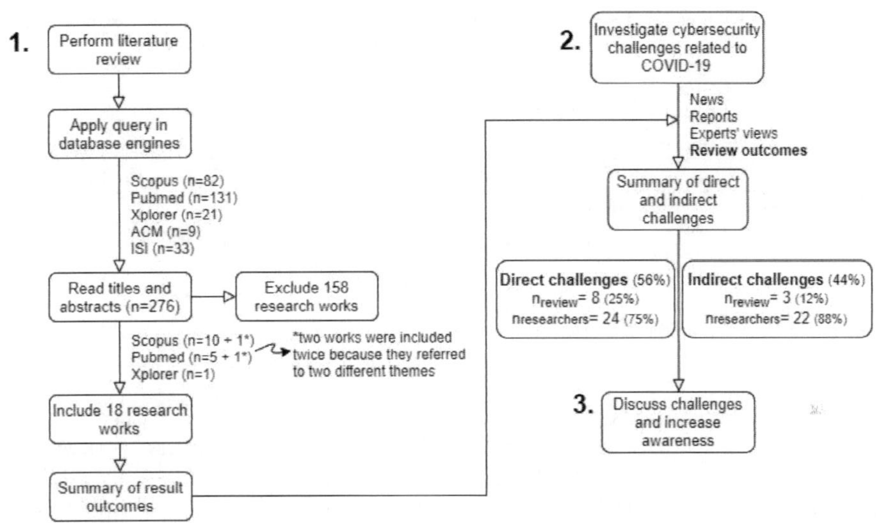

Figure 3. Illustration of the methods performed in this work

3. RESPONSE FROM THE RESEARCH COMMUNITY

Eighteen research works (10 articles; 8 comment/letters/opinions) were included in the review. These focus mostly on sharing data to find out more about the disease (8), provide means to trace infected people (7), better understand how the disease propagates and the implications in terms of privacy and data protection. In terms of consequences of COVID-19 to cybersecurity, 83% (n=15) of the works focus on **direct** (i.e., data sharing and contact tracing)**,** while 17% (n=3), on **indirect** consequences (i.e., mental health (2) and legal procedures (1)). Common **problems** and proposed **solutions** are highlighted in Tables 1 and 2, respectively.

Table 1. Cybersecurity problems and concerns found in the review, directly or indirectly associated with COVID-19

Areas of research	Cybersecurity problems/concerns found in the review	References
DATA SHARING*		
Contact Tracing	Breach of privacy can result in decreased compliance, public trust and adherence	(Pesce) (Nature) (Ienca)
	How to wind back to a world with greater privacies after the crisis abates?	(Pesce) (Kummitha)
	Countries are developing apps independently, and there are no global standards	(Nature)

	Scant evidence on how effective these apps are/will be	(Nature)
	Contact data such as location in tracing apps present significant privacy concerns	(Yasaka) (Finch)
	It is becoming easier to identify individuals from anonymized data sets	(Nature)
Other purposes	Virologists/epidemiologists are consulted, and other health/non-health experts?	(Rajan)
	Health professionals should share their knowledge/suspicion a patient is infected to others when 'distancing'/'shielding' is impracticable, ineffective or refused	(Finch)
MANAGE OTHER DISEASES REMOTELY⁺		
Mental health	Social isolation in seniors has been linked to increased depression, suicidality, proinflammatory and decreased immune responses - privacy and safety concerns	(Jawaid)
Impact on diseases	Epidemics do not affect all population equally, which can contribute to the spreading of infections - privacy and safety concerns	(Yao)
Legal procedures	Laws to arrest people infected to limit the spread of disease – privacy concerns	(Finch)

*DIRECT CONSEQUENCES; +INDIRECT CONSEQUENCES

Table 2. Cybersecurity solutions found in the review, directly or indirectly associated with COVID-19

Areas of research	Cybersecurity solutions found in the review	References
DATA SHARING*		
Contact Tracing	Use anonymized graph of interactions to conduct contact tracing apps	(Yasaka)
	Apps should not be rolled out without pilot studies or risk assessments	(Nature)
	Apps that link to official validated tests are more likely to give accurate results	(Nature)
	The need for data minimisation for tracing apps	(Mirco)
	Critical to balance the need for information with legitimate privacy concerns	(Park)(Abeler)
	Contact tracing is possible without collecting a lot of data in central databases	(Abeler)
Other purposes	Provide insights to guide behaviour of individuals, public policy and dialogue - provide transparency on how, when, and why data are being used	(Pesce) (Rajan) (Mirco) (Jawaid) (Kummitha)
	More involvement from researchers, also from ethics and law, and commitment from governments that personal data is secure and used only for requested reasons	(Nature)
	Create a unified framework for data exchange about the epidemic and research is an essential step towards effective responses to the challenges we currently face	(Lenert)
	Use these data and algorithms in a responsible manner, in compliance with data-protection regulations and due respect for privacy and confidentiality	(Ienca) (Jawaid)
	Balance between data privacy, public health and rigorous human-AI interaction	(Naudé)
	Build community-level resilience in order to cope with the crisis	(Kummitha)
	Cities need to partner with each other and actively coordinate with governments	
	Need to share best practices among countries and encouraging different key stakeholders to invent/implement technologies to detect/isolate infected patients	
MANAGE OTHER DISEASES REMOTELY⁺		
Mental health	Keep elderly engaged/motivated with volunteers, hotlines or online resources	(Yao)(Jawaid)

*DIRECT CONSEQUENCES; +INDIRECT CONSEQUENCES

4. PANDEMIC CHALLENGES FOR CYBERSECURITY

A brainstorm was performed by the authors, experts in cybersecurity, to identify complementary cybersecurity challenges associated with the COVID-19, not yet identified by the research community, DIRECTLY (Table 3) and INDIRECTLY (Table 4) related to COVID-19.

Table 3. Privacy/security challenges DIRECTLY associated to COVID-19, not presented in Table 1

Categories	Cybersecurity issues directly related to COVID-19 - NOT found in the review
DATA SHARING	
Contact Tracing	EU member states need to comply with GDPR (e.g., right to be forgotten, access, etc)
	Need for agreements from authorized Ethics Commissions (same for other purposes)
	Lack of IT and cybersecurity literacy can compromise development/use of such apps
	Triangulation of tracing data, with data from the same device or service provider has to be avoided. The device is not anonymous and other apps may share those data
	Bluetooth has its known vulnerabilities depending on the device and protocol version
	If these apps work, is there an effective return to the individual or the public health?
	Anonymized data: more susceptible to wardriving, jamming and sabotage, leading to poor data quality, with no proper means available to guarantee data integrity
	Total anonymity is never possible because when people are in isolation, it can be easy to find out who is infected or not, due to low interaction activities
	Many false positives can be obtained when Bluetooth/GPS traverse walls, cars, etc
Other purposes	Mandatory measure of body temperature or immunity tests of employees/students before entering the premises. Need for proper national regulations for these issues
	Data sharing from government institutions for research purposes may be adequate and justified. For example: GDPR has such provisions (Articles 6(e) & 9(g)(i))
	Feeding on the current fear of the population, authorities may act with impropriety to the privacy of citizens without proper justification on why the decisions were made
	Integrity of research results needs to be checked by other impartial research groups to make sure important decisions are taken based on the most current/correct data
FRAUD & THEFT	
Theft of health emergency material	International espionage and sabotage, like health material held or confiscated by another country or region for which it was supposed to be delivered in the first place
Ransomware	Messages associated to COVID-19, with an increased degree of persuasion due to fear and uncertainty, to block access to data and steal money by asking for a ransom
Phishing	Messages associated to COVID-19, with an increased degree of persuasion due to fear and uncertainty, to steal credentials from online or banking accounts Fake messages with online orders with an increased degree of persuasion since people are in isolation and, therefore, buying many products online
False campaigns	False fundraising campaigns associated to COVID-19 with the false aim to help people in distress or buy specific resources
Uncertified material	Non-certified or false health material that is sold (such ventilators, masks) which may be very dangerous to whomever uses them (both patients and professionals)
VULNERABLE SYSTEMS	
Unpatched systems	Healthcare systems likely to be under budgeted and not with the latest patches/versions
Lack of resources	Not enough technology/devices to monitor a high number of hospitalized patients
Stress use	Health systems are diverse and not interoperable, which can make them less resilient
Access control	Lack of access control updating and monitoring may compromise patients' privacy, with no control of the high turnover and additional professionals in times of pandemic
Vulnerable communications	Lack of systems protection can lead to more leakages in terms of communication via many devices/technologies that can be used (smartphones, email, social networks, etc)

Table 4. Privacy/security challenges INDIRECTLY associated to COVID-19, not presented in Table 1

Catgories	Cybersecurity issues indirectly related to COVID-19 - NOT found in the review
MANAGE OTHER DISEASES	
Telemedicine/online advice	Unsecure home technology for health communications - privacy/physical security*
EXACERBATION OF CONDITIONS	
Mental health, obesity, etc	Emergency/continued support via unsecure technologies -privacy/physical security*
Lack of timely resources - medication/treatments	Stress or lack of security in communication infrastructures can end in resources' deviation or treatments' delay - privacy/physical security*

PRIVACY & PHYSICAL SECURITY*	
Constantly online	Increased risky behaviours such as "click on anything" or no conscious assessment or time to correctly assess each message and action performed online
Shared contexts (personal, professional, familiar)	The use of the same accounts on different devices, or different accounts on the same device – prone to confusion and mistakes. No strong different passwords in them
Unlocked sessions/devices	This situation, in shared spaces/contexts increases risk of attacks/mistakes
Children and teenagers	Increased time online in various activities (homeschooling, connect with friends, etc) can increase risky behaviours and cyberbullying Overburdened/distracted parents do not always monitor/control children's activities There is a lack of IT and cybersecurity literacy from both parents and children
Unsecure home channels	No custom use of secure communication channels, such as VPNs
Equipment configurations/patches	Not updated anti-virus or anti-malware software, lack of timely patching, no passwords, weak passwords or same passwords, for different equipment/devices
Regular Video calls	Privacy threats with the use of personal spaces, images of children or other personal information in the background
Shared spaces	Increased frequency, due to quarantine, of discussing/disclosing personal, private or professional matters in shared spaces such as garden, balcony or living room
Lack of backups	Lack of regular data backups, in at least 2 different devices, in different locations
Targets always home	Easier to target cybersecurity and physical attacks to people confined in their home
Cameras and microphones	With people constantly online, cameras and microphones are on, most of the time. They can still be on even when not needed and attackers can take advantage of this
No/improper access control	Lack of access control measures for all the people using shared devices and accounts
Fake deliveries and access to the house	With people in insolation and constantly online buying food and other products, fake deliveries can easily deceive people and attackers can rob/harm people
Fear and discrimination	People infected may be discriminated/mistreated and targeted online
Masks	With masks it is even easier to impersonate someone and enter a health institution, it is difficult to quickly ascertain someone's identity
Gloves	Cannot use biometric fingerprint authentication to access health spaces/systems
Break work policies/rules	In mixed contexts people should be aware of the same regulations and policies of their organization to always apply, when working from home
Stress use	Home systems/infrastructures may not be regularly updated and their continued use on periods such as this, can make them less resilient

5. DISCUSSION AND RECOMMENDATIONS

This work identifies and discusses very relevant cybersecurity issues in times of pandemic, and strives to make the research community, and the public in general, aware, so that adequate means are developed to face them. Due to space constraints, the authors will highlight and discuss only some of the identified issues.

In terms of review results, works mostly focus on direct challenges and much less on indirect. In fact, maybe due to the short time that has elapsed since the beginning of the pandemic, is still early for researchers and people to think, and even realize, the wider range of consequences that COVID-19 has brought to our lives. For one of the most focused direct cybersecurity challenges in the literature (i.e., contact tracing), many more issues need to be discussed and addressed. In fact, the balance between sharing and the privacy of personal data is really crucial but very hard to achieve, since the need for a high degree of anonymity makes it much harder to control or audit, and if we cannot trust data integrity and quality, it will not be useful for the proposed goals. Data integrity can only be achieved by letting some degree of privacy slip away, ideally to a trusted entity. It is advised that apps should not be rolled out without pilot studies or risk assessments being published (Nature, 2020), yet this has not been the case, as various countries have already released this kind of technology without proper transparency measures (Ekong, 2020). Means are needed to wind back to a world with greater privacies, after the crisis declines (Pesce, 2020) and to provide transparency and continuous risk assessment, including, the certification of the used technology, the provision of links with proper regulatory and cybersecurity authorities, together with external auditing to keep track of what is being done and how.

In terms of data sharing, the authors will highlight two issues. Now that some countries are starting to deconfine, there have been some measures to monitor workers' and students' body temperature. These, or similar measures, can be invasive to the privacy and physical space of the individuals, and may not even have a legal basis, so need to be properly addressed and discussed. Another issue relates to a wider subject, also important in non-pandemic times, but that may negatively exacerbate in pandemic times, which is research integrity. Many governments and health authorities may try to base their decisions on epidemiological and research results, without proper verification/certification by other research groups or entities (although time is of essence). Such important life/death decisions must be taken on the most updated and corrected data results.

One final direct issue, with no solution at sight but very relevant, is the great increase of social engineering attacks and fake campaigns associated with COVID-19 (WHO, 2020). Fakenews, fear, isolation and lack of awareness, can turn the confined population into a highly vulnerable target to those attacks (INTERPOL, 2020). This issue needs addressing as no proper solutions exist, and this is only bound to exponentiate.

On the indirect cybersecurity issues related to COVID-19, these are very relevant to address, since other fatalities have and will happen as a consequence of the pandemic, which could probably have been avoided. The stress on remote support and consultation (e.g., telemedicine) for chronical, mental diseases or other conditions, such as pregnancy, is now a reality, but no proper cybersecurity standards are available to protect these connections. The subject has been studied for several years but now that is really needed, pressure is put on home, health and professional's infrastructures, without proper testing and monitoring, and without proper study on its impact to patients' privacy, physical integrity and safety. For cybersecurity, the most important asset are people, so the connection between people's health, the pandemic and cybersecurity, is close indeed.

One more word in relation to children and adolescents and how they are also affected by cybersecurity challenges. They are, together with elderly people, more prone to attacks and persuasion, so privacy and security measures in times of pandemic (but not only) need to adapt to various types of populations and their specific characteristics and needs. Current infrastructures are not adequate for kids to perform secure and safe homeschooling, sports, music and communication activities online, with teachers or friends and family. Out of the box and innovative solutions are required. Maybe transforming peoples' identities into high-fidelity digital humans could prevent, at least, some of the privacy and physical issues that arise from online conference calls.

Finally, many new behaviours that have been introduced quickly to face emergency states may not have been tested accordingly and may also generate undesired consequences. Examples include material such as masks, gloves and even full protection suits in areas with infected patients. These are certainly very important equipment to be used by health professionals and other types of workers. In some countries, regulations are imposed for the public to wear masks in public confined places. However, all these facilitate identity hiding and impersonation, which in healthcare related situations (but not only) can be life threatening. Maybe this is not yet the time to deal with these specific problems as attending to patients is certainly more important, still, attackers and fraudsters are always alert and take advantage of these holes in the system.

Cybersecurity as well as any other recommendations, measures, regulations or tools, can be used to perform the required goals, but also be subverted and used to perform less appropriate actions. Researchers, developers and everyone in general, need to be more aware of this to better balance these dual risks and consequences, especially crucial in life threatening situations.

Limitations: the sample size found in the literature was small probably due to the short period of time analysed and the time needed to publish those works; also, the review was performed by only one researcher. The identified complementary issues rely mostly on authors' expertise and a few on other sources such as news, technical magazines and cybersecurity sites. However, those data are scattered around, and this study helps to provide a unique/comprehensive way to make them available for future research and improvement. Due to space constraints, it was not possible to lengthily discuss recommendations to mitigate all presented challenges.

6. CONCLUSION

None of the identified cybersecurity challenges is new, however, the current pandemic has stressed the need for a more comprehensive analysis and adequate response to both direct and indirect issues related to COVID-19, very much intertwined into our daily lives. This work discusses those cybersecurity issues that require more resilient solutions from multidisciplinary sources that range from physical, social, ethical, behavioural, psychological to regulatory and technical knowledge; and stresses the necessity to take back our fundamental right for privacy after this unusual situation ends, as the impact of COVID-19 in cybersecurity and in the privacy and physical integrity of individuals will have to be handled, much after the pandemic subsides.

ACKNOWLEDGEMENT

This work has been funded by TagUBig - Taming Your Big Data (IF/00693/2015) from Researcher FCT Program funded by National Funds through FCT.

REFERENCES

Abeler, J., Bäcker, M., Buermeyer, U. and Zillessen, H., 2020. COVID-19 Contact Tracing and Data Protection Can Go Together. *In JMIR Mhealth Uhealth*, 8(4):e19359.

COVID-19 Cyberthreats, 2020. INTERPOL. Available at: https://www.interpol.int/en/Crimes/Cybercrime/COVID-19-cyberthreats. Accessed in May 2020.

Ekong, I., Chukwu, E. and Chukwu, M., 2020. COVID-19 Mobile Positioning Data Contact Tracing and Patient Privacy Regulations: Exploratory Search of Global Response Strategies and the Use of Digital Tools in Nigeria. *In JMIR Mhealth Uhealth*, 8(4):e19139.

Finch. J, 2020. Legal aspects of COVID-19 pandemic management for community nurses. *In British Journal of Community Nursing*, 25(4): 196-199.

Ienca, M. and Vayena, E., 2020. On the responsible use of digital data to tackle the COVID-19 pandemic. *In Nat Med*, 26(4):463-464.

Lenert, L. and McSwain, BY., 2020. Balancing Health Privacy, Health Information Exchange and Research in the Context of the COVID-19 Pandemic [published online ahead of print, 2020 Mar 31]. *In J Am Med Inform Assoc*.

Jawaid, A., 2020. Protecting older adults during social distancing. *In Science*, 368(6487):145.

Kavoor, A.R., Chakravarthy, K., and John, T., 2020. Remote consultations in the era of COVID-19 pandemic: Preliminary experience in a regional Australian public acute mental health care setting. *In Asian Journal of Psychiatry*, 51.

Kummitha, RKR, 2020. Smart technologies for fighting pandemics: The techno- and human- driven approaches in controlling the virus transmission [published online ahead of print, 2020 Apr 20]. *In Gov Inf Q*, 101481.

Mirco, N. et al, 2020. Give more data, awareness and control to individual citizens, and they will help COVID-19 containment. *In Transactions on Data Privacy*, 13(1): 61-66.

Nature, 2020. COVID-19 digital apps need due diligence. *In The international journal of science*. Springer Nature, Vol. 580, 563.

Naudé, W., 2020. Artificial intelligence vs COVID-19: limitations, constraints and pitfalls [published online ahead of print, 2020 Apr 28]. *In AI Soc*, 1-5.

Now, You Can Consult a Doctor on WhatsApp: Govt Issues Telemedicine Guidelines, 2020. india.com. Available at: https://www.india.com/news/india/now-you-can-consult-a-doctor-on-whatsapp-govt-issues-telemedicine-guidelines-3981377/. Accessed in May 2020.

Park, S. and Choi GJ, Ko. H, 2020. Information Technology-Based Tracing Strategy in Response to COVID-19 in South Korea-Privacy Controversies [published online ahead of print, 2020 Apr 23]. *In JAMA*.

Pesce, M., 2020. Privacy in the time of Covid-19. *In IEEE Spectrum*, vol. 57, no. 5, pp. 23-23.

Rajan, D., Koch, K., Rohrer, K., Bajnoczki, C., Socha, A., Maike, V., Marjolaine, N., Valéry, R. and Justin, K., 2020. Governance of the Covid-19 response: a call for more inclusive and transparent decision-making. *In British Medical Journal Global Health*, 5, 2655.

Rajkumar, P., 2020. COVID-19 and mental health: A review of the existing literature. *In Asian Journal of Psychiatry*, 52.

WHO reports fivefold increase in cyber-attacks, urges vigilance. 2020. WHO news release. Available at:

https://www.who.int/news-room/detail/23-04-2020-who-reports-fivefold-increase-in-cyber-attacks-urges-vigilance. Accessed in May 2020.

Yao H, Chen JH, Xu YF., 2020. Patients with mental health disorders in the COVID-19 epidemic. *In Lancet Psychiatry*, 7(4): e21.

Yasaka, TM., Lehrich, BM. and Sahyouni, R., 2020. Peer-to-Peer Contact Tracing: Development of a Privacy-Preserving Smartphone App. *In JMIR Mhealth Uhealth*, 8(4):e18936.

IDENTIFYING AND UNDERSTANDING OPIOID USE DISORDER IN CLINICAL NOTES

T. Elizabeth Workman[1], Joel Kupersmith[2], Joseph L. Goulet[3], Christopher Spevak[4], Cynthia Brandt[5], Friedhelm Sandbrink[6], Marc R. Blackman[7], Nawar M. Shara[8] and Qing Zeng-Treitler[1]

[1]*Biomedical Informatics Center, The George Washington University, Washington, D.C., U.S.A.*
[2]*Veterans Initiatives, Georgetown University, Washington, D.C., U.S.A*
[3]*VA Connecticut Healthcare System, West Haven, CT, U.S.A*
[4]*Prescription Medication Misuse Program, Georgetown University, Washington, D.C., U.S.A*
[5]*Biostatistics, Yale University, New Haven, CT, U.S.A*
[6]*Pain Management, VA Medical Center, Washington, D.C., U.S.A*
[7]*Medicine and Rehabilitation Medicine, Georgetown University, Washington, D.C., U.S.A*
[8]*Biostatistics and Biomedical Informatics, MedStar Health Research, Hyattsville, MD, U.S.A*

ABSTRACT

Opioid use, abuse and misuse afflicts many populations, including Veterans. The objective of this ten-year retrospective study was to identify documentation of potential opioid abuse, both treated and untreated, in clinical notes, by developing and applying a natural language processing tool to a corpus of clinical notes documenting the healthcare of U.S. Veterans. To better understand the issue of opioid abuse among Veterans, we also extracted descriptive data on prescription counts, patient demographics, and diagnoses. The natural language processing tool we developed achieved F1 scores of 88% and 91% in identifying opioid abuse with treatment, and without treatment, respectively, among U.S. Veterans receiving healthcare in the Baltimore, Maryland and Washington DC VA service regions. This resulted in identifying 809 additional patients experiencing opioid abuse. The descriptive data give insight by elucidating trends that enhance understanding of opioid abuse among Veterans receiving healthcare in these service regions, and suggest future research.

KEYWORDS

Opioid Abuse Disorder, Machine Learning, U.S. Veterans

1. INTRODUCTION

The misuse of prescribed opioids is a global problem affecting numerous populations (Martins et al., 2015). In 2017, there were 47,600 deaths the United States attributed to opioid overdoses (Scholl et al., 2018). The Diagnostic and Statistical Manual of Mental Disorders, 5th Edition (Association, 2013) defines Opioid Abuse Disorder (OUD) as a problematic pattern of opioid use leading to clinically significant impairment or distress. Its characteristics include clinical heterogeneity (Kendler et al., 2013), having diverse risk factors, including past substance abuse, age, untreated psychiatric disorders, family situation, traumatic brain injury, and pain (Webster, 2017, Cottler et al., 2011, Edlund et al., 2007). Variance in geographic prescribing patterns affects the complexity of OUD (Garcia et al., 2019). Facets involving gender and race also present parallel considerations in studying OUD, particularly in pain management. African Americans and other ethnic minorities are consistently under-accessed and undertreated for pain (Green et al., 2003). Women experiencing chronic pain and opioid abuse tend to report more pain-related impairment and psychiatric co-morbidity, while their male counterparts report more aberrant behaviors (Manubay et al., 2015). Women may also demonstrate lower pain thresholds, and higher pain ratings than men (Chartoff and Mavrikaki, 2015). Opioid abuse also affects Veterans of the United States Military, for whom male sex, young age, and mental health disorders are notable risk factors (Edlund et al., 2007). Several policies and practice guidelines have attempted to address the problem of opioid abuse among U.S. Veterans, including the Opioid Safety Initiative (OSI) in 2013 and the stratification tool for opioid risk management (STORM) in 2018 (Management, Gellad et al., 2017, Lin et al., 2017, Wentling, 2018).

In electronic medical records (EMRs), opioid use, abuse, and misuse is recorded as structured data, in the form of diagnostic codes, or in the unstructured data found in clinical notes. While clinicians often record information as structured data, sometimes some information may be found only in a patient's notes (Moss et al., 2007). Using only ICD codes in structured data as an information source may underestimate the incidences of opioid abuse and result in errors in subsequent analyses. It is therefore desirable to identify clinical notes specifically addressing opioid abuse along with information addressing treatment. This can potentially be facilitated through natural language processing (NLP) and machine learning by preparing the notes using standard NLP techniques such as ngram assessment, where an ngram is an instance of a single word (unigram), or instances of adjacent multiple words in a corpus. This type of assessment includes capturing instances of two adjacent words (bigrams) throughout a corpus of clinical notes, and using the counts of each bigram as features for each clinical note, in training a machine learning model.

Other studies have applied machine learning applications to gain insight into opioid use, abuse and misuse. Several of these focus on social media content (Sarker et al., 2019b, Sarker et al., 2019a) or structured EMR data (Hastings et al., 2020, Workman et al., 2019b), or structured data and clinical notes, with NLP (Lingeman et al., 2017, Sharma et al., 2020). Sharma et al. identified a random case-control cohort of 1000 patients with or without ICD-9/ICD-10 OUD related diagnosis or positive urine drug screen for opioids, subjected to chart review (Sharma et al., 2020). They experimented with both ngram and UMLS concept-mapped representations of the data, and several machine learning models, using a 63,301 note corpus; in their experiments, a convolutional neural network achieved the best performance with an F1 score of 0.84. Wadekar (Wadekar, 2020) also experimented with multiple machine learning models, using 42,324 observations from the 2016 National Survey on Drug Use and Health, along with patient demographic, vital, and health information. The best model was a random forest, achieving a specificity (i.e., recall) of 0.81 and sensitivity of 0.82 (F1 was not disclosed). Prieto et al. used a document term frequency matrix derived from a corpus of 1166 paramedic trip reports containing keywords 'heroin' or naloxone', or both. Of several models tested, a logistic regression model achieved an F1 score of 0.93, based on reported data (Prieto et al., 2020).

The Veterans Health Administration (VHA) within the Department of Veterans Affairs (VA), a branch of the United States Federal Government, directs one of the world's largest healthcare systems, providing care to over 9 million Veterans at 1255 in-patient and out-patient facilities, throughout the 50 U.S. states, 4 U.S. territories, and the Philippines (Administration). It is divided into regional units known as Veterans Integrated Service Networks (VISNs), which in turn consist of regional service districts known as stations. A station generally consists of one or more hospitals, and other facilities such as outreach clinics, substance abuse care centers, and domiciliary care programs for homeless Veterans. The Washington DC and Baltimore VHA stations are part of the VISN 5, VA Capitol Health Care Network, consisting of 6 stations based in Washington DC, Maryland, and West Virginia. The EMR developed by the VA, called the Veterans Health Information Systems and Technology Architecture (VistA)(Brown et al., 2003), is one of the earliest developed and widely used EMRs in the world. These efforts have resulted in a vast clinical data resource, housed in the VHA's corporate data warehouse, and made available through the Veterans Affairs Informatics and Computing Infrastructure (VINCI), a secure research platform.

In this retrospective study, we sought to develop an NLP tool to identify opioid abuse and its treatment as documented in clinical notes, using a corpus of notes recorded between 2010 and 2019 that were drawn from the Washington DC and Baltimore stations. We applied NLP analysis methods in data preparation, and a subsequent support vector machine (SVM) model. To place these findings in context, we independently examined structured and unstructured data from the two stations, within the same ten-year period, to extrapolate on commonalities and differences in terms of patient demographics, prescribing patterns, and opioid use, abuse and misuse rates. All data retrieval, and NLP and machine learning processes were performed within the VINCI platform. This project was carried out in support of the VA Clinical NLP Ecosystem study, which was approved by the VA Central IRB #CIRB 13-17 Zeng. No patients were contacted, only data from the EMR were used.

2. METHODS

2.1 Data for NLP and Machine Learning

We extracted 495 clinical notes for patients in the Washington DC and Baltimore VA stations that contained key phrases associated with opioid and other forms of substance abuse. A committee of clinicians familiar with OUD documentation in clinical notes in the VA identified these phrases. The key phrases were "opioid", "opioids", "dependence", "addiction", "detox", "withdrawal", "taper", and "tapered". We then extracted snippets, i.e. segments of text including the key phrase used to retrieve the given note, within a contextual window of the 15 terms preceding the key phrase, and the 15 terms following it in the note. One author labeled each snippet with one of three categories: 1) no evidence of opioid abuse, 2) evidence of opioid abuse without treatment, and 3) evidence of opioid abuse with treatment. To validate this process, a master set of 50 randomly selected snippets was drawn, and then reviewed by a second author. Where there were differences in labeling, the two annotators reached a consensus and made changes in the classification of the full body of snippets as dictated.

After converting text to lower case, removing punctuation and common markup such as tabs, carriage returns, and excessive whitespace, and removing non-alphanumeric terms, we identified bigrams that occurred more than four times throughout the corpus. The features for each clinical note were the occurrence frequencies of each bigram within the note, and the character offset at which the key phrase used to retrieve the note occurred within its content. The data, consisting of these features for each note, were divided into training (80%) and test (20%) sets.

2.2 SVM

The SVM implemented a radial basis function (RBF) kernel, regularization parameter C of 1.0, scaled gamma, a stopping criterion tolerance of 1e-3, and the one-vs-rest decision function shape (all default values for the RBF kernel). We used the python scikit-learn svm SVC package. The SVM model was trained using 80% of the data, and tested using the other 20% that had been reserved for that purpose. There were 137, 53, and 80 support vectors for the respective classes no evidence of opioid abuse, evidence of opioid abuse without treatment, and evidence of opioid abuse with treatment. We evaluated the model's performance for precision (true positives/(true positives + false positives)), recall (true positives/(true positives + false negatives)), F1 score (2*((precision * recall)/(precision + recall))), and total accuracy.

2.3 Descriptive Data Extraction

To establish context for the extent of opioid abuse within these two VA stations, we analyzed structured and unstructured data in terms of opioid prescriptions issued between 2010 and 2019 (retrieval variables in Table 1). We extracted all prescriptions for eight opioid drugs, using their generic names as recorded in structured VA data, issued the same day as an outpatient visit at the Washington DC or Baltimore station. We then extracted counts of opioid abuse diagnoses, using ICD-9-CM and ICD-10-CM codes, where a diagnosis was recorded sometime in the study period after a prescription was issued. Because structured data is sometimes incomplete, we also searched for term patterns explicitly indicating the abuse of opioid drugs in notes recorded after receiving an opioid prescription the day of an outpatient visit, for patients who did not have a subsequent ICD-9-CM or ICD-10-CM opioid abuse diagnosis. Since notes are not limited to generic names, we also searched for common trade names of these drugs. These term patterns consisted of "opioid abuse", "opiate abuse", "abusing opiates", "abusing opioids", and the terms "abuses" and "abusing" directly preceding a specific drug (e.g., "abuses hydrocodone", "abusing oxycontin"). To gauge the utility of this last step, we reviewed 100 randomly selected notes that contained the term patterns. Of the 100 notes, 77 provided clear evidence that the associated patient was experiencing current or past opioid abuse at the time the note was written.

We also extracted demographic data for these patients, including age (at time of prescription), marital status, gender, race, and ethnicity. We also extracted co-diagnosis information for traumatic brain injury, mental health issues, and pain diagnosis.

Table 1. Structured and unstructured data retrieval variables

Structured Data	Type	Example Values
Drug	Categorical	Morphine, Codeine, Tramadol, Hydrocodone, Oxycodone, Hydromorphone, Oxymorphone, Fentanyl (Patch or Lozenge)
Patient Demographics	Numeric & Categorical	Age, Gender, Ethnicity, Race, Marital Status
OUD Diagnosis (outcome variable)	Binary	ICD-9-CM codes (304.0, 304.00-304.03, 304.7, 304.70-304.73, 305.50-305.53, 965.00, E850.2, E935.2); ICD-10 codes (F11.10-F11.19, F11.20 – F11.29);
Pain Diagnosis	Binary	ICD-9-CM codes (715.16, 715.96, 719.40-43, 719.45-49, 723.1, 724.2, 724.4, 729.5, 524.60-64, 524.69); ICD-10 codes (G43-, G44-, G50-, M25.5, M26.6, M54-, M79.1-2, M79.5-7, R51-);
TBI Diagnosis	Binary	ICD-9-CM codes (310.2, 800-, 801-, 803-, 804-, 850-, 852-, 853-, 854-, 907.0, 959,01); ICD-10 codes (F07.81, S01-, S02-, S06-, S09.8, Z87.820);
Mental Health Diagnosis	Binary	ICD-9-CM codes (296.2-3, 300.00-02, 300.09, 309.81-82, 309.89, 311-); ICD-10 codes (F32-, F33, F41-, F43-);
Unstructured Data	**Type**	**Example Values**
Pattern phrases	Binary	"opioid abuse" "abusing oxycodone" "abuses oxycontin"; generic drug names used in structured data; additional trade names: Oxycontin, Lortab, Percodan, Percocet, Vicodin, MS Contin, Roxanl, Dilaudid

3. RESULTS

3.1 SVM

The SVM achieved 92.9% accuracy, with weighted averages of 93% for precision, recall, and F1 score (Table 2). The SVM, with the hyperparameters described in the methods section, achieved this performance using a relatively small corpus of clinical notes.

Table 2. SVM results

Class	Precision	Recall	F1	Support (Class Occurrence)
1: no evidence of opioid abuse	0.90	1.00	0.95	55
2: evidence of opioid abuse without treatment	0.92	0.85	0.88	13
3: evidence of opioid abuse with treatment	1.00	0.84	0.91	31
Weighted on Support/Total Support	0.93	0.93	0.93	

3.2 Descriptive Data Analysis

Within the study period a total of 609777 prescriptions were issued to 69398 unique patients, as itemized in Table 3. Because many patients received prescriptions for multiple years, there is a difference between the cumulative patient counts and the sums of the year-by-year patient counts. Calculating the cumulative patient count (i.e., unique patients served in the ten-year period) provides an overview of the patient population receiving opioid prescriptions during the study. Over the ten-year period, 69398 patients received opioid prescriptions, with an average of 8.8 prescriptions per patient, and an average of 768 days between the first and last prescription.

Table 3. Descriptive data analysis, prescriptions issued and patients receiving prescriptions during the study period

	All Prescriptions	All Patients	DCVA Prescriptions	DCVA Patients	Baltimore Prescriptions	Baltimore Patients
Cumulative (unique to study period)	609774	69398	287553	35353	322221	34045
2010	66834	17981	32394	8935	34440	9046
2011	73159	19356	35202	9530	37957	9826
2012	74434	19758	35436	9624	38998	10134
2013	75079	19984	35915	9768	39164	10216
2014	74658	20101	34613	9751	40045	10350
2015	69780	19058	31919	9318	37861	9740
2016	60660	16382	26915	7940	33745	8442
2017	48532	13796	21876	6701	26656	7095
2018	37382	11067	17953	5512	19429	5555
2019	29256	9187	15330	4837	13926	4350

Of the 69398 patients receiving at least one opioid prescription, 4209 received at least one opioid abuse diagnosis (by ICD code) after receiving a prescription. These patients received an average of 22.3 opioid prescriptions, with an average of 1388 days between the first and last prescription. After applying the NLP tool to the clinical notes of the patients receiving prescriptions but not an opioid diagnosis, we identified 809 additional patients with opioid abuse documented in their clinical notes within the study's time period. Table 4 itemizes these results by year and station. Again, comparing the cumulative totals to the year-by-year figures reveals that there are select patients counted for multiple years, as these patients received a prescription and a later diagnosis, or alternatively a prescription and documentation indicating opioid abuse issues, for multiple years.

Table 4. Descriptive data analysis, patients suspected of experiencing OUD; findings presented side-by-side by data analysis type

	Structured Data Analysis			Clinical Notes Analysis		
	Both Stations	DCVA	Baltimore	Both Stations	DCVA	Baltimore
Cumulative (unique to study period)	4209	1758	2451	809	203	606
2010	676	234	442	61	18	43
2011	985	352	633	87	23	64
2012	1241	416	825	111	29	82
2013	1344	474	870	135	34	101
2014	1307	492	815	118	23	95
2015	1321	491	830	139	36	103
2016	1152	439	713	136	31	105
2017	1181	430	751	141	34	107
2018	1190	385	805	91	20	71
2019	1145	344	801	121	32	89

There are notable trends within these results, by station and by year. While the cumulative number of patients receiving prescriptions over the study period by station is comparable (slightly more for Washington DC than Baltimore) more prescriptions were issued by the Baltimore station; however, the decrease in prescriptions issued in the last three years is especially notable in the Baltimore data (Table 3). In addition, there are more patients in the Baltimore data suspected of experiencing OUD, in both the structured data and clinical notes analyses (Table 4).

The patient demographic and co-diagnosis data (Table 5) are itemized according to percentage and data type. Data points are separated by data type groups, including gender, race, ethnicity, marital status, and co-diagnosis.

Table 5. Patient demographic and co-diagnosis percentages, by data type

Data Point	Structured Data	Clinical Notes
Patient Demographics		
Male	92.8%	90%
Female	7.2%	10%
Mean Age	55.1 years	57.8 years
Patient Race		
Black or African American	67.6%	58.6%
White	29.5%	37.3%
Declined to Answer	1%	1.1%
Native Hawaiian or Pacific Islander	<1%	<1%
Unknown by Patient	<1%	<1%
Asian	<1%	<1%
American Indian or Alaska Native	<1%	<1%
Ethnicity		
Not Hispanic or Latino	97.8%	96.4%
Hispanic or Latino	1.3%	1.2%
Unknown by Patient	<1%	1.4%
Declined to Answer	<1%	1%
Marital Status		
Married	24.7%	22.6%
Divorced	32.7%	37.2%
Never Married	25.5%	25.3%
Separated	12.7%	10%
Widowed	4.4%	4.7%
Unknown	<1%	<1%
Co-diagnosis		
Traumatic Brain Injury (ICD-9-CM / ICD-10-CM)	5.7% / 5.7%	5.4% / 6.1%
Mental Health Issues (ICD-9-CM / ICD-10-CM)	47.4% / 58.8%	44.4% / 57.5%
Pain (ICD-9-CM / ICD-10-CM)	82.6% / 68.4%	79.6% / 65.1%

Percentage breakdowns for the demographic data are inclusive by data type group, and differences are due to rounding. In the co-diagnosis data, percentages are indicative of the number of diagnoses for unique patients, for the given co-diagnosis in terms of the primary diagnosis of OUD, according to ICD code type.

4. DISCUSSION

Empirically, clinicians and clinical researchers believe that the ICD coding of opioid abuse does not capture all known cases. This study demonstrates that NLP and machine learning methods can identify additional patients who are potentially experiencing opioid related problems, (with and without treatment), and differentiate them from those experiencing other forms of substance abuse, using clinical notes. Many of the notes classified as showing no evidence of opioid abuse did document other forms of substance abuse. In our experiment, the NLP tool identified over 800 patients who did not have ICD codes for OUD (4209 patients had been identified by ICD codes). F1 scores of 0.88 and 0.91 for evidence of opioid abuse without treatment and with treatment, respectively, demonstrate that these types of documents (and the associated patients) can be identified with considerable exactitude, even using a relatively small corpus. The weighted averages of 0.93 for precision, recall, and F1 are comparable to the similar projects described in the Introduction, but through leveraging less data. Thus, healthcare institutions and researchers can identify patients potentially suffering from opioid abuse through clinical note analysis, in addition to coded data, with less annotation burden.

The descriptive data analysis demonstrates geographical differences in terms of opioid prescriptions and opioid abuse. More prescriptions were issued by the Baltimore station, despite there being slightly more cumulative patients served by the Washington DC station during the study period (Table 3). More Baltimore station patients exhibited evidence of opioid abuse than did those of the Washington DC station, particularly in the analysis of clinical notes (Table 4). The dynamics driving these effects warrant further study.

There is notable variation in the number of opioid prescriptions issued by year (Table 3). These relate to some extent to the policies and practice guidelines outlined in the Introduction, suggesting that these measures have had a positive effect in reducing overall prescription frequencies. In contrast, we do not see the same effect in OUD (Table 4). The study's cumulative patient counts (Table 3) for each station are comparable, and their surprisingly small numbers (35353 for Washington DC and 34045 for Baltimore) compared to the amount of prescriptions issued over the ten-year study period, and the amount of patients receiving a prescription each year, suggest that opioid use, abuse and misuse are ongoing issues for many Veterans.

There is notable variation within the OUD patient demographic data (Table 5), in terms of comparing OUD documentation in structured and unstructured data. This is especially seen in regard to race, and to a lesser extent in co-diagnosis. In the co-diagnosis breakdowns, percentages represent the total of OUD patients receiving a specific type of co-diagnosis code, which depends on year. The VA phased out ICD-9-CM and began using ICD-10-CM in 2015, which was the sixth year of this ten-year retrospective study. While these percentages are derived using the OUD totals for the ten-year study, they are comparable to findings in a similar 2012-2018 study (Workman et al., 2019b), thus further demonstrating that OUD is a problem framed by an extended, multiyear viewpoint. Also, the differences in average prescriptions (8.8 vs. 22.3) and average days between first and last prescription (768 vs. 1388) between all patients and those receiving an OUD diagnosis reveal OUD is an issue framed by time as well as drug quantity. In this current study, the differences in percentage, by type of code, suggest tendencies for clinicians to note OUD for patients with the given co-diagnosis over time. They suggest that clinicians are more likely to note OUD, in both structured and unstructured mediums, for patients experiencing traumatic brain injury and mental health issues after 2015, and less likely to do so for patients with pain diagnoses. The collective reasons behind these differences also warrant further analysis.

4.1 Future Work

The future work inspired by this study addresses additional efforts in machine learning-enabled clinical note identification, and data analysis. We intend to develop NLP and machine learning methods to distinguish notes documenting OUD by station. This can be seen as a sublanguage task (Doing-Harris et al., 2013, Workman et al., 2019a, Zeng et al., 2011) benefiting from sublanguage-oriented techniques. We also intend to further study the dynamics behind the differences in data, in terms of station prescription issuance, and demographic and co-diagnosis data, particularly in terms of structured and unstructured data types.

5. CONCLUSION

We developed an NLP tool to identify instances of opioid abuse receiving treatment, and instances of opioid abuse not receiving treatment, among Veterans served by two VA stations, using a corpus of clinical notes. The tool achieved significant performance, demonstrating that opioid abuse can be detected by healthcare institutions through NLP and machine learning, even using a small corpus. To better understand the results, we also retrieved and analyzed data for opioid prescriptions, opioid abuse, patient demographics, and co-diagnoses, from structured and unstructured data (i.e., clinical notes). This resulted in notable trends and differences among the two VA stations.

ACKNOWLEDGEMENT

The views expressed are those of the authors and do not necessarily reflect those of the Department of Veterans Affairs, the United States Government, or the academic affiliate organizations. This work was supported by VA Projects SIP 18-329 Utilization of Opioids in Post 9/11 Veterans in the VA Community, HSR&D CRE 12-315 A VHA NLP Software Ecosystem for Collaborative Development and Integration, and NIH grant UL1TR001876 from the National Center for Advancing Translational Sciences.

REFERENCES

Administration, V. H. *About VA* [Online]. Available: https://www.va.gov/about_va/ [Accessed 04 May 2020].

Association, A. P. 2013. *Diagnostic and statistical manual of mental disorders (DSM-5®)*, American Psychiatric Pub.

Brown, S. H., Lincoln, M. J., Groen, P. J. & Kolodner, R. M. 2003. VistA--U.S. Department of Veterans Affairs National-scale HIS. *Int J Med Inform,* 69, 135-56.

Chartoff, E. H. & Mavrikaki, M. 2015. Sex Differences in Kappa Opioid Receptor Function and Their Potential Impact on Addiction. *Front Neurosci,* 9, 466.

Cottler, L. B., Ben Abdallah, A., Cummings, S. M., Barr, J., Banks, R. & Forchheimer, R. 2011. Injury, Pain, and Prescription Opioid Use among Former National Football League (NFL) Players. *Drug Alcohol Depend,* 116, 188-94.

Doing-Harris, K., Patterson, O., Igo, S. & Hurdle, J. 2013. Document Sublanguage Clustering to Detect Medical Specialty in Cross-institutional Clinical Texts. *Proc ACM Int Workshop Data Text Min Biomed Inform,* 2013, 9-12.

Edlund, M. J., Steffick, D., Hudson, T., Harris, K. M. & Sullivan, M. 2007. Risk Factors for Clinically Recognized Opioid Abuse and Dependence among Veterans Using Opioids for Chronic Non-cancer Pain. *Pain,* 129, 355-62.

Garcia, M. C., Heilig, C. M., Lee, S. H., Faul, M., Guy, G., Iademarco, M. F., Hempstead, K., Raymond, D. & Gray, J. 2019. Opioid Prescribing Rates in Nonmetropolitan and Metropolitan Counties Among Primary Care Providers Using an Electronic Health Record System - United States, 2014-2017. *MMWR Morb Mortal Wkly Rep,* 68, 25-30.

Gellad, W. F., Good, C. B. & Shulkin, D. J. 2017. Addressing the Opioid Epidemic in the United States: Lessons From the Department of Veterans Affairs. *JAMA Intern Med,* 177, 611-612.

Green, C. R., Anderson, K. O., Baker, T. A., Campbell, L. C., Decker, S., Fillingim, R. B., Kalauokalani, D. A., Lasch, K. E., Myers, C., Tait, R. C., Todd, K. H. & Vallerand, A. H. 2003. The Unequal Burden of Pain: Confronting Racial and Ethnic Disparities in Pain. *Pain Med,* 4, 277-94.

Hastings, J. S., Howison, M. & Inman, S. E. 2020. Predicting High-risk Opioid Prescriptions before they are Given. *Proc Natl Acad Sci U S A,* 117, 1917-1923.

Kendler, K. S., Ohlsson, H., Sundquist, K. & Sundquist, J. 2013. A Latent Class Analysis of Drug Abuse in a National Swedish Sample. *Psychol Med,* 43, 2169-78.

Lin, L. A., Bohnert, A. S. B., Kerns, R. D., Clay, M. A., Ganoczy, D. & Ilgen, M. A. 2017. Impact of the Opioid Safety Initiative on Opioid-related Prescribing in Veterans. *Pain,* 158, 833-839.

Lingeman, J. M., Wang, P., Becker, W. & Yu, H. 2017. Detecting Opioid-Related Aberrant Behavior using Natural Language Processing. *AMIA Annu Symp Proc,* 2017, 1179-1185.

Management, V. P. *Opioid Safety Initiative (OSI)* [Online]. Available: https://www.va.gov/painmanagement/opioid_safety_initiative_osi.asp [Accessed 04 May 2020].

Manubay, J., Davidson, J., Vosburg, S., Jones, J., Comer, S. & Sullivan, M. 2015. Sex Differences among Opioid-abusing Patients with Chronic Pain in a Clinical Trial. *J Addict Med,* 9, 46-52.

Martins, S. S., Sampson, L., Cerda, M. & Galea, S. 2015. Worldwide Prevalence and Trends in Unintentional Drug Overdose: A Systematic Review of the Literature. *Am J Public Health,* 105, e29-49.

Moss, J., Andison, M. & Sobko, H. 2007. An Analysis of Narrative Nursing Documentation in an otherwise Structured Intensive Care Clinical Information System. *AMIA Annu Symp Proc*, 543-7.

Prieto, J. T., Scott, K., Mcewen, D., Podewils, L. J., Al-Tayyib, A., Robinson, J., Edwards, D., Foldy, S., Shlay, J. C. & Davidson, A. J. 2020. The Detection of Opioid Misuse and Heroin Use From Paramedic Response Documentation: Machine Learning for Improved Surveillance. *J Med Internet Res,* 22, e15645.

Sarker, A., Gonzalez-Hernandez, G. & Perrone, J. 2019a. Towards Automating Location-Specific Opioid Toxicosurveillance from Twitter via Data Science Methods. *Stud Health Technol Inform,* 264, 333-337.

Sarker, A., Gonzalez-Hernandez, G., Ruan, Y. & Perrone, J. 2019b. Machine Learning and Natural Language Processing for Geolocation-Centric Monitoring and Characterization of Opioid-Related Social Media Chatter. *JAMA Netw Open,* 2, e1914672.

Scholl, L., Seth, P., Kariisa, M., Wilson, N. & Baldwin, G. 2018. Drug and Opioid-Involved Overdose Deaths - United States, 2013-2017. *MMWR Morb Mortal Wkly Rep,* 67, 1419-1427.

Sharma, B., Dligach, D., Swope, K., Salisbury-Afshar, E., Karnik, N. S., Joyce, C. & Afshar, M. 2020. Publicly Available Machine Learning Models for Identifying Opioid Misuse from the Clinical Notes of Hospitalized Patients. *BMC Med Inform Decis Mak,* 20, 79.

Wadekar, A. S. 2020. Understanding Opioid Use Disorder (OUD) Using Tree-based Classifiers. *Drug Alcohol Depend,* 208, 107839.

Webster, L. R. 2017. Risk Factors for Opioid-Use Disorder and Overdose. *Anesth Analg,* 125, 1741-1748.

Wentling, N. 2018. Watchdog: VA Cuts Opioid Prescription Rate by 41 Percent in Past 5 Years. *Stars and Stripes*.

Workman, T. E., Divita, G. & Zeng-Treitler, Q. Discovering Sublanguages in a Large Clinical Corpus through Unsupervised Machine Learning and Information Gain. 2019 IEEE International Conference on Big Data (Big Data), 2019a. IEEE, 4889-4898.

Workman, T. E., Shao, Y., Kupersmith, J., Sandbrink, F., Goulet, J. L., Shaar, N. M., Spevak, C., Brandt, C., Blackman, M. R. & Zeng-Treitler, Q. Explainable Deep Learning Applied to Understanding Opioid Use Disorder and Its Risk Factors. 2019 IEEE International Conference on Big Data (Big Data), 2019b. IEEE, 4883-4888.

Zeng, Q. T., Redd, D., Divita, G., Jarad, S., Brandt, C. & Nebeker, J. 2011. Characterizing Clinical Text and Sublanguage: A Case Study of the VA Clinical Notes. *J Health Med Informat S,* 3, 2.

HOW ACCURATE DO YOU WANT IT? DEFINING MINIMUM REQUIRED ACCURACY FOR MEDICAL ARTIFICIAL INTELLIGENCE

Federico Sternini[1], Alice Ravizza[1] and Federico Cabitza[2]
[1]*USE-ME-D srl, I3P Politecnico di Torino*
Corso Castelfidardo 30/a, Torino, Italy
[2]*Dipartimento di Informatica, Sistemistica e Comunicazione, Università degli Studi di Milano-Bicocca*
Viale Sarca 336, 20126, Milano, *Italy*

ABSTRACT

Artificial intelligence (AI) is becoming a more and more common component of biomedical engineering solutions, and these latter systems are getting promising results in terms of diagnostic and prognostic accuracy. Medical AI (MAI) is then reaching the maturity level for its appropriate use in clinical practice, but to this end, its efficacy needs to be demonstrated first. Currently, this efficacy is proven in terms of the reported accuracy of the algorithm, especially for diagnostic tasks. But also in this case, how much accurate is "enough" accurate? To address this question means to define the minimum required accuracy for a system to be valid, that is fit to its intended use. To this aim, we propose a risk-based approach to the definition of adequate accuracy, in accordance with a risk-based regulatory classification. We investigated whether the current state of the art is already compliant with this standard-based approach, by performing a literature review in four application domains, one for each of the four risk classes we identified: the diagnosis of psoriasis, of knee osteoarthritis, the screening of breast cancer screening, and the detection of influenza outbreaks. The evaluation of the literature review highlighted that this approach is still not widely adopted, but that there is a partial presence of an implicit, conventional scheme that is similar to our proposal, especially in the high-impact literature. We also provide some guideline to assess the minimum required accuracy but also sheds light on the need for further official guidelines that ensure the wider application of the regulatory risk-based approach by the scholarly community of MAI.

KEYWORDS

Accuracy, Performance, Validation, Machine Learning, Medical Artificial Intelligence, Standards

1. INTRODUCTION

Artificial Intelligence (AI) is acquiring increasing importance in all fields of technology and industry, including the healthcare domain. When AI is used to diagnose, prevent, monitor, predict, treat or alleviate diseases or injuries, it should be classified as a software as a medical device (FDA, 2019; GHTF, 2012) and we denote it as Medical AI (MAI) (Cabitza and Zeitoun, 2019). For the delicate and sensitive context in which is applied, MAI needs a regulatory path intended to prove compliance with the safety, effectiveness and quality requirements (i.e., verification), as defined by the relevant regulatory system and in many cases also prove some form of clinical effectiveness and actual benefit (i.e., validation). Regulations enacted worldwide all follow the World Health Organization recommendation (World Health Organization, 2011) and share a classification approach, which typically divides medical devices into different classes depending on the risk to the patient (European Parliament and European Council, 2017a, 2017b). Each class is then associated with different requirements, which increase as patient risk increases. The main goal of this approach is the efficient distribution of resources from the competent authorities, where the most critical (and impactful) cases receive greater attention and resources.

Grounding on international standards and regulations, we decided to propose a classification of AI in terms of different levels of minimum required safety depending on the risk profile of the algorithm. We propose that those requirements should be defined for the whole software life cycle, from the training phase to the end of service, including all updates and further training phases. Our proposal is based on two regulatory references:

the identification of the requirement for the pre-market assessment of the In Vitro Diagnostic devices drafted by the Global Harmonization Task Force (GHTC), and the framework for risk Categorization of Software as a Medical Device (SaMD) proposed by the International Medical Device Regulators Forum (IMDRF) (GHTF, 2008; IMDRF, 2014) on one hand, and the European Regulations 2017/745 and 2017/746, respectively on Medical Devices and InVitro Diagnostic medical devices, on the other hand (European Parliament and European Council, 2017a, 2017b). The choice of these references is backed up by consistency of them with the ones used by the Food and Drug Administration to propose a Regulatory Framework for modifications to MAI (FDA, 2019).

In this study, we propose an interpretation of the classification to analyze requirements for AI intended to complete diagnosis automatically (e.g., in screening ambit) or, more commonly, support the clinicians in their diagnostic tasks. The classification is defined as follows:

- I: Non-serious condition: A condition for which wrong or delayed diagnosis can cause minor injuries if any (e.g. acne diagnosis, psoriasis diagnosis).
- II: Serious condition: when wrong or delayed diagnosis can cause mild injuries to the patient (e.g. osteoarthritis).
- IIIa: Critical condition: when wrong or delayed diagnosis can cause death or severe injury (e.g. cancer, heart attack).
- IIIb: Public health threat: when wrong or delayed diagnosis can expose public health to harm (e.g. epidemic influenza, HIV).

In the specific context of AI embedded in diagnostic support systems, the safety is tightly correlated with the accuracy of the algorithm's output. All accuracy-related metrics are aimed at representing the extent a classifier makes mistakes, from a quantitative and probabilistic point of view. Specificity and sensitivity (and all their averages and combinations, like f- and g-scores) also express a qualitative, and yet crucial, aspect, by distinguishing between false negatives and false positive mistakes. However, to this respect, the most common approach is to weight both classes (i.e., positive and negative cases) equally, and this is clearly unrealistic in many cases. Moreover, to our knowledge, only one accuracy metrics takes the severity of condition into account, by weighting mistakes according to the complexity of the cases in the test set when this is annotated with this additional information (Cabitza and Campagner, 2019): however, this method has not yet found widespread application in medical ground truthing.

Our proposal takes into account the current regulation, which we here quote: "Devices shall achieve the performance intended by their manufacturer and shall be designed and manufactured in such a way that, during normal conditions of use, they are suitable for their intended purpose" (European Parliament and European Council, 2017a), and it grounds on an intuitive, if not obvious, notion: not all mistakes (of a classifier) are made equal, and the higher their potential impact on the patients' health, the lower the probability of their occurrence should be. In other words, we claim that: first, scholars should consider a classifier skill *acceptable* (and hence worthy of publication and dissemination) when it is better than previous already-released systems (being considered acceptable at their time); moreover, they should demand higher accuracy for more impactful errors and more serious conditions than for lighter ones. At the moment, we are not aware of similar proposals for the definition of the accuracy requirements for MAI.

In this paper, we will pursue then two research questions: first, we will report for a single prototypical health condition extracted from each one of the risk classes considered above, what the average accuracy of existing models is; in doing so, we can consider such an estimate a sort of lower bound threshold of acceptable accuracy for that condition (and, more loosely, for the corresponding risk class). Secondly, we will report whether the above idea to require higher accuracy for higher criticality is already an established and common practice in the specialist community of ML researchers.

To either the above aims, in light of the definition of the risk classes proposed above, we performed a concise, systematic review for one representative application of each risk class. The applications are the following ones:

- Non-serious condition: psoriasis;
- Serious condition: knee osteoarthrosis;
- Critical condition: Breast cancer;
- Public health threat: Influenza;

2. METHODS

To perform the literature review, four queries were executed on the article repository indexed by Pubmed on the 22nd of January 2020. The queries were designed to include only results published in English in the last five years. The following keywords were also used:

- "(((((artificial intelligence) OR deep learning) OR machine learning) AND diagnosis) AND [condition])", with [condition] being one of the prototypical conditions mentioned above.

2.1 Eligibility Criteria and Evaluation

All abstracts were manually reviewed, and the sources that were found not related to the specific disease or an AI application were excluded. Subsequently, the whole manuscript of the remaining articles was manually reviewed, and the articles found relevant to the research questions were included. In addition, the manuscripts were assessed in accordance with Meddev 2.7/1 rev 4 in terms of relevance to the research question (EUROPEAN COMMISSION, 2016). The assessment procedure requires to assign a score to each source, and it is composed of three parts. The first one regards the level of evidence reported in the manuscript, defined in accordance with the Oxford evidence scale (OCEBM Levels of Evidence Working Group, 2011); as shown in Table 1, the second part regards the impact factor of the journal, while the third contribution is determined by the application of the appraisal plan as described in Meddev 2.7/1 rev 4, as described in Table 2. The three sections maximum score are respectively 6, 6 and 18. The total result maximum score is 30. It should be noted that regarding the second element of the procedure described in Table 2, the required performance should not be defined by simply comparing similar studies of similar applications.

Table 1. Scores based on the evidence level

Description	Score
Systematic review of cross sectional studies with consistently applied reference standard and blinding	6
Individual cross sectional studies with consistently applied reference standard and blinding	4
Non-consecutive studies, or studies without consistently applied reference standards	2
Case-control studies, or "poor or non-independent reference standard	1
Mechanism-based reasoning	0

Table 2. Score on the base of the Meddev appraisal plan

Description	Examples	Score
To what extent are the data generated representative of the device under evaluatior	The device is an AI whose output is a diagnosis or a diagnostic suggestion	6
	The device is an AI whose output is additional information to aid the diagnosis	3
	Other devices	Not Applicable
What aspects are covered?	Required performances	8
	Performances	4
	Other aspects	Not Applicable
Are the data relevant to the intended purpose of the device or to claims about the device?	Searched disease	4
	concerns specific models/ sizes/ settings, or concerns specific aspects of the intended purpose or of claims	0
	Other diseases	Not Applicable
If the data are relevant to specific aspects of the intended purpose or claims, are they relevant to a specific - type and severity of the medical condition?	Specific type OR severity of the medical condition	2
	Specific type AND severity of the medical condition	0

Finally, the scores of the different categories are evaluated to understand if the found evidence is compatible with our proposal or not.

In addition, during all the literature review process, information regarding the performance of the applications in terms of accuracy, AUC (c-statistics), f-score, sensitivity and sensibility have been collected. These values have then been used for a twofold aim coherent with our two research questions: first to propose a level of minimum necessary accuracy for each risk level; then to understand if the literature supports our proposal. To this aim, we adopt the following approach: the average value of the metric per application is calculated, and then, if the relationship of the average values with the average value of the other applications is consistent with the proposed approach (i.e. the accuracy is higher as the risk level is higher), the papers reporting the metric are considered as supporting our framework; in the case of metric values not consistent with the proposed scheme, the negative evidence is calculated as the average of the scores of studies presenting those metrics. In addition to data related to Meddev evaluation and accuracy data, we analysed how many of the papers compared the MAI accuracy with accuracy performed by human operators

As a last task, we also collected the qualitative terms associated with each paper to evaluate how these attributes are related to different accuracy levels. To assign a score to each qualitative value, three coders rated all of the qualitative attributes associated with the judgments of performance found in the reviewed articles on a 4-value scale, with no neutral code to minimize central-tendency bias: strong negative, weak negative, weak positive and strong positive.

3. RESULTS

The research performed on PubMed identified 124 articles. After the abstract and full-text evaluation, 59 articles were identified as applicable, divided as follows between the different applications:

- Psoriasis diagnosis: 7 papers;
- Knee Osteoarthritis diagnosis: 25 papers;
- Breast cancer screening: 17 papers;
- Influenza detection: 10 papers;

All of the papers were evaluated, and were associated with at least 12 out of 30 points as their evidence score; no article got a score higher than 25 out of 30. The average evidence score was 18. A vast majority of studies were not compatible with our proposal (52 vs 19), while the evidence scores were almost evenly matched with a slight primacy of the studies backing up our proposal (18.5 vs 17.9). The details of all the papers evaluated are listed in the appendix to this paper, which is available at shorturl.at/GKO19[1].

52 out of 59 papers (88%) included at least one value of the accuracy metrics listed above, thus leading to the collection of 22 values for the accuracy, 28 AUC values, 12 f-score and 21 values of both sensitivity and specificity. The average scores are reported in table 3. Moreover, 13 studies out of 59 (22%) compared the MAI accuracy with the accuracy obtained by human operators.

Table 3. Average accuracy metrics grouped by application

Application	Accuracy (Mean St.Dev.)	AUC (Mean ± St.Dev.)	f-score (Mean ± St.Dev.)	Sensitivity (Mean St.Dev.)	Specificity (Mean St.Dev.)
Psoriasis diagnosis	0.86 ± 0.21	0.87 ± 0.00	0.59 ± 0.00	0.75 ± 0.10	0.70 ± 0.00
Osteoarthritis diagnosis	0.77 ± 0.10	0.83 ± 0.09	0.78 ± 0.04	0.81 ± 0.11	0.80 ± 0.11
Breast cancer screening	0.76 ± 0.28	0.84 ± 0.11	0.88 ± 0.03	0.87 ± 0.08	0.74 ± 0.31
Influenza detection	0.87 ± 0.00	0.89 ± 0.05	0.84 ± 0.08	0.93 ± 0.04	0.90 ± 0.09
Breast cancer screening and Influenza detection	0.78 ± 0.26	0.85 ± 0.11	0.86 ± 0.07	0.88 ± 0.08	0.78 ± 0.29

[1] The original URL is: https://drive.google.com/file/d/1ZXf2Mr-nbvhY-JPTvquV0vje013MHkBI/view

Regarding the qualitative attributes characterizing the MAI performance, three coders rated 26 expressions and their agreement, computed in terms of Krippendorff's Alpha, was found "substantial" (0.62) (Landis and Koch, 1977); this allowed us to proceed with majority voting and assign each accuracy score with an appreciation level. This classification yielded to 19% strong positive appreciations, 68% weak positive ones, 17% weak negative ones, and no strong negative ones. We anticipate here that this latter result was expected, as models that perform so badly to be connotated in strongly negative terms hardly are published in impacted journals (by either file-drawer effect or review filtering). Thus, the performance of the vast majority (more than two thirds) of the predictive models reviewed was characterized in weakly positive terms by their developers or authors: this class was associated with average and minimum scores listed in Table 4.

Table 4. Average, median and minimum accuracy of models characterized by weakly positive terms

	Psoriasis diagnosis	Knee osteoarthrosis diagnosis	Breast cancer screening	Influenza detection	Breast cancer screening and influenza detection	Global
Average	48.75%	79.17%	94.55%	86.70%	92.59%	82.07%
Median	48.75%	82.50%	97.91%	86.70%	92.71%	86.35%
Minimum	48.75%	69.00%	87.50%	86.70%	86.70%	48.75%

Table 5. Evaluation of the qualitative judgements expressed by the authors. The values from 1 to 4 correspond to the grades Strong Negative, Weak Negative, Weak Positive, Strong positive

Judgement of performance	Coder 1	Coder 2	Coder 3	Majority vote
Highest accuracy	4	4	4	4
Appreciable	3	3	3	3
Good agreement	3	3	3	3
Outperforms human	4	4	4	4
Considerably high	4	4	4	4
Robust	3	3	3	3
Success	3	3	3	3
relatively high	3	3	2	3
Satisfactory	3	3	2	3
Fair	3	3	2	3
High	3	4	3	3
Promising	2	2	1	2
Good	3	3	3	3
Reliable	3	3	3	3
Encouraging	2	2	2	2
well-performing	3	3	2	3
Accurate	3	3	3	3
high level accuracy	4	4	4	4
potential effectiveness	3	3	2	3
comparable with experts	4	4	4	4
Acceptable	3	3	3	3
Accurately	3	3	3	3

4. DISCUSSION

As a first consideration, let us consider a parameter that was not reported in the Results Section but nevertheless can inform the interpretation of the findings: the percentage of papers that included an explicit requirement of the accuracy beyond the simple comparison with the performance of similar applications. This parameter was not used to evaluate our proposal because the requirement definition is part of the Meddev: therefore, it would have inflated our results on the positive side. The percentage of papers that explicitly declared the requirement in the context of the psoriasis diagnosis was 14.3%, while in the context of osteoarthritis was 28% and in the context of breast cancer screening 29.4%. None of the papers concerned with flu detection described the required accuracy for the system, probably due to the difficulty and novelty of the application domain. Indeed, while the other applications are consolidated in the medical practice and are usually performed by specialists, the detection of a flu outbreak is not an activity that is part of the common clinical practice completed by single physicians. In addition, but not less importantly, many of the current flu applications are designed to early detect flu outbreaks by means that are out of the usual scope of routine clinical evaluations, like the use of social media and search query analysis. These peculiarities make understanding what the adequate threshold could be of performance acceptability difficult for these applications.

Another parameter that shall be considered is the percentage of studies that involved the comparison with the human operator. Even in this case, the percentage increases as also the risk level increases (psoriasis diagnosis: 14.3%; osteoarthrosis diagnosis: 24%; breast cancer screening: 35.3%), but this also drops to zero for flu detection. All the possible reasons identified above can also be used in this case: the proposed systems are not intended to give a second opinion to the physician but are intended to detect situations that are defined as dangerous for the public health to epidemiologists and health policymakers.

An additional parameter that could be considered is the percentage of studies that compared the results with already published literature. In this case, the percentage decreases as the risk level increases (psoriasis diagnosis: 57.14%, osteoarthrosis diagnosis: 52%, breast cancer screening: 35,29%, flu detection: 30%). Even if this trend could seem to be related to the risk level of the different applications, we decided to exclude this term from further analysis due to the high number of factors that could affect the availability of a literature evaluation (e.g. novelty of the technology) and that are out of the scope of this review.

For the above reasons, in what follows we will focus on the first three risk levels (I, II, IIIa), as those concerning individual clinical cases (indeed, we recall here that the public health case is associated with a higher magnitude of risk, IIIb, not for the disease criticality per se but for the multitude of patients that can potentially be affected by it). In this case, accuracy and AUC present an opposite trend with respect to our proposal: the average performance reported in the reviewed studies decrease as the risk class these studies belong to increase, with the last two risk levels being almost evenly matched. In fact, the only two metrics that are consistent with our proposal are the sensitivity (which is so in regard to all four risk levels) and the f-score. However, as counterintuitively as it might seem, these findings can be read as a confirmation of the reasonability of our framework and as an indication of what difficulties the scientific community is currently facing. In fact, we recall that sensitivity regards the capability of the MAI to avoid false-negative errors: intuitively, the more serious the condition, the higher impact the consequences of not detecting and treating the condition timely, and the harm for the patient. This intuition must have passed from the medical community to the MAI practitioners, who, in trying to optimize their predictive models, must have favoured sensitivity for potentially serious conditions. This is also mirrored by the finding of the f-score, which is the (harmonic) mean of the model's sensitivity and its capability not to raise false positive alarms (cf. the positive predictive value of the model).

The opposite trend for the accuracy metrics does not come totally unexpected: lower risk conditions are such also for their higher recognizability and their relationship with a smaller number of physiological states and causal links; this could be mirrored by perceptual tasks that are easier to automate and smaller predictor spaces to be computationally represented. State of the art is more than adequate to detect their signs and identify powerful predictive features. The same does not hold for most serious and complex health conditions.

Also, the quantitative analysis of the reviewed studies confirms this reading of the results: Whilst more studies were not compatible with our proposal (52 vs 19), the evidence score associated with the compatible studies is slightly higher (18.5vs 17.9). Intuitively this means that the higher-impact resources are already more aligned with our proposal, while more resources contradict it but on (relatively) lower-value sources. Therefore, for the moment, we can claim that the majority of recent studies on diagnostic MAI do not implicitly adopt our approach to self-validate their results, but the most impacted (and higher-value) contributions show a tendency to its adoption: that notwithstanding, validation is commonly based on the comparison with the performance of similar previous algorithms or the subjective evaluation of the authors, without considering the risk class of the intended use.

From the analysis of the qualitative terms, we here focus on the accuracy values that we associated with the weak positive scale. The intuitive idea is that the average value of the accuracy reported by the studies falling in this category would provide a threshold for acceptability that is based on the implicit consensus of the scholarly community. Moreover, the weak positive class was the more equally distributed among all risk classes (Class I: 12,5%, Class II: 50%, Class IIIa: 25%, Class IIIb: 12,5%), thus providing mean scores that are internally more representative. The total average across all risk classes, as reported in Table 4, is 82%, which is nevertheless a pessimistic (low) estimate, also for the minimum extreme that is even lower than the chance guess (i.e., 50%). The average found for each class also provides an intuitive criterion for the specific health conditions we chose to focus on, as it represents the average performance of comparable applications whose performance was found to be at least weakly positive by their developers. Similar consideration should be done for each single health condition, as reasoning for risk class would likely provide too coarse-grained indications, for the high heterogeneity of diagnostic class in the same risk class. All things considered, a gross, but easy-to-remember, rule of thumb to set a minimum acceptable for diagnostic MAI is what we call the "die-roll" threshold.: Every side of dice has one sixth probability of getting out (ca. 17%); this could be the acceptable error rate for decent decision support, which also finds an interesting parallel with recent estimates of human diagnostic error, that is, 13%-15%, (Graber, 2013). The die-roll threshold that is substantially confirmed by our bibliographic study would then be more or less a "psychological" level, set at approximately 83% (accuracy).

5. CONCLUSION

In this work, we have proposed a general approach for the definition of requirements of minimum accuracy for AI applications in medicine: a risk-based approach with a classification of the application risk that is consistent with the current regulatory framework worldwide. We performed a literature survey to understand if this approach is already applied in the scientific community of those who develop and deploy medical AI systems.

The above literature survey was completed in four application domains of increasing severity and impact to be loosely representative of the whole medical domain. The literature evaluation showed that, for the moment, the most used approach is an "opportunistic" approach that does not consider the risk of the application but relies on the judgement of the authors or comparison with similar algorithms. Even though, the literature analysis suggests that a similar validation approach is getting wider and wider application, especially in the most impacted specialist literature, and in regard to the risk of committing a false negative mistake, that is the sensitivity metric.

After the collection of all the accuracy score for the different applications, we propose these values as state-of-the-art baseline and acceptability threshold for future studies that are in the scope of our literature review.

This study also suggests that providing a clear guideline may encourage future studies to take regulatory requirements more seriously, thus setting higher expectations for those models that are intended to treat or diagnose more critical conditions.

REFERENCES

Cabitza, F., Campagner, A., 2019. Who wants accurate models? Arguing for a different metrics to take classification models seriously. ArXiv191009246 Cs Stat.

Cabitza, F., Zeitoun, J.-D., 2019. The proof of the pudding: in praise of a culture of real-world validation for medical artificial intelligence. 2019 7, 1.

European Commission, 2016. Clinical Evaluation: A Guide For Manufacturers and Notified Bodies Under Directives 93/42/EEC and 90/385/EEC, MEDDEV 2.7/1 revision 4.

European Parliament, European Council, 2017a. Regulation (EU) 2017/745 of the European Parliament and of the Council of 5 April 2017 on medical devices, amending Directive 2001/83/EC, Regulation (EC) No 178/2002 and Regulation (EC) No 1223/2009 and repealing Council Directives 90/385/EEC and 93/42/EEC (Text with EEA relevance.), OJ L.

European Parliament, European Council, 2017b. Regulation (EU) 2017/746 of the European Parliament and of the Council of 5 April 2017 on in vitro diagnostic medical devices and repealing Directive 98/79/EC and Commission Decision 2010/227/EU (Text with EEA relevance.), OJ L.

FDA, 2019. Proposed Regulatory Framework for Modifications to Artificial Intelligence/Machine Learning (AI/ML)-Based Software as a Medical Device (SaMD) - Discussion Paper and Request for Feedback.

GHTF, 2008. Principles of In Vitro Diagnostics (IVD) Medical Devices Classification.

GHTF, 2012. Definition of the Terms 'Medical Device' and 'In Vitro Diagnostic (IVD) Medical Device.'

Graber, M.L., 2013. The incidence of diagnostic error in medicine. BMJ Qual. Saf. 22, ii21–ii27.

IMDRF, 2014. "Software as a Medical Device": Possible Framework for Risk Categorization and Corresponding Considerations.

Landis, J.R., Koch, G.G., 1977. The measurement of observer agreement for categorical data. Biometrics 33, 159–174.

OCEBM Levels of Evidence Working Group, 2011. The Oxford 2011 Levels of Evidence.

World Health Organization (Ed.), 2011. Development of medical device policies, WHO medical device technical series. World Health Organization, Geneva, Switzerland.

ALL YOU NEED IS HIGHER ACCURACY? ON THE QUEST FOR MINIMUM ACCEPTABLE ACCURACY FOR MEDICAL ARTIFICIAL INTELLIGENCE

Federico Cabitza[1], Andrea Campagner[1], Francesco Del Zotti[2], Alice Ravizza[3] and Federico Sternini[3]
[1]University of Milano-Bicocca, Viale Sarca 336, 20126, Milano, Italy
[2]NetAudit, Verona, Italy
[3]USE-ME-D srl, I3P Politecnico di Torino, Corso Castelfidardo, 30/a, 10129, Torino, Italy

ABSTRACT

In this paper we will discuss the concept of the minimum level of accuracy an Artificial Intelligence system must exhibit in medical settings to be fit to its intended use and improve the daily practice of its intended users, the medical doctors. We will consider simple binary classification tasks in both diagnostic and prognostic ambit (like to discriminate between normal/abnormal case, and improvement/no improvement prospects). We will make the point that the common ways to determine this minimum acceptable accuracy are fraught with many conceptual and practical troubles. We will report about a small user study conducted to elicit the discriminative requirements from a sample of medical doctors, stratified both in general practitioners and specialists. Finally, we will present a simple nomogram by which to determine the minimum accuracy of a technological aid, once the human average performance and the desired level of accuracy are known. The nomogram is to be intended as a provocative simple tool to recognize that the technological tool is less important than a sound protocol in which to use it, responsibly and paying due attention to the role of the human decision makers.

KEYWORDS

Medical Artificial Intelligence, Minimum Acceptable Accuracy, Validation, Healthcare

1. INTRODUCTION

In this paper we focus on a concept that is seemingly straightforward to understand, the "Minimum Acceptable Accuracy" (MAA) that a classification system must exhibit in medical settings, but that nevertheless is difficult to define. Setting a specific MAA can be important for at least three broad purposes: for verification and validation of a Medical Artificial Intelligence (MAI); for certification purposes of the products embedding such a MAI; and also to allow for a reasonable accuracy – generalization trade-off, which parallels the known bias – variance trade-off at the point-of-care (Belkin et al., 2019). In particular, this latter concern refers to the fact that, once a system has achieved a required MAA, it could be counterproductive to train it further, as increases in accuracy could actually be associated with some sort of statistical overfitting, and hence with lower performance whenever the "best" model were fed with instances that are different from the ones used for training and in the validation phase. Not to mention that a MAI that greatly exceeds the required level of accuracy could induce stronger over-reliance on its capabilities and mislead its users (paradoxically) more often than a less accurate system of which the end users are a little wary of (we will come back to this point later on, when speaking of automation bias and complacency).

A MAA threshold can be set in mainly three ways, of which the first two are the most common ones: first, by asking the experts (including the prospective users, that is the medical doctors who are supposed to make decision with the MAI support); then, by considering previous performance as a benchmark (both of human decision makers and previous existing MAI systems). We will briefly discuss these two broad methods, and refer to two studies that we performed in this mold. Lastly, we will also propose a third approach, grounding on a new analytical way to assess MAA, and we will introduce a nomogram to help set its value graphically. This nomogram is a two-dimensional diagram that allows to calculate a probabilistic function yielding the required MAA of a MAI in a prototypical business use case.

2. A SURVEY OF METHODS AND THEIR SHORTCOMINGS

In this section we will briefly review three ways to focus on the concept of acceptability regarding the predictive performance of a Machine Learning (ML) model to be deployed in a medical context.

2.1 Benchmarking with Similar Systems

As we mentioned above, a common, bottom-up, approach to define acceptability in regard to the predictive performance of a Machine Learning (ML) model is to rely on the community of ML practitioners, and take them as sort of arbiters: according to this common-sense view, it is acceptable the performance that the community (including both the authors of a study and its peer reviewers) consider to be worthy of publication in the specialist literature. This translates in the comparing of the performance of a new ML model to the performance of models that had been previously developed for the same medical task: the MAA is then associated with a performance that is at least equal, if not superior, to that of existing decision support aids, the so called "state of the art".

In (Sternini et al., 2020), we reviewed 49 recent articles reporting the MAI performance for three diagnostic tasks for health conditions of different severity, namely: psoriasis, knee osteoarthritis and breast cancer. As expected, we found great differences in the average scores among the medical tasks (from breast cancer applications, average accuracy 76%, to the psoriasis detection ones, 86%) but, more surprisingly yet, also high variability within the single tasks (e.g., 28% and 21% were the observed standard deviations of the average scores for breast cancer and psoriasis, respectively). Although of limited scope, our review shows how nearly 2 applications out of 3 could exhibit performances as good as that of oracles (nearly 100%) and as poor as that of systems that just guess (nearly 50%).

Thus, although this comparative approach is straightforward and seemingly effective, here we make the point that it is also inadequate. First of all, accuracy-related measures (ranging through sensitivity, specificity, area under the curve, accuracy, F_1 score, et cetera) are only seemingly comparable with each other across different studies: strictly speaking these metrics can be compared only if related to the same "test set", so that it is possible to speak of models that support the same discriminative task. If we intended that different test sets for the same discriminative task have to be considered random samples drawn from the same population of similar cases (an assumption that is difficult to consider fully applicable), then comparisons can yield statistically significant results only if confidence intervals are provided and these are computed verifying the underlying assumptions on the nature of the distribution: however, reporting performance metrics with their confidence intervals is still not a widespread and common practice.

Thus, even if we assumed two studies comparable when pertaining to the same "medical task" (like "detecting diabetic retinopathy"), the current lack of reporting standards (the guidelines known as TRIPOD-ML, CONSORT-AI and SPIRIT-AI are still in development (Liu et al., 2019a)) often makes the comparison of models' skills an arbitrary and vague endeavour. Moreover, even if we accepted that the literature could provide reliable benchmarks, comparison with other tools cannot tell whether the new system is really *acceptable* in absolute terms, but only that it is better than other, possibly inadequate, tools.

2.2 Benchmarking with Humans

To overcome the objection just mentioned, another common approach to setting MAA requires to relate it to the average accuracy of human doctors performing the same discriminative task: the reasoning here is that if doctors, after extensive training and sufficient experience, are considered adequate to the task, no one should expect an automatic system perform better than them: and when it does (as often reported in the recent specialist literature, e.g., (Liu et al., 2019b)) big headlines on the news are guaranteed too.

In the literature review mentioned above (Sternini et al., 2020), we also observed that comparing the machine performance with that of humans on the same instance is relatively, and surprisingly, common (especially considering the additional effort required to involve professionals with usually very busy schedules, like doctors are), and it was done in 13 studies out of 49 (27%), i.e., in slightly more than one quarter of the survey sample.

More generally speaking, we could be interested in understanding what the average human baseline is, so that we can set acceptability right at that level. To this aim, in the literature there are many studies aimed at assessing "the incidence of error rate" in medicine. Among the most-cited studies, (Graber, 2013) reviews several previous contributions and estimates diagnostic human accuracy at 85%-87%, while in the radiological domain estimations ranges from approximately 80% (Bruno et al., 2015) to even more pessimistic estimations, up to just 70% (Berlin, 2014).

However, although setting human performance as a benchmark could sound reasonable, estimating any average rate can be difficult, or even unfeasible: first of all, human accuracy does not only depend on the specific medical task, which in its turn depends on the health condition at hand and the investigation tool (e.g., MRI, CTs, X-rays, lab exams); but also on idiosyncratic aspects, like the complexity, severity and rarity of the case (for instance, (Semigran et al., 2016) report that doctors are better in high-acuity (vs low-acuity cases) and uncommon cases (vs common cases)); and on characteristics of the work setting (like throughput, time constraints (ALQahtani et al., 2016), work shift policies (Hanna et al., 2018), modality brand and age), not to mention the doctors involved, their expertise mix, and even their states of mind. Thus, relying on the thaumaturgical value of averages on a sufficiently large sample (which is seldom the case), could not be enough to assess an average accuracy threshold above which to set the MAA in a less than highly situated and local manner.

Furthermore, the merely discriminative task for which the error rate is measured could be just an abstraction of a more complex interpretative task, from which it cannot be decoupled unless in a fictional, laboratory setting (e.g., radiologists do not only look at and classify images, but also consult the patient record and consider the needs and hypotheses of the referring specialist). In particular, also a "laboratory effect" has been observed (Gur et al., 2008): this is a somehow counterintuitive phenomenon by which doctors are more likely to perform worse in controlled experimental conditions, even if these create an ideal situation of uninterrupted concentration, than when they actually work at the hospital, and their decision can affect the health of real patients. If this effect is common as conjectured, researchers performing MAI-humans comparisons on retrospective data could seriously underestimate the actual doctors' performance.

That said, even assuming the estimate correct, and setting the MAA slightly above the average human performance, we should consider the following point: can the introduction of a decision aid in a hospital be assimilated to the entry of a new member into an existing team of doctors? If this were the case, would it be right to bear the investment necessary for acquiring a technological member, and the related operating costs, when this latter one is roughly worse than half the other members?

2.3 Asking the Experts

Another simple way to set the MAA threshold requires to rely on the experts' opinion. As arbitrary as this approach might seem, we also will give our opinion on what the MAA should be at the end of this contribution. However, few of these opinions can be found in the specialist literature. Combining the two approaches mentioned above, (Floares, 2009) sets MAA at about 80% grounding on both "the performances of the best commercial systems" and the fact that "the majority of physicians [involved in his study] will agree that the prediction accuracy should be at least 80%".

We wanted to replicate this claim and performed a simple user study by asking a convenience sample of specialists and family doctors what the MAA for a MAI should be in both a generic diagnostic and prognostic task (i.e., respectively, in predicting the patient's health problem and in predicting the patient's outcome after treatment). Forty-eight (48) General Practitioners and 31 specialists (mainly orthopedists) participated, by filling-in a two-question online questionnaire where we explicitly asked "What is, in your opinion, the maximum error rate (in percentage) of the system that you believe is tolerable for a MAI system to be *practically useful to you* in your daily work?": on average the general practitioners indicated a diagnostic MAA of 88% (median: 93%) and a prognostic MAA of 83% (median: 88%). On the other hand, the specialists were harder to please (not totally surprisingly) and expressed a MAA of 95% (median: 94%) and 90% (median: 90%) for diagnostic and prognostic purposes, respectively.

In particular, with reference to Figure 1, we can notice that the differences between general practictioners (GPs) and specialists are not statistically significant, while the difference between the minimum acceptable diagnostic and prognostic performance reported by the specialists is statistically significant (i.e., no overlap occurs between the confidence intervals of the medians).

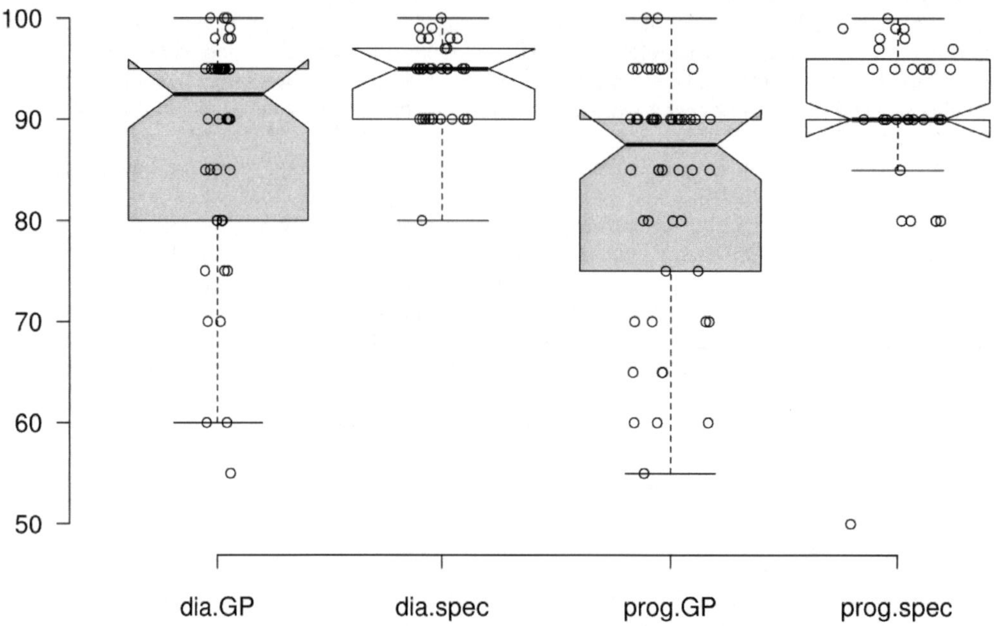

Figure 1. Minimum Acceptable Accuracy, as it is perceived by a sample of General Practitioners (GP, N=48) and Specialists (spec, N=31) for diagnostic (dia) and prognostic (prog) purposes

Asking the key opinion leaders, or just the prospective users of the system to be validated, is a sound principle of the requirement engineering practice. However, just asking "how much accurate users want their decision support to be" could be not sufficient to improve their practice. First, doctors could overestimate their own accuracy, and hence be too much demanding with respect to their decision aids, and end up by asking for unrealistically (and unnecessarily) accurate tools if their main aim is the improvement of doctors' performance, especially in terms of effectiveness (rate of correct and appropriate decisions) and ultimately, the improvement of the patients' outcome. Moreover, even if doctors were warned of their average performance (at least in terms of approximate estimate, as this can be drawn from the literature as mentioned above), one should not expect they can understand how the accuracy of a decision tool could impact the overall team performance, as this aspect has still to be investigated in the specialist literature.

2.4 Looking at How Teams Work (or the Protocol-Based Approach)

All of the above methods, although very common and sounding perfectly reasonable at first glance, suffer from both minor and major flaws. The most important flaw, in our view, is that they subsume an idea of decision support that is disembodied and abstracted away from the situated ways in which a decision aid is actually used in real-world practice. In the next future, and probably even in the long term, no MAI will be supposed to provide a diagnosis or formulate a prognosis alone, since these tasks will continue to be well under the human responsibility, also (but not only) because these latter ones will get access to many more elements (including the psychological, emotional, social ones) than those can be properly digitized, and hence made available to the machine's scrutiny.

Likewise, users could find difficult to understand what kind of support they need to improve their own performance, as they could underestimate how their expectations, biases and prejudices could affect their decisions, when these are made in light of the machine's recommendations: suffice it here to mention the effects known as "automation bias" (Goddard et al., 2012), that is over-relying on the machine's advice believing that it will never (or very seldom) be wrong, "automation complacency" (Lyell and Coiera, 2016), that is when one overestimate their capability to understand when the machine is wrong or it should raise an alarm, and the "prejudice against the machine"(Cabitza, 2019), which is a sort of opposite attitude observed in a cardiological study, when the participants trusted more a simulated human colleague than a MAI although they were proposing the same wrong options.

For all these reasons, we propose a further approach to determine MAA, which can be called analytical but also protocol-based, because we can trace it back to a reasonably common, and realistic, business scenario of MAI use. In this case, one could think of a typical point-of-care use case, where the medical AI is used by one doctor to be advised about a case. This setting is coherent with the so called centaur model (Goldstein et al., 2017), that is a model proposed in medical decision making in which humans and machines team up in sort of augmented, hybrid teams that collaborate to make a decision and where every member of the team does what they excel at the most: intuition, empathy, communication the human part (represented as the head and heart of the centaur), big data mining, logical reasoning, trend detection and pattern matching the computational part (i.e., the muscles and legs of the centaur). In most of the cases, the machine would just confirm and corroborate the doctor's interpretation, even in those (hopefully few) cases that both the human and the MAI were wrong (in this case, the final decision by the doctor could easily be affected by a technologically-augmented, if not induced, confirmation bias). However, in a certain number of cases, doctors and AI will disagree. In that case, a safe protocol would then require the involvement of a third agent, a second human doctor acting as an arbiter, who could re-read the available imaging (or evidence) available for the case at hand and express a sort of second (or actually third) opinion. In the simplest case, that is a binary task, the third opinion would just confirm one diagnostic hypothesis and automatically outvote the other. Although a final decision could be taken by majority vote, a deeper interaction between the human raters is probably as much an inefficient further step as an advisable one in the protocol to achieve full accountability.

If we take this simple case as a basis for our centaur protocol, we can show that the analytical expression of the formula giving the accuracy of the technology-supported agency can easily be derived. First of all, we adopt the simplifying assumption that the accuracy of the human raters is known and is equal to a constant number acc_i. Then, the formula that represents the probability that the majority decision of the group of three raters (thus, including the AI) is indeed the correct one (under the assumption that the raters' opinions are independent or can be requested to be so) is given by the following formula:

$$\sum_{k=2}^{3} \sum_{R \epsilon R_k} \prod_{r_i \epsilon R} acc_i \prod_{r_j \epsilon R} (1 - acc_j)$$

In this formula (which follows from a simple combinatorial argument) R_k is the set of groups of raters of size k, acc_i is the accuracy of rater i.

As we previously mentioned, this formula can then be used to determine an analytical expression for the MAA of a MAI which is to support the other 2 human raters. Indeed, let us suppose that the two human doctors have an average accuracy of 85% and we want to establish a team that is augmented with a MAI and we want to determine the MAA for this MAI so that the overall accuracy of the augmented team is higher than the accuracy of the single human doctors, and is at least 90%. The MAA can then be found by solving an inequality in which the previous formula is set as greater than our desired group accuracy (I.e. 90%) and the only parameter to be determined is the accuracy of the MAI. It could then seem counterintuitive to notice that to satisfy such a requirement and guarantee an overall accuracy of 90%, only a 70%-accurate AI is needed!

For the sake of reference, we translate this analytical expression into a diagram, or better yet, a nomogram. By means of such a nomogram (depicted in Figure 2) it is possible to graphically determine the MAA of a MAI, given the accuracy of the two human doctors, and the desired level of accuracy of the whole "hybrid team", which is augmented by the MAI and adopts the simple majority voting strategy to settle case of discordance. The example depicted in the nomogram reported in Figure 2 represents the above described example scenario.

Obviously, the nomogram can be read also the other way round, by drawing a broken line from the vertical axis on the right up to the colored segment and then down up to the horizontal axis: so, if we use an 90% accurate MAI within a team where it is used under the engagement protocol described above, the performance of a team of doctors who are 85% accurate, would be improved up to 95%.

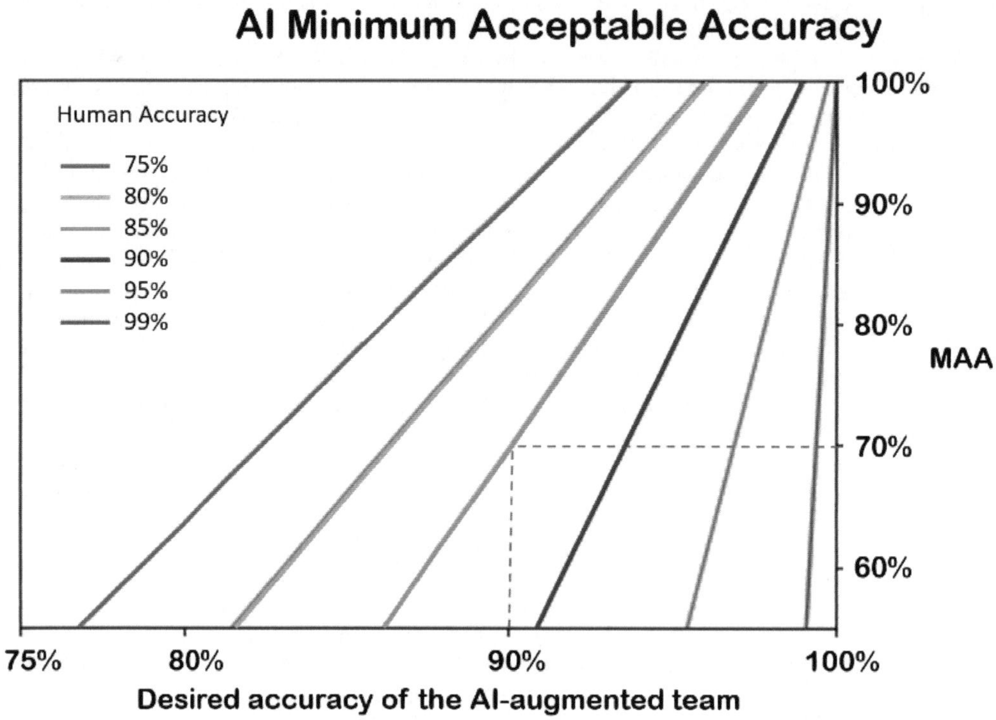

Figure 2. The nomogram to determine the minimum acceptable accuracy (MAA) of an AI in medical context. The dashed line suggests how to use the nomogram to read the MAA on the vertical line, once the desired accuracy of the medical team has been set, knowing the accuracy of the human members (colored segment)

Thus, this analytical approach allows to the determine of the MAA of a MAI (and the related illustrative nomogram) and also shows how the technological tool (I.e. the MAI), and its accuracy, in itself could be less important than the interaction protocol in which the MAI is considered: indeed, as we previously shown, a 70% accurate MAI may be in some cases sufficient to set up an interaction protocol that is more accurate than the sum of its parts.

3. CONCLUSIONS

In this paper we have outlined the case of determining a minimum threshold according to which to consider the performance of an Artificial Intelligence - AI (applied to discriminative and classification tasks) sufficiently valid in medical settings: we called such a threshold, the Minimum Acceptable Accuracy (MAA) a valid medical AI (MAI) system must exhibit to be such. Although it is easy to convince ourselves that such a concept is important and defining a MAA in Medical decision supports (one for each medical task or health condition) is the first step to understand if (buying) and adopting such an aid is necessary to improve medical practice, we also shed light on the fact that the common ways to understand if a MAI is accurate enough to be fit to its intended use (that is to support and improve doctors in their daily tasks) are fraught with simplistic assumptions or over-ambitious estimates.

We also reported the results of a small user study, in which we asked 79 medical doctors "how much accurate" they would require their MAI to be to perceive it as useful for their practical aims; and then we presented a simple nomogram by which to determine the MAA in a task-agnostic manner, when the accuracy of the human doctors is known (or assumed) and the accuracy requirement is adopted at the higher level, that is the level of the hybrid human-machine team where the MAI is adopted within a usage protocol that is aimed at finding the appropriate trade-off between efficiency (e.g., faster decisions), effectiveness (e.g., higher accuracy) and accountability (e.g., humans have the final say and can discuss controversial cases). This contribution is also aimed at addressing the provocative question mentioned in the title with a reflective attitude. In fact, we have also shown how doctors, if invited to express their requirements, would probably exaggerate the importance of relying on accurate decision aids to have their practice improved, and would push for more and more accurate tools (which nevertheless could facilitate the emergence of harmful attitudes like automation bias and complacency).

The nomogram we devised is then to be intended more as an informative simple tool, than an actual decision aid, to recognize that the technological tool is less important than a sound protocol in which to use it, both responsibly and by paying due attention to the role of the human decision makers. In fact, medicine has always been a collaborative practice, and a simple protocol can make up the shortcomings of technology and the limits of any single human by simply fostering discussion and making a technologically-augmented support better than the sum of its parts and better than each of the single parts.

ACKNOWLEDGEMENT

The authors acknowledge that the GP study was conducted on the basis of a sample of data coming in part from a survey conducted by INTRAMED COMMUNICATIONS SRL on a sample of doctors who had been invited to participate in an ECM event organized by this company. The authors are also grateful to the anonymous General Practitioners who accepted the invitation to participate in the survey sent by FDZ, as well as the anonymous specialists who participated to the same survey from the IRCCSS Istituto Ortopedico Galeazzi, Milano, Italy.

REFERENCES

Al Qahtani, D.A., Rotgans, J.I., Mamede, S., Al Alwan, I., Magzoub, M.E.M., Altayeb, F.M., Mohamedani, M.A., Schmidt, H.G., 2016. Does Time Pressure Have a Negative Effect on Diagnostic Accuracy?. Acad. Med. 91, 710-716. https://doi.org/10.1097/ACM.0000000000001098

Belkin, M., Hsu, D., Ma, S., Mandal, S., 2019. Reconciling modern machine-learning practice and the classical bias–variance trade-off. Proc. Natl. Acad. Sci. 116, 15849–15854. https://doi.org/10.1073/pnas.1903070116

Berlin, L., 2014. Radiologic errors, past, present and future. Diagnosis 1, 79–84. https://doi.org/10.1515/dx-2013-0012

Bruno, M.A., Walker, E.A., Abujudeh, H.H., 2015. Understanding and Confronting Our Mistakes: The Epidemiology of Error in Radiology and Strategies for Error Reduction. RadioGraphics 35, 1668–1676. https://doi.org/10.1148/rg.2015150023

Cabitza, F., 2019. Biases Affecting Human Decision Making in AI-Supported Second Opinion Settings, in: Torra, V., Narukawa, Y., Pasi, G., Viviani, M. (Eds.), Modeling Decisions for Artificial Intelligence. Springer International Publishing, Cham, pp. 283–294.

Floares, A.G., 2009. Using computational intelligence to develop intelligent clinical decision support systems, in: International Meeting on Computational Intelligence Methods for Bioinformatics and Biostatistics. Springer, pp. 266–275.

Goddard, K., Roudsari, A., Wyatt, J.C., 2012. Automation bias: a systematic review of frequency, effect mediators, and mitigators. J. Am. Med. Inform. Assoc. 19, 121–127. https://doi.org/10.1136/amiajnl-2011-000089

Goldstein, I.M., Lawrence, J., Miner, A.S., 2017. Human-Machine Collaboration in Cancer and Beyond: The Centaur Care Model. JAMA Oncol. 3, 1303. https://doi.org/10.1001/jamaoncol.2016.6413

Graber, M.L., 2013. The incidence of diagnostic error in medicine. BMJ Qual. Saf. 22, ii21–ii27. https://doi.org/10.1136/bmjqs-2012-001615

Gur, D., Bandos, A.I., Cohen, C.S., Hakim, C.M., Hardesty, L.A., Ganott, M.A., Perrin, R.L., Poller, W.R., Shah, R., Sumkin, J.H., Wallace, L.P., Rockette, H.E., 2008. The "Laboratory" Effect: Comparing Radiologists' Performance and Variability during Prospective Clinical and Laboratory Mammography Interpretations. Radiology 249, 47–53. https://doi.org/10.1148/radiol.2491072025

Hanna, T.N., Lamoureux, C., Krupinski, E.A., Weber, S., Johnson, J.-O., 2018. Effect of Shift, Schedule, and Volume on Interpretive Accuracy: A Retrospective Analysis of 2.9 Million Radiologic Examinations. Radiology 287, 205–212. https://doi.org/10.1148/radiol.2017170555

Liu, X., Faes, L., Calvert, M.J., Denniston, A.K., 2019a. Extension of the CONSORT and SPIRIT statements. The Lancet 394, 1225. https://doi.org/10.1016/S0140-6736(19)31819-7

Liu, X., Faes, L., Kale, A.U., Wagner, S.K., Fu, D.J., Bruynseels, A., Mahendiran, T., Moraes, G., Shamdas, M., Kern, C., others, 2019b. A comparison of deep learning performance against health-care professionals in detecting diseases from medical imaging: a systematic review and meta-analysis. Lancet Digit. Health 1, e271–e297.

Lyell, D., Coiera, E., 2016. Automation bias and verification complexity: a systematic review. J. Am. Med. Inform. Assoc. ocw105. https://doi.org/10.1093/jamia/ocw105

Semigran, H.L., Levine, D.M., Nundy, S., Mehrotra, A., 2016. Comparison of Physician and Computer Diagnostic Accuracy. JAMA Intern. Med. 176, 1860. https://doi.org/10.1001/jamainternmed.2016.6001

Sternini, F., Ravizza, A., Cabitza, F., 2020. How accurate do you want it? Defining Minimum Required Accuracy for Medical Artificial Intelligence, in: To Be Published.

PRIMARY USE CASE IMPLEMENTATION OF INTERNATIONAL PATIENT SUMMARY ON openEHR PLATFORM

Evgeniy Krastev[1], Petko Kovatchev[1], Dimitar Tcharaktchiev[2] and Simeon Abanos[1]
[1]Faculty of Mathematics and Informatics, Sofia University St. Kliment Ohridsky, Bulgaria
5 James Bouchier blvd., 1164 Sofia, Bulgaria
[2]Department of Medical Informatics, Medical University, Sofia, Bulgaria
University Hospital of Endocrinology, 2 Zdrave str., 1431 Sofia, Bulgaria

ABSTRACT

The International Patient Summary (IPS) comprises the essential clinical data that must be available in case an European citizen needs unscheduled or urgent health care while abroad. The implementation of cross- border exchange of IPS is among the highest priorities of the European commission. Detailed functional requirements and use cases have been elaborated in addition to a new IPS standard in recent years as part of the efforts for enabling interoperability in information exchange and continuity of care. This paper presents a practicable approach and a solution to the problem of achieving semantic interoperability among the IPS sources in the European union. A novelty approach in this paper is first to design and analyze the business process of the primary use case and after that to propose a solution. It allows to identify several activities that are out of the scope in the existing documents and at the same time require special attention at the stage of software implementation. The proposed solution makes use of the openEHR information model to represent the IPS standard dataset in terms of archetypes. This way we preserve the semantic context in clinical concepts as well as introduce interoperability among data providers. Moreover, it overcomes the severe drawback of the message paradigm, where every two communicating parties must apply transformation of existing data that is specific to their needs. The discussion considers a computer experiment with real data employing a fully functional web application for IPS cross-border exchange.

KEYWORDS

International Patient Summary, Use Case, Business Process Model and Notation, openEHR Information Model, Web Application, Semantic Interoperability

1. INTRODUCTION

The free movement of people poses several challenges to continuity of care and coordination of activities in cases of health accidents or emergency situations. Essential clinical data is urgently needed in understandable form so that the patient receives timely and adequate medical therapy. Reliable information like the patient's history of past illnesses, allergies and intolerances is available in the country of permanent stay and quite often it is not accessible at the time of hospitalization of a foreign citizen. Prevention the spread of infectious diseases such as COVID-19 requires travelers to provide evidence about their immunization status. Knowledge about the past illnesses of a foreign employee is an important factor in preventing workplace accidents in executing duties incompatible with his health status. These are only few of the typical healthcare problems that people might encounter while abroad. The seriousness of the problems becomes obvious once we take in consideration the large number of tourists in addition to people who migrate seeking better employment in developed countries (Footman, et al., 2014).

The European commission realized long ago the need for establishing a standard for the patients' minimal dataset of health information that is necessary to ensure continuity and coordination in administering medical treatment activities. The European Patients– Smart Open Services (epSOS) project proposed a pilot infrastructure for sharing and exchanging patient summary records (European Commission, 2012). The interoperability of health information systems (HIS) has become an obligatory requirement in cross- border exchange of patient summary. It sparked the development of interoperable solutions based on best practices in implementing eHealth standards and specifications (European Commission, 2015). An important step in this direction is the specification of detailed functional requirements and use case specifications (European Commission, 2018) applicable to the current standard EN 17269:2019 for International Patient Summary (IPS) (CEN/TC 251, 2019a). Nowadays special attention is dedicated to standards and specifications that enable semantic interoperability in exchange of clinical data. It allows to use information models and technologies for preserving the semantic context in which clinical data has been generated and make it equally understandable in the language of the receiving country. Typical representatives of such standards and specifications correspondingly, EN 13606:2019 (ISO/TC 215, 2019b) and openEHR (openEHR, 2020a), make use of the Archetype information model (openEHR, 2020b).

The review of existing literature sources shows that important implementation issues are usually not taken in consideration. For example, there is no publication that attempts to create a business process model and analyze the activities in cross- border exchange of IPS according to the use cases outlined by the European Commission. The primary use case considers a peer- to- peer message exchange of IPS clinical data between two countries. At the same it is recognized that European countries employ different eHealth information models for representing IPS entities, while some of them are even incompatible with internationally approved standards and specifications. Therefore, these use case do not support direct connectivity between major healthcare information system in Europe and semantic interoperability or even interoperability in cross- border exchange of IPS remains a serious challenge in practice.

A novel approach is here proposed for the development of a solution to this problem based on the analysis of business process model. The business process model makes use of Business Process Model and Notation 2.0 (BPMN) (Freund & Rücker, 2020). This model elaborates the primary case study with the objective of achieving semantic interoperability among all countries participating in IPS data exchange. A novelty from practical point of view is the web application we have developed on openEHR platform to demonstrate the execution activities (Gutiérrez, 2015). Both the business process model and the web application for IPS exchange have no analogs in public available sources. The here reported results are part of ongoing eHealth research program (Krastev, et al., 2019), (Tcharaktchiev, et al., 2020), (Krastev, et al., 2020).

The main results in this paper are presented in the following four sections. Section 2 formulates the problem in terms of functional requirements and participants in the primary use case. Section 3 presents a BPMN model of the primary use case and elaborates on the web application that implements the activities in that business process. Section 4 evaluates the results obtained from the computer experiments and Section 5 summarizes the advantages of the business process model for executing the primary use case and the potential applications of the web application that implements it.

2. PROBLEM STATEMENT

The current IPS standard (CEN/TC 251, 2019a) has been developed with the initial objective to support the use case of cross- border exchange of essential patient's clinical data in situations of unscheduled care. During its development process the scope of application has been extended to cover scenarios of both scheduled and unscheduled care that involve the patient while he is abroad or in the country of permanent stay (CEN/TC 251, 2018). For clarity in this paper we refer to the primary use case as the use case of IPS cross- border exchange abroad related to unscheduled or scheduled care.

The IPS standard determines a well- structured model of the clinical document and comprises 6 required sections, namely, Patient Attributes, Allergies and Intolerances, Medication Summary, Problems, Provenance and a conditional Cross- Border section. The conditional section is ignored in the secondary use cases dealing with scheduled and unscheduled care in the country, where the patient lives permanently. The first three IPS sections are mandatory, while the standard allows the list required sections to be extended with additional "*required if- known*" sections. This way the IPS standard can be accommodated for exchange of clinical data in other use cases like for example exchanging clinical data between registers for rare diseases (Tcharaktchiev, et al., 2020).

The official definition of the primary use case involves the following actors, where for uniformity we preserve in bold their original notations (European Commission, 2018) (European Commission, 2020):

- ✓ Human actors:
  - o **Patient** is the citizen from **country A** seeking health care service in country B. Although a **Patient** may have different IPS documents
  - o **HCP B** is the healthcare professional providing health care service in country B
- ✓ Institutional actor:
  - o **HCPO B** is the healthcare service providers organization in country B. All healthcare professionals belong to a specific **HCPO.**
- ✓ System actors:
  - o **NCP A** denotes the national contact point in country A. Respectively, **NCP B** denotes the national contact point in country B. The national contact point provides visibility to a single IPS copy to the outside world. Hence, a **Patient** could have at most one IPS document in each country.

For clarity of presentation we omit semantic and business rule services that are part of the complete description of the System actors.

A description of the flow of events in the primary use case is presented in relation to key functional requirements in (European Commission, 2018), where the communication between the above identified actors is visualized in a UML Sequence diagram. In short, a **Patient** needs health assistance in country **B,** he visits **HCP B** from **HCPO B** and **HCP B** wants to display the patient's IPS that is maintained by **NCP A. HCP B** establishes connection with **NCP A** through **NCP B,** the **Patient** identity is confirmed and after that the **Patient** authorizes **HCP B** to extract his IPS from **NCP A.** Finally, **HCP B** displays the IPS document and executes the requested health assistance taking in consideration the clinical data.

Although this description sounds complete enough it appears to be very difficult to implement it in practice because it avoids answering several important questions. For simplicity, assume both national contact points implement the IPS standard. First, this use case considers data exchange between two countries that in general use different information models to represent the same IPS standard. Note, that the IPS standard does not impose constraints on the information model that could be used to implement its data structures. The primary use case leaves out of scope the event of transformation needed between these two different models. Moreover, if we extend the primary use case and assume that country **A** could be any other European country, then we find out more complications in the practical implementation of this use case. For example, then **NCP B** must invest a lot in support for transformation between the information model adopted in country **B** and any other information model employed by the other national contact points. Besides, most of the time such transformations are "*one- way*". Another important issue is how to preserve the semantic context in the original IPS document during its digital transformation. This issue underpins the importance of using information models that support semantic interoperability. However, experience in using such models shows that interoperability even between such information models is difficult to achieve (Krastev, et al., 2020). Finally, the primary use case assumes **HCP B** can display the IPS. It means each **HCP B** must use a tool for IPS visualization compatible with the information model and data structures employed by **NCP B**. The implementation of the IPS in different information models and enabling support for digital transformation ("*two-way*") between each two countries is the top of the iceberg, because for shortness here we leave out of scope other issues like information security and confidentiality.

All these problems and more become visible when we attempt to create a model of the business process model pertinent to the primary use case. A business process model allows not only to analyze and understand the process activities, but also evaluate key performance indexes of the process execution through simulation (Shahinyan & Krastev, 2013). Therefore, the main objective in this paper is to design a business process model employing BPMN and use this model to analyze the implementation of the primary use case in a fully functional web application on an openEHR platform with real- life clinical data. A secondary objective is to summarize the advantages and disadvantages in the proposed business process model and the implementation of the primary use case.

3. PRIMARY USE CASE IMPLEMENTATION

A detailed analysis of existing descriptions of the primary use case for cross- border IPS exchange allows to identify the sequence of activities and events that execute each one of the actors involved in that use case.

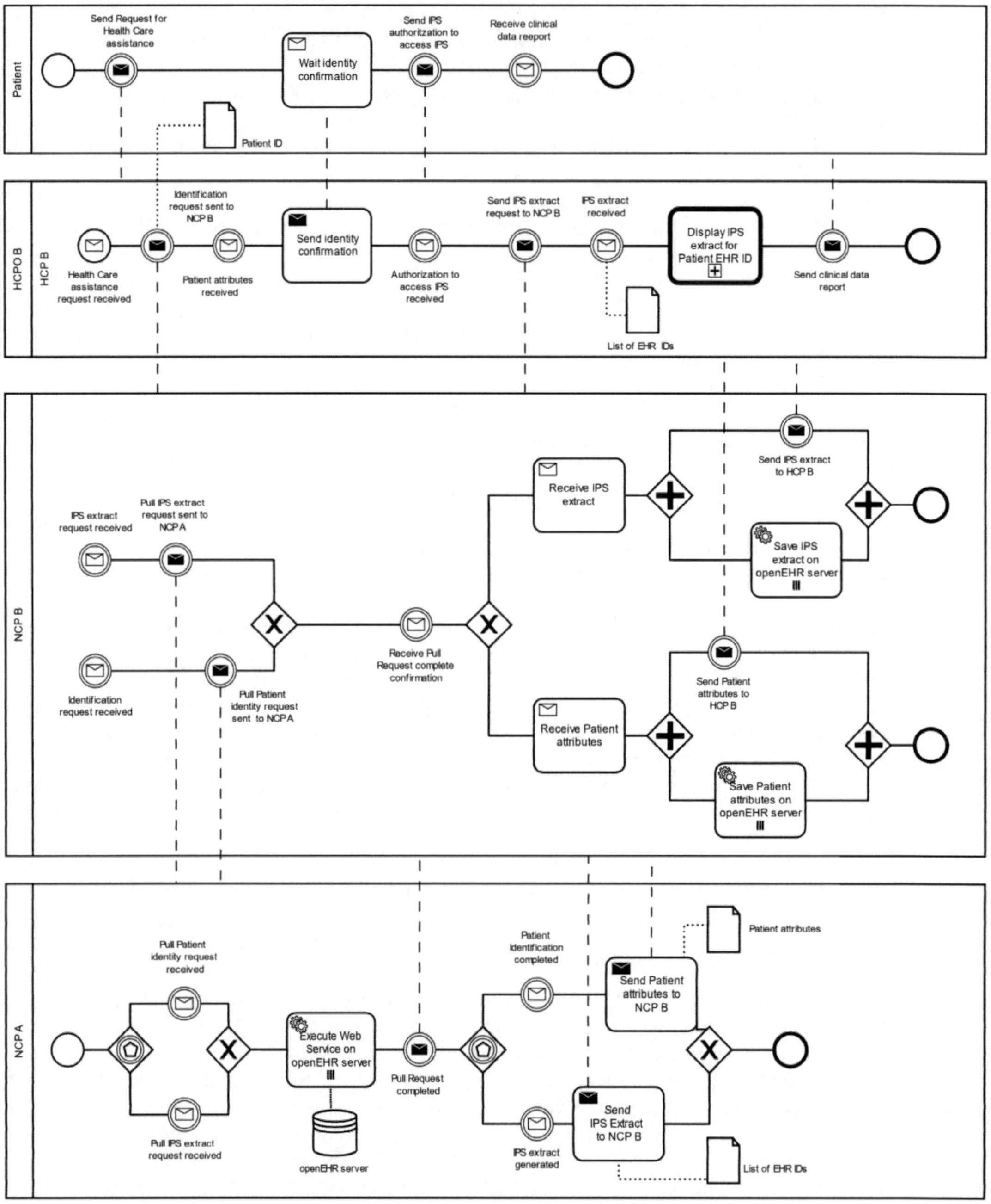

Figure 1. Diagram of the business processes in the Primary Use case of the IPS using BPMN 2.0

Following the syntax requirements in BPMN we design the business process models using open source Business Process Modeler (Camunda, 2020). Figure 1 displays the business processes executed respectively by the **Patient**, **HCP B**, **NCP B** and **NCP A**. In addition to the visual interpretation each one of these diagrams has XML representation that can be executed on a BPMN Engine for process simulation and evaluation (Rademakers & van Liempd, 2012).

The business processes on collaborate by means of exchanging message events or executing *Send* and *Receive* tasks. The dashed lines between the BPMN swimlanes depict the communication between such events and tasks.

An important element in the primary use case communications is the communication between **NCP B** and **NCP A**. Unlike the description of the flow of events in a UML Use case diagram at this stage of business process modeling we must clarify the type of activity **NCP B** and **NCP A** execute. Here we have chosen to use an openEHR server and a stack of web services published by each server to enable the persistence of IPS standard documents and exchange of these documents between **NCP B** and **NCP A**. The choice of openEHR platform is not accidental (Pazos, 2020) (openEHR, 2020c). Note activity "*Execute Web Services on openEHR Server*" executing Service task in the business process owned **NCP A**. Similar Service tasks are used in the business process owned by **NCP B** to save the patient attributes and IPS document extracted from **NCP A.** This choice is motivated by the fact that this platform employs an Archetype information model and this way it satisfies the nonfunctional requirement for semantic interoperability in cross- border IPS exchange. Moreover, openEHR identifies a patient by a unique Electronic Health Record (EHR) ID and this way supports confidentiality and security requirements in accessing sensitive patient data. Hence, in his communication with **HCP B** the **Patient** identifies himself with personal attributes, while **HCP B** receives access to the IPS document through the EHR ID linked to these attributes. In related research (Krastev, et al., 2019) (Krastev, et al., 2020) we have described the representation of an IPS standard document in terms of an openEHR archetype. Once the archetype IPS standard document is shared on public repositories, it will eliminate duplication of efforts and costs to implement IPS standard data structures in heterogenous and usually non- standard information models. Besides, it facilitates smooth and effortless digital sharing of IPS content from **NCP A** into the openEHR server on **NCP B**.

Another key contribution in the proposed business model is the introduction of a BPMN ***Call*** activity to display the IPS document in the business process of **HCP B** (see activity "*Display IPS extract for EHR ID*" in Figure 1). Designed this way the activity is interpreted as a global task in sense that it will be implemented in software once and reused by any other instance of **HCP B**. Finally, note that at the end of the business process of **NCP B** makes a copy of the patient attributes and his IPS on the openEHR platform managed by country **B**. Thus, the IPS copy may be used next time the **Patient** needs health assistance in country B.

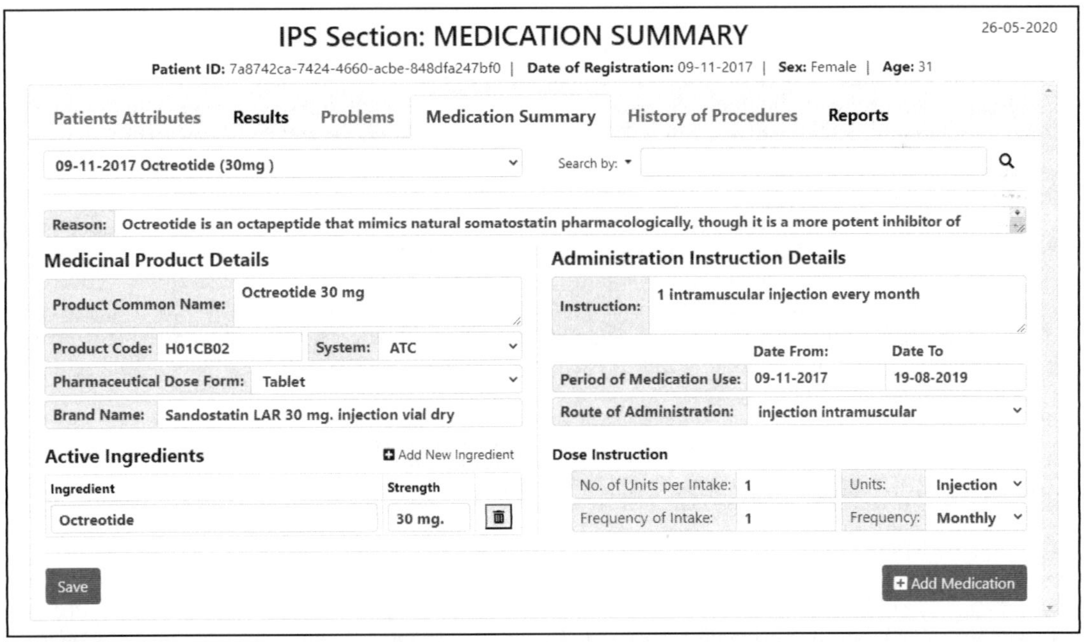

Figure 2. Execution of activity "*Display IPS extract for Patient EHR ID*" by **HCP B**

Once the business process model is clarified we attempt to implement it in a real-life web application. To simulate communication between **NCP B** and **NCP A** we have created a cluster of two virtual machines running openEHR servers. Additionally, we have developed a PHP client application to execute the business process of **HCP B**. In this scenario we assume the **Patient** is a human that collaborates with **HCP B** in executing validation of patient identity and authorizing the access to the patient's EHR ID. The web application is presented in a related research paper (Krastev, et al., 2020). Detailed demonstration in the execution of the web application is provided in (Kovachev, 2020). A sample snapshot of the web page with required *Medication summary* section of the extracted IPS is shown in Figure 2. The *Medication summary* section content entirely satisfies the IPS standard, while representing an instance of the openEHR archetype model. For clarity we have preserved in this web page the same label notations and data layout as those used in the IPS standard. This web page allows the **HCP B** to view previously prescribed medications with detailed semantic context describing each medication. For example, each medication is uniquely identified by its ATC code (WHO Collaborating Centre for Drug Statistics Methodology, 2020), instructions for administering the medication and the active ingredients are part of this description.

4. DISCUSSION

The design and the analysis of the business process model for cross- border exchange of IPS documents contributed significantly in identifying missing activities that are required in the software implementation. This model elaborates the existing descriptions of the primary use case. Therefore, its execution is entirely compatible with the functional requirements and the flow of events in that use case description. Moreover, the business process model adds semantic interoperability to the set of non- functional requirements of the originally use case description. Another advantage of this model is the introduction of openEHR as a common platform for providing persistence and web service communication between the actors involved in a real- life scenario (Gutiérrez, 2015). The above presented web application has no analog in the existing literature. It proves that a standard based IPS document can be represented as openEHR archetype. Moreover, instances of this archetype can be used in cross- border exchange (read and write operations) of IPS electronic health records.

Without loss of generality the openEHR platform can be replaced by another platform employing Archetype information model like EN 13606 to represent the IPS clinical document with respect to the existing standard in the European Union. In a related research paper, we have shown the standard IPS document representation in terms of an EN 13606 archetype as well (Krastev, et al., 2019). This approach eliminates duplication of considerable amounts costs and efforts that otherwise are necessary to implement the existing primary use case in practice. The development of archetypes and transformation of existing data into instances of these archetypes is not a straightforward process. It represents one of the disadvantages in implementing the Archetype information model. Employing open source code such as the freely available openEHR platform is acceptable for research purposes (Pazos, 2020). The development of the PHP client of the demonstrated web application needs a lot of programming experience as well. In general, the development of standard- based eHealth applications require a synergy of knowledge in health informatics, healthcare organization and medicine. However, the efforts invested in development of archetypes based on international standards payback and at the end cost less than the implementation of many heterogeneous and incompatible with each other eHealth systems. Moreover, such systems will not be semantically interoperable either.

5. CONCLUSION

This paper considers the primary use case for cross- border exchange of clinical documents satisfying the current IPS standard EN 17269. A detailed description of this use case is available in the existing literature, while its implementation in real- life application remains a challenge. This paper presents a novel approach to analyze this use case by designing a BPMN business process model employing a Camunda Modeler. The obtained BPMN model is executable on a BPMN engine and allows to discover missing activities and events of practical importance in the existing scenarios for IPS implementation. For example, scenarios in existing use case descriptions do not take in consideration the activities for transformation between the different IPS information models employed in EU countries. Open source tools are used to design and evaluate the execution of the BPMN process with clinical data. An openEHR platform is introduced in the execution of the process employing the openEHR Archetype information model and adds semantic interoperability capabilities of the software implementation. A sample web application proves the correctness of the obtained business process model. The computer experiments show that the web application precisely executes the sequence of activities in the BPMN model. Moreover, the application reads and writes electronic health records of a patient's IPS, whose data structure follows the current EN 17269:2019 standard for this document. The experimental results demonstrate that the electronic health records satisfy both the dual information model of openEHR and the IPS standard EN 17269:2019. We show that the Medication summary section on these records encapsulates detailed semantic context describing each previously prescribed medication.

Both the here developed BPMN model and the web application for cross- border management of IPS records based on the openEHR dual information model and the EN 17269 standard have no analogs in the existing literature. In addition to cross- border exchange of clinical documents this web application can be accommodated locally in each EU country and serve secondary scenarios in FprCEN/TS 17288 comprising both unplanned and planned cases of medical treatment in the healthcare domain. Our current research work is focused on adapting the web application to the requirements of the national legislation for monitoring of patients suffering severe infectious diseases like COVID-19 and prevention the spread of such diseases. In our future work we plan to extend the usability of the web application by implementing the IPS standard EN 17269 in terms of the EN 13606:2019 dual information model.

ACKNOWLEDGEMENT

This research is supported by the National Scientific Program eHealth in Bulgaria.

REFERENCES

Camunda, 2020. *Modeler. The developer-friendly desktop app for editing BPMN process diagrams and DMN Decision Tables.* [Online] Available at: https://camunda.com/products/modeler/ [Accessed 22 May 2020].

CEN/TC 251, 2018. *FprCEN/TS 17288. Health informatics - The International Patient Summary: Guidance for European Implementation,* Brussels: Eurpoean Committee for Standadization.

CEN/TC 251, 2019a. *EN 17269. Health informatics - The Patient Summary for Unscheduled, Cross- border Care,* Brussels: Eurpoean Committee for Standadization.

European Commission, 2012. *European Patients Smart Open Services (epSOS). D3.2.2 Final definition of functional service requirements- Patient Summary.* [Online] Available at: https://ec.europa.eu/digital-single-market/en/news/cross-border-health-project-epsos-what-has-it-achieved [Accessed 5 June 2019].

European Commission, 2015. *The Antilope project.* [Online] Available at: https://www.antilope-project.eu/ [Accessed 22 May 2020].

European Commission, 2018. *eHealth DSI Patient Summary and ePrescription. PS Functional requirements.* [Online] Available at: https://ec.europa.eu/cefdigital/wiki/download/attachments/35210463/PS%20functional%20requirements_v2.2.2.pdf [Accessed 20 May 2020].

European Commission, 2020. *Use Case Repository.Patient Summary Sharing On A Cross-Border And International Scale.* [Online] Available at: https://usecase-repository.ihe-europe.net/content/patient-summary-sharing-cross-border-and-international-scale [Accessed 20 May 2020].

Footman, K. et al., 2014. *Policy Summary 14. Cross-border health care in Europe. Copenhagen, Denmark: World Health Organization. Regional Office for Europe.* [Online] Available at: https://www.euro.who.int/_data/assets/pdf_file/0009/263538/Cross-border-health-care-in-Europe-Eng.pdf [Accessed 20 June 2020].

Freund, J. & Rücker, B., 2020. *Real-Life BPMN: With introductions to CMMN and DMN. Analyze, improve and automate your business processes.* 4 th ed. s.l.:Camunda.

Gutiérrez, P. P., 2015. *Towards the Implementation of an openEHR-based Open Source EHR Platform (a vision paper).* São Paulo, Brazil, IOS Press, pp. 45-49.

ISO/TC 215, 2019b. *ISO 13606-1:2019 Health informatics -- Electronic health record communication -- Part 1: Reference model.* [Online] Available at: https://www.iso.org/standard/67868.html [Accessed 6 June 2019].

Kovachev, P., 2020. *openEHR implementation of International Patient Summary based on the Fpr CEN 17269 standard.* [Online] Available at: https://www.youtube.com/watch?v=lDReBt9DsAs [Accessed 19 May 2020].

Krastev, E. et al., 2019. *Software Implementation of the EU Patient Summary with Archetype Concepts.* Porto, Portugal, IARIA, pp. 8- 13.

Krastev, E., Tcharaktchiev, D., Kovatchev, P. & Abanos, S., 2020. International Patient Summary Standard Based on Archetype Concepts. *International Journal On Advances in Life Sciences,* 12(1-2), p. (in print).

openEHR, 2020a. *The EHR Information Model.* [Online] Available at: https://specifications.openehr.org/releases/RM/latest/ehr.html#_the_ehr_information_model_[Accessed 4 May 2020].

openEHR, 2020b. *Archetype Model (AM) Component - latest.* [Online] Available at: https://specifications.openehr.org/releases/AM/latest/index [Accessed 22 May 2020].

openEHR, 2020c. *EHRServer. Platform. Free to use & open source.* [Online] Available at: https://www.openehr.org/downloads/platform/_[Accessed 10 May 2020].

Pazos, P., 2020. *Service-oriented openEHR repository for clinical data with composition commit, query and retrieve capabilities.* [Online] Available at: https://github.com/ppazos/cabolabs-ehrserver [Accessed 20 May 2020].

Rademakers, T. & van Liempd, R., 2012. *Activiti in Action.* s.l.:Manning Publications.

Shahinyan, K. & Krastev, E., 2013. *Extending a BPMN Engine with Evaluation Metrics for KPIs.* Sofia, Sofia University Press, pp. 102- 109.

Tcharaktchiev, D. et al., 2020. Cross-border Exchange of Clinical Data using Archetype Concepts Compatible with the International Patient Summary. In: L. Pape-Haugaard, et al. eds. *Digital Personalized Health and Medicine. Proceedings of MIE 2020.* Geneva: Studies in Health Technology and Informatics, IOS Press, p. (in print).

WHO Collaborating Centre for Drug Statistics Methodology, 2020. *Anatomical Therapeutic Chemical code classification system (ATC).* [Online] Available at: https://www.whocc.no/atc_ddd_index/

APPLYING DEEP LEARNING METHODS TO THE AI EXPERT SYSTEM ON ALZHEIMER`S DISEASE FOR THE ELDERLY

Chun-Yang Chang[1], You-Hsun Wu[2], Wei-Pin Hong[3], Chien-Hsu Chen[4] and Yang-Cheng Lin[4]
[1]*National Cheng Kung University, Dept. of Industrial Design, Taiwan (R.O.C), Graduate Student*
[2]*National Cheng Kung University, Dept. of Techno Art, Taiwan (R.O.C), Graduate Student*
[3]*National Cheng Kung University Hospital, Dept. of Neurology, Taiwan (R.O.C), Attending Physician*
[4]*National Cheng Kung University, Dept. of Industrial Design, Taiwan (R.O.C), Professor*

ABSTRACT

With the advent of an aging society, the number of people who are being afflicted with Alzheimer's disease is also on a gradual rise. However, an effective medical treatment to contain the contraction of the disease still does not exist. There is a widespread lack of understanding about dementia among the general populace, and the amount of available research into early pathological prediction of Alzheimer's disease is also quite low. This study proposes a testing system to detect cases of dementia, which is designed to assist doctors in diagnosing the disease. The system can be applied to embedded devices and mobile devices in the hospital in the future to promote the development of artificial intelligence in the medical field and improve diagnosis efficiency. On the basis of minuscule size and a small number of parameters, lightweight convolutional neural networks can be deployed on devices with finite memory and computing without connecting any cloud platform to avoid the breach of image data and ensure the quality of its security. For this purpose, three common lightweight convolutional neural networks are used in this study — MobileNet V2, NASNetMobile, and ShuffleNet V2. In cases where the parameters are identical to those corresponding to other conditions, the Alzheimer's disease predictive identification is applied to use open-source magnetic resonance imaging (MRI) scans obtained from the Kaggle platform. The results of the study indicate that MobileNet V2 exhibits the highest prediction accuracy (80.78%). Additionally, the system proposed in this study can be integrated into physicians' workflows during the diagnosis of Alzheimer's disease, thereby making their medical judgments more accurate. It can, therefore, address the tedious and time-consuming nature of the current methods of diagnosis of the disease, improve the efficiency of the medical treatment of patients, and improve the possibility of early detection of the disease.

KEYWORDS

Alzheimer Disease, Image Recognition, MobileNet V2, NASNetMobile, ShuffleNetV2

1. INTRODUCTION

With the gradual increase in the life expectancy of the world's population, the number of patients suffering from dementia has also been observed to grow proportionally (Prince et al., 2013). Globally, approximately 9.9 million people suffer from dementia each year, implying that a new patient is diagnosed with the disease once every 3.2 s. The number of affected people is expected to exceed 100 million by 2050. In particular, the number of people suffering from dementia in Taiwan is expected to increase by an average of 35 per day over the next 45 years, amounting to 1 person being afflicted with dementia every 40 minutes (Figure 1.). Currently, no concrete cure exists for dementia (Prince et al., 2014), and owing to a general lack of understanding and awareness about dementia, the disease is often perceived as a natural and unavoidable companion of old age rather than a treatable disease.

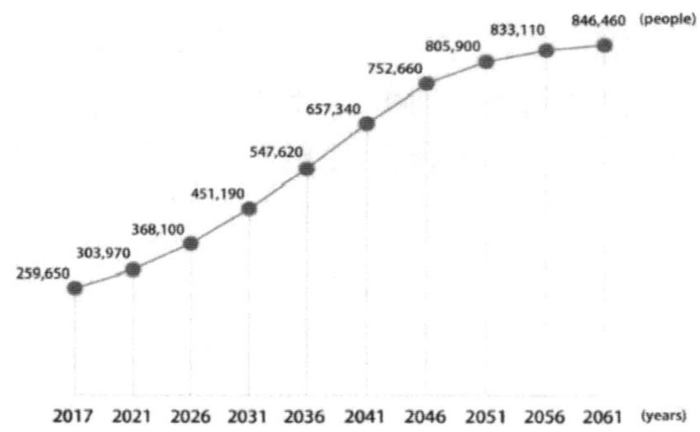

Figure 1. Estimation of the total population of Dementia in Taiwan

Dementia is the common name for memory loss. Depending on the severity of the disease, patients' daily lives are fundamentally affected including problems with their mental abilities. Unlike normal aging, Dementia is a neurodegenerative disease that tends to occur in the elderly. It can be classified as one of degenerative, vascular, or mixed dementia based on the causes and symptoms of the disease. Alzheimer's disease is one type of degenerative dementia, accounting for 50% to 60% of all cases of dementia (Cummings & Cole, 2002).

Several different biomarkers based on different medical imaging techniques can be used to provide additional information on pathological testing of Alzheimer's disease to improve the accuracy of detection (Zhang et al., 2011). Compared to other imaging methods, magnetic resonance imaging (MRI) is the most widely used in clinical detection, which enables the detection of different clinical stages during the diagnosis (Jack Jr et al., 2013). Hence, we use MRI images as the dataset in the experiments in this study. For patients suffering from the Alzheimer's disease, pathological changes often occur in the brain before any clinical symptoms are exhibited. If the diagnosis is Alzheimer's, then beginning a treatment in the early phases of the disease holds the promise of slowing the rate of disease degradation and improving quality of life for the patient (Brookmeyer et al., 2007).

In the era of big data, artificial intelligence (AI) has become a developmental trend in various fields. AI-based methods have emerged as powerful tools that have transformed medical care. In particular, they have the potential to achieve early diagnosis of dementia (Lei et al., 2016). Medical imaging data account for approximately 80% of all medical information, and their proper use and analysis can significantly enhance medical efficiency. Additionally, with the rapid development of cloud services and the Internet of Things, non-information-related industries such as medical and health information are increasingly dependent on information systems and networks,thereby leading to increased risks and threats of information security. The protection of information security and the transmission cost of the network cannot be ignored. In this study, we use three lightweight convolutional neural networks (CNNs) and evaluate their respective efficiencies and accuracies in terms of their performance and speed. With Alzheimer's disease being our primary object of research, we develop an AI expert system that is capable of diagnosing Alzheimer's disease at a stage as early as possible. This could help to improve the efficiency of doctors' diagnoses and allow early detection of Alzheimer's disease in a patient. The goals of this study can be summarized as follows:

1. Design a lightweight CNN-based diagnostic model for Alzheimer's disease that is capable of providing instant diagnosis and can predict the patient's condition.

2.Analyze and compare the three kinds of lightweight CNNs (i.e. MobileNet V2, NASNetMobile, and ShuffleNet V2).

3.Serve as a reference for the research and development of deep-learning-based platforms for the detection of Alzheimer's disease.

2. LITERATURE

This section focuses on the medical knowledge of Alzheimer's disease and the discussion of deep learning knowledge. This section has been classified into the following topics: 1. Alzheimer's disease 2. Deep Learning 3. Application of deep learning in medical field.

2.1 Alzheimer's Disease

Dementia is often incorrectly referred to as senility or senile dementia, but, in reality, it is very different from normal aging. Dementia is caused by abnormal brain changes, systemic diseases of the body, and inappropriate use of drugs or substances. Alzheimer's disease is the most common type of dementia, accounting for 60 percent or more of all cases, and it is an age-dependent neurodegenerative disorder (Tanzi, 2012). Especially during the early stages of the disease, the symptoms of Alzheimer's disease vary from person to person. It is a progressive disease that begins with mild memory loss, and, with time, it can possibly lead to loss of the ability to carry on a conversation and respond to the environment, besides also seriously affecting the patient's ability to carry out daily activities (Serrano-Pozo et al., 2011).

From a clinical perspective, there is a critical period of time between the moment at which it becomes possible to diagnose the disease and the moment when it is actually diagnosed. Therefore, taking more practical precautions for Alzheimer's disease is imperative to reduce this period as much as possible, enabling the swift introduction of treatment during this period.

2.2 Deep Learning

Deep learning is an artificial intelligence function that imitates the workings of the human brain during the processing of data and creation of patterns for use in decision making. Owing to its ability to detect abstract and complex patterns, deep learning has been applied in neuroimaging studies of psychiatric and neurological disorders, which are characterized by subtle and diffuse alterations from the norm (Vieira et al., 2017). The architecture of deep learning is composed of multiple processing layers of modules that retain a lot of flexibility to enhance the computing power of the model. Deep learning can be applied in different fields, and many neural networks dedicated to image recognition have been developed.

A CNN comprises one or more convolutional layers followed by one or more fully connected layers. It contains neurons that self-optimize through learning and imitating the cognitive methods of the human brain, providing an optimal architecture for image recognition and pattern detection. Therefore, many image recognition-related models are extended and applied based on this architecture (O'Shea & Nash, 2015). In this study, we used three lightweight CNNs as deep learning model architectures.

2.3 Application of Deep Learning in the Medical Field

With the rapid development of science and technology, medical breakthroughs continue to be made every day all around the world. In the field of medicine, the novel use of deep learning methods to analyze the relationship between technology and disease prediction has positive implications in clinical testing. The application of artificial intelligence in medical imaging is primarily meant to help hospitals or physicians provide better medical care for patients, rather than to completely replace physicians. Deep learning has been widely used to predict missing information, including the search for and identification of target objects in images (Simonyan & Zisserman, 2014). The core concept of deep learning is to analyze and summarize historical data, and to judge new image data on this basis. Deep learning has exhibited significant progress in the medical field, especially in the development of models for chronic diseases (Green & Zhang, 2016).

Currently, breast ultrasound systems trained by CNNs have been successfully used to assist physicians in diagnosis of breast tumors. By capturing the subtle features of medical image, a tumor and its severity can be simultaneously detected at an early stage of its development. The identification rate of this method has been observed to be higher than 90% (Fakoor et al., 2013). On the other hand, by extracting physiological signals during operations related to heart diseases, it is possible to identify patients with acute renal failure early after surgery to reduce the occurrence of other subsequent problems. Therefore, with the maturity of image recognition technology, deep learning exhibits increasing potential to bring clinical breakthroughs in the diagnosis of dementia by analyzing complex brain imaging data and identify early subtle changes in them (Lei et al., 2016).

3. RESEARCH METHODS

In this study, we evaluate the performance of three lightweight CNNs for Alzheimer's disease detection, namely, MobileNet V2, NASNetMobile, and ShuffleNet V2. A MRI dataset is obtained from the Kaggle platforms model training data. The data in this dataset contains 6,400 MRI images related to Alzheimer's disease. The dataset is confirmed and evaluated by the attending physician of the Department of Neurology at the - National Cheng Kung University Hospital, Taiwan.

The data in the MRI database had been pre-categorized into 4 levels, namely, NonDemented, VeryMildDemented, MildDemented, and ModerateDemented (Figure 2.). As our research was at an initial phase, and overly complex data were expected to adversely affect the effectiveness of the neural networks being assessed, thus compromising their comparison, we recategorized the data into three classes of different disease levels (VeryMildDemented, MildDemented, and ModerateDemented) with respect to one attribute of the disease (Dementedness). Following this process, we evaluated the three lightweight neural networks by comparing relevant data corresponding to two distinct attributes and analyzed their respective levels of applicability in the identification of Alzheimer's disease. The results obtained in this study could serve as references for future research. For any subsequent research, four levels of disease severity will be restored and analyzed to estimate the effectiveness of models under consideration on the database.

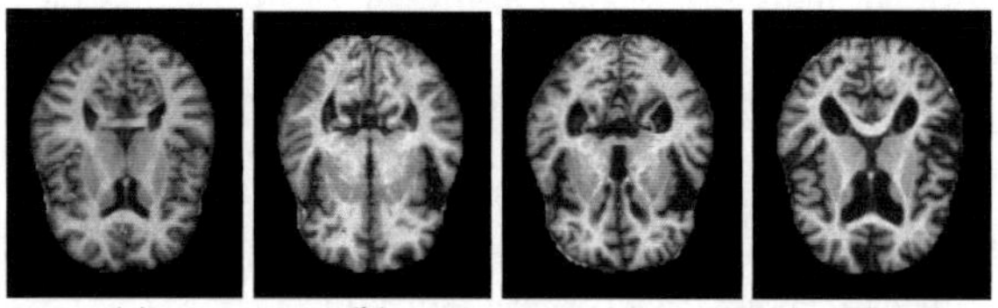

(a) (b) (c) (d)

Figure 2. MRI image dataset of Alzheimer's disease (a) NonDemented (b) VeryMildDemented (c) MildDemented (d) ModerateDemented(taken from Kaggle:https://www.kaggle.com/tourist55/alzheimers-dataset-4-class-of-images)

3.1 Fine-Tuning

On the first step in the identification process, we modified the structure of the pre-trained network model via fine-tuning and by altering the load model weights to enable its use as a model initialization for new tasks. To accomplish this, we first made consistent adjustments to the training models of the three lightweight neural networks. To prevent overfitting caused by too many parameters, GlobalAveragePooling2D was added at the end of the model to replace the fully connected layer. The average number corresponding to each feature map was designated as the associated output. Two layers of Batch Normalization were then added to ensure that the input corresponding to each layer of the network had the same distribution as those corresponding to the others. Besides reducing the occurrence of overfitting, this process was also observed to accelerate the convergence speed of the training process. Finally, a Dropout layer was added to improve the generalization of the model and avoid overreliance on certain regional features (Table 1.).

In this study, AMSGrad was used as an optimizer of the model to avoid the non-convergence or convergence of the training results on the local optima. The optimizer was used to identify the most suitable learning rate while maintaining superior convergence. AMSGrad was used to ensure that the learning rate always remained non-negative and to update the learning rate with the maximum among all the gradient values (Reddi et al., 2019).

Table 1. Layers added at the end of three lightweight networks after fine-tuning

A Lite Convolutional Neural Network (MobileNet V2 / NASNetMobile / ShuffleNetV2)	
(Layer)	**(type)**
global_average_pooling2d_1	**GlobalAveragePooling2D**
batch_normalization_1	**BatchNormalization**
dense_1	**Dense**
batch_normalization_2	**BatchNormalization**
dropout_1	**Dropout**
dense_2	**Dense**

3.2 Dataset

We used the public dataset obtained from the Kaggle platform as training data for the models. This dataset contained 6400 samples of MRI images of Alzheimer's patients and were divided into three folders, training, validation, and testing. Each folder contained subfolders for images that corresponded to normal patients and those suffering from the Alzheimer's disease. First, MRI data of each classification dataset was randomly split into a large training and validation set (80% of images) and a testing set (20% of images). Data augmentation was then applied on images selected for training and validation (not testing) in order to generate additional artificial images and consequently prevent overfitting. Second, each augmented dataset was randomly split into two subsets (80% for training and 20% for validation).

As an insufficient quantity of data could lead to poor performance and overfitting problems after training, the training data were increased via the data augmentation method to horizontally flip, randomly scale, and fill the original image (Figure 3.) before executing the analysis of the data. Further, new images that were completely different from the original data were generated to solve the problems of data imbalance and insufficient data.

```
horizontal_flip=True
zoom_range=0.2
rescale=1/255
fill_mode="nearest"
```

Figure 3. Preprocessing for datasets

3.3 MobileNet V2

MobileNet V2 is an effective feature extractor used for object detection and segmentation. It introduces the Inverted Residual block, which enables features in an image to be used repeatedly and improves the performance of the network. To address the prospective problem of feature destruction, MobileNet V2 uses Linear Bottlenecks to remove ReLU from the network to reduce ReLU's damage to features. Pointwise convolution is added before Depthwise convolution to increase the dimension, which addresses the problem that the number of channels could not be changed to obtain a greater number of features (Figure 4.) (Sandler et al., 2018). MobileNet V2 uses lightweight convolution to filter features, increasing the efficiency of its Random-Access Memory. MobileNet V2 operates with 3.4 million parameters and sets the initial value of the learning rate to be equal to 0.001 (lr = 1e-3)\. Colab with Tesla P100 Video Graphic Array Card is then used to train MobileNet V2 until the loss is observed to stabilize (1e-4).

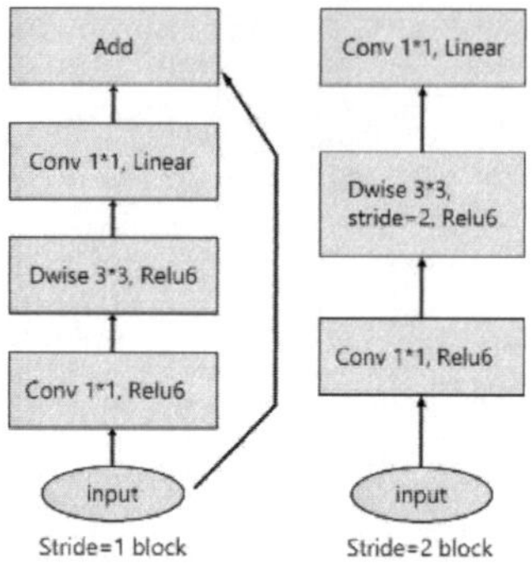

Figure 4. The convolutional block of MobileNet V2

3.4 NASNetMobile

NASNet is a scalable CNN structure inspired by the Neural Architecture Search (NAS) framework (Zoph et al., 2018) that consists of modules optimized for reinforcement learning. In NAS, a controller recurrent neural network samples child networks with different architectures to enable the generation of better architectures over time. In NASNet, the most suitable architecture identified on the CIFAR-10 dataset would scale to the ImageNetdataset. Finally, NASNet consists of modules that perform well on both CIFAR-10 and ImageNet datasets (Figure 5.). The NASNet used in this study possesses 4.8 million parameters, and as in the training of MobileNet V2, the initial value of the learning rate is set to be equal to 0.001 (lr = 1e-3). Colab with Tesla P100 Video Graphic Array Card is subsequently used to train NASNet until the loss is observed to stabilize (1e-4).

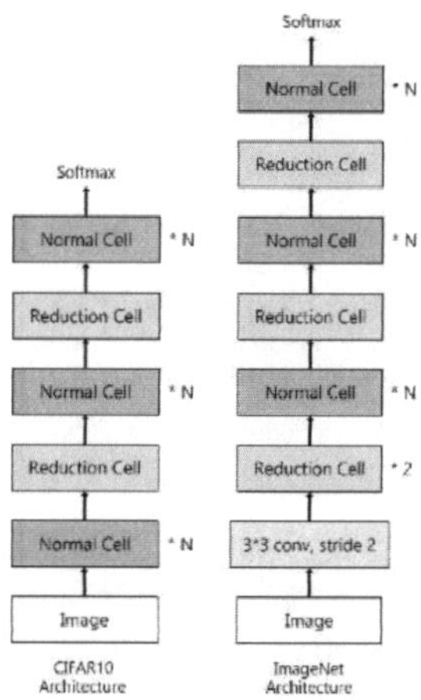

Figure 5. The cells for CIFAR-10 and ImageNet

3.5 ShuffleNet V2

To achieve increased capacity and efficiency, ShuffleNet V2 uses Channel Split, which is designed to reduce the number of branches of neural networks. In this architecture, after each split, one branch is retained as the identity while the other branch is made to consist of three convolutions with the same input and output channels, thereby minimizing multiply-accumulates (MACs). The two 1 × 1 convolutions are not oriented group-wise to avoid increase in MAC. Following the convolution, the two branches are stitched together to keep the number of channels constant, followed by the execution of Channel Shuffle to ensure the undisturbed exchange of information between the two branches (Figure 6.) (Ma et al., 2018). The ShuffleNet V2 used in this study possesses 6.4 million parameters, and, as in the previous two models, the initial value of the learning rate is set to be equal to 0.001 (lr = 1e-3). Colab with Tesla P100 Video Graphic Array Card is subsequently used to train ShuffleNet V2 until the loss is observed to stabilize (1e-4).

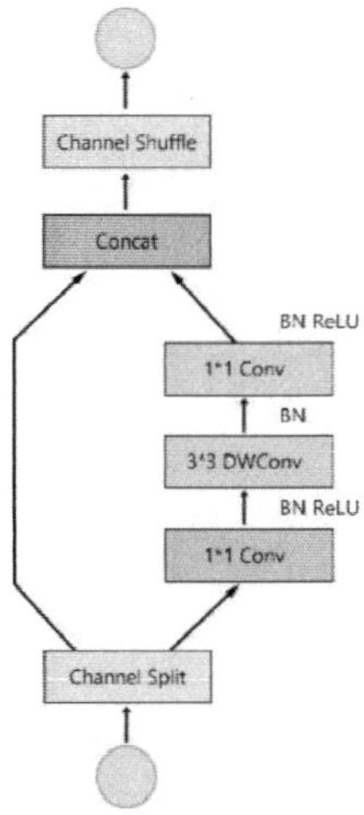

Figure 6. Structure of ShuffleNet v2

4. RESULTS

The MRI dataset used in this study was obtained from the Kaggle platform and contained 6,400 pieces of data related to Alzheimer's disease. Eighty percent of the data were used as training and validation data, and the remaining twenty percent were used as testing data. The data included 3200 normal people and 3200 patients suffering from Alzheimer's disease. We used three common lightweight CNNs (MobileNet V2, NASNetMobile, ShuffleNet V2) and evaluated the possibility of predicting Alzheimer's disease with each. Before training the models, all three neural networks were unified under the following parameter settings: the initial Learning rate was set to be equal to 0.001 (lr = 1e-3), Epoch was set to be 50, Batch-size was set to be 32, and the loss function was set to be binary_crossentropy.

Table 2 depicts the comparison of the performances of the three lightweight CNNs. It is concluded that ShuffleNet V2 exhibits the best training speed (35 min 54 s) of the three, followed by MobileNet V2 (43 min 25 s) and NASNetMobile (1 h 11 min 41 s). With respect to prediction, MobileNet V2 exhibits the highest accuracy of 80.78%. Additionally, on the medium datasets (6400 data points) used in this study, ShuffleNet V2 is observed to be significantly less accurate than the other two networks. This demonstrates that ShuffleNet V2 performs poorly in prediction, i.e., the compatibility of ShuffleNet V2 with this data type is not satisfactory.

Table 2. Comparison of efficiencies and accuracies of three lightweight convolutional neural networks

Model	CPU times (total)		Loss fuction	Accuracy (%)
MobileNet V2	43min 25s	Train	0.0076	99.73
		Test	0.5912	**80.78**
NASNetMobile	1hr 11min 41s	Train	0.0153	99.59
		Test	0.8296	74.61
ShuffleNet V2	35min 54s	Train	6.2956	52.41
		Test	0.7441	50.00

5. CONCLUSIONS

Since the emergence of LeNet, CNNs have promoted the development of deep learning. With problems caused by the deepening of the network, the idea of simplifying the model has begun to prevail, and more excellent networks have been created. In this paper, we have outlined how to use image recognition in deep learning to detect whether a subject is suffering from Alzheimer's disease in real-time, thereby promoting the development of a greater number of deep learning solutions for image recognition in the medical field. Further, we have evaluated the performance of three lightweight CNNs with respect to Alzheimer's disease detection (i.e. MobileNet V2, NASNetMobile, and ShuffleNet V2), and trained the models using a dataset of 6400 samples. Our experiments show that NASNetMobile is not as fast as MobileNet V2 and ShuffleNet V2 with respect to training speed. However, with respect to accuracy of prediction, MobileNet V2 has the highest value of 80.78%, while ShuffleNet V2 exhibits considerable disparity among the three CNNs. The experimental method used in this study is fast, accurate, and easy to implement. It will contribute to provide better calculation methods in the current era of big data.

REFERENCES

Brookmeyer, R., Johnson, E., Ziegler-Graham, K., & Arrighi, H. M. (2007). Forecasting the global burden of Alzheimer's disease. Alzheimer's & dementia, 3(3), 186-191.

Cummings, J. L., & Cole, G. (2002). Alzheimer disease. Jama, 287(18), 2335-2338

Fakoor, R., Ladhak, F., Nazi, A., & Huber, M. (2013, June). Using deep learning to enhance cancer diagnosis and classification. In Proceedings of the international conference on machine learning (Vol. 28). New York, USA: ACM.

Green, C., & Zhang, S. (2016). Predicting the progression of Alzheimer's disease dementia: a multidomain health policy model. Alzheimer's & Dementia, 12(7), 776-785.

Jack Jr, C. R., Knopman, D. S., Jagust, W. J., Petersen, R. C., Weiner, M. W., Aisen, P. S., ... & Lesnick, T. G. (2013). Tracking pathophysiological processes in Alzheimer's disease: an updated hypothetical model of dynamic biomarkers. The Lancet Neurology, 12(2), 207-216.

Lei, B., Chen, S., Ni, D., & Wang, T. (2016). Discriminative learning for Alzheimer's disease diagnosis via canonical correlation analysis and multimodal fusion. Frontiers in aging neuroscience, 8, 77.

Ma, N., Zhang, X., Zheng, H. T., & Sun, J. (2018). Shufflenet v2: Practical guidelines for efficient cnn architecture design. In Proceedings of the European Conference on Computer Vision (ECCV) (pp. 116-131).

O'Shea, K., & Nash, R. (2015). An introduction to convolutional neural networks. arXiv preprint arXiv:1511.08458.

Prince, M., Guerchet, M., & Prina, M. (2013). The global impact of dementia 2013-2050. Alzheimer's Disease International.

Prince, M., Albanese, E., Guerchet, M., & Prina, M. (2014). Dementia and risk reduction: an analysis of protective and modifiable factors. World Alzheimer Report, 66-83.

Reddi, S. J., Kale, S., & Kumar, S. (2019). On the convergence of adam and beyond. arXiv preprint arXiv:1904.09237.

Sandler, M., Howard, A., Zhu, M., Zhmoginov, A., & Chen, L. C. (2018). Mobilenetv2: Inverted residuals and linear bottlenecks. In Proceedings of the IEEE conference on computer vision and pattern recognition (pp. 4510-4520).

Serrano-Pozo, A., Frosch, M. P., Masliah, E., & Hyman, B. T. (2011). Neuropathological alterations in Alzheimer disease. Cold Spring Harbor perspectives in medicine, 1(1), a006189.

Simonyan, K., & Zisserman, A. (2014). Very deep convolutional networks for large-scale image recognition. arXiv preprint arXiv:1409.1556.

Tanzi, R. E. (2012). The genetics of Alzheimer disease. Cold Spring Harbor perspectives in medicine, 2(10), a006296.

Vieira, S., Pinaya, W. H., & Mechelli, A. (2017). Using deep learning to investigate the neuroimaging correlates of psychiatric and neurological disorders: Methods and applications. Neuroscience & Biobehavioral Reviews, 74, 58-75.

Zhang, D., Wang, Y., Zhou, L., Yuan, H., Shen, D., & Alzheimer's Disease Neuroimaging Initiative. (2011). Multimodal classification of Alzheimer's disease and mild cognitive impairment. Neuroimage, 55(3), 856-867.

Zoph, B., Vasudevan, V., Shlens, J., & Le, Q. V. (2018). Learning transferable architectures for scalable image recognition. In Proceedings of the IEEE conference on computer vision and pattern recognition (pp. 8697-8710).

EVALUATION OF COST-SAVING MACHINE LEARNING METHODS FOR PATIENT BLOOD MANAGEMENT

Davide Brinati[1], Andrea Seveso[1], Paolo Perazzo[2], Giuseppe Banfi[2] and Federico Cabitza[3]
[1]*Dipartimento di Informatica, Sistemistica e Comunicazione, Università degli Studi di Milano-Bicocca Viale Sarca 336, 20126, Milano, Italy*
[2]*Istituto Ortopedico Galeazzi, Via Riccardo Galeazzi, 4, 20161 Milano, Italy*
[3]*Dipartimento di Informatica, Sistemistica e Comunicazione, Università degli Studi di Milano-Bicocca Viale Sarca 336, 20126, Milano, Italy*

ABSTRACT

Our objective is the development of cost-saving methods for the patient blood management in Galeazzi Orthopedic Institute, a large Italian hospital. The methods have been developed in relation to the known costs of the hospital, both in terms of unused blood bags and drugs. Observational data about 4593 patients have been retrieved, with anagraphical and pre-operational clinical features. Model's performances have been compared to an existing baseline in terms of both accuracy measures (F1, recall, AUC) and saved costs per patient. The proposed methods recorded an enhancement of performances for the adopted measures, demonstrating a possible useful application of machine-learning-based methods for the patient blood management task.

KEYWORDS

Patient Blood Management, Machine Learning, Sensitivity Analysis, Operation Costs

1. INTRODUCTION

Blood transfusion is a widely used procedure in hospitalized patients, used in many different clinical conditions, such as traumas, infections, hemorrhages. Notwithstanding this therapy represents a lifesaving treatment for some patients, it is also one of the most overused procedures in hospitals. Unnecessary transfusions often lead to increased risks and costs, due to the suboptimal allocation of such an important good. An efficient strategy aims to improve the transfusion practice by avoiding unhandy operations for the patient and saving blood costs. This is an effective method of promoting high-value cares. Some of these costs could be saved by treating the patient with medicines that reduce the risk of transfusion, thus reducing the costs, since medicines have a lower price than blood bags.

In addition to the economic aspect, a successful patient blood management system diminishes unnecessary allocations of blood bags, which remain available for the patients that actually need transfusions.

Some previous works (Sadana et al. 2000, Hayn, 2017) focused their attention on the optimization of blood transfusion. The latter describes how predictive modelling and machine learning tools applied on pre-operative data can be used to predict the amount of red blood cells to be transfused during surgery and to prospectively optimize blood ordering schedules.

This work aims for evaluating the current patient blood management system of the Italian hospital *Galeazzi Orthopedic Institute (Istituto Ortopedico Galeazzi - IOG),* a center for medical research in the orthopedic field, based in Milan, and investigate if and how it can be improved, in terms of saved costs. Furthermore, a machine learning method will be proposed and compared with the previous methods.

IOG already has a method to deal with the patient blood management, based on a heuristic approach dictated by clinical expertise. Patients hemoglobin levels are evaluated, and a threshold is set: if the patient presents a hemoglobin value less than 12.9, then he is likely to undergo a transfusion, and so he is given a set of erythropoietin-stimulating agents in order to decrease the transfusion risks.

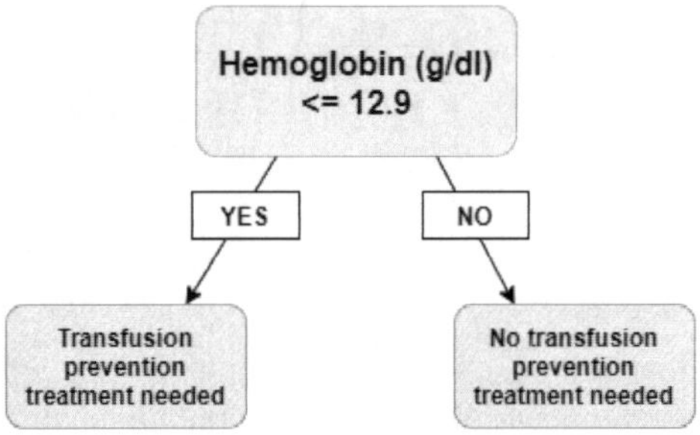

Figure 1. Current strategy at IOG: if the patient hemoglobin level falls below 12.9, transfusion prevention treatment is deemed required, meaning the patient will receive erythropoietin-stimulating drugs

This method seems preventive because it is overly cautious, causing the rise of *false positive* patients, e.g. those would not require a transfusion, e.g. the drugs administered would be an unnecessary cost. This situation can be visualized in *figure 2.*

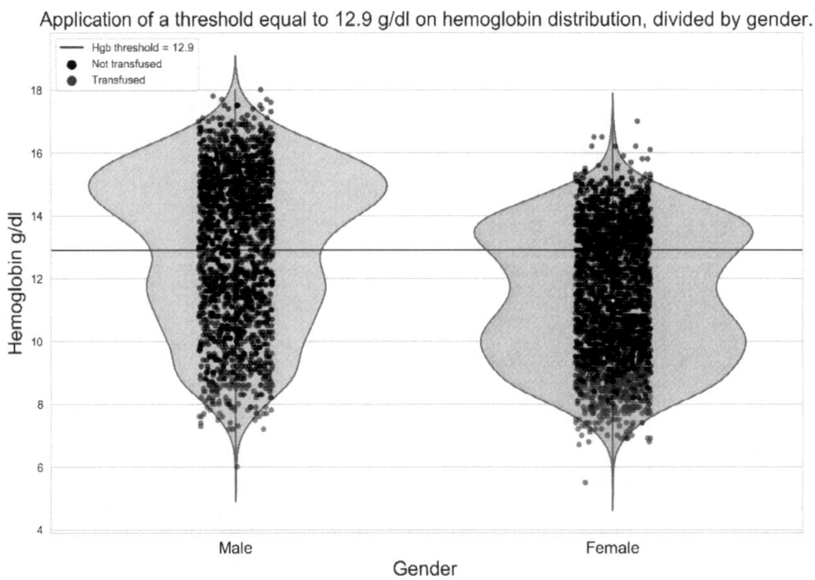

Figure 2. Violin plots showing the distribution of hemoglobin, divided by gender. The red horizontal line denotes the current preemptively method: observations below this threshold are predicted as transfusion requiring and those above as non-transfusion requiring. Points marked in red are patients who have been transfused, while black points are patients who have not received a transfusion. A large portion of patients are classified by the simple existing method as likely to be transfused, even if they will not require it

Our research activity has two goals: the first one concerns if an improvement over the heuristic method based on hemoglobin thresholds is possible by performing a sensitivity analysis on the threshold, using a data-driven approach, also keeping the method as simple as possible. Secondly, we propose a blood-transfusion machine learning classification system to decrease the costs attributable to blood bags.

2. METHODS

2.1 Dataset Description

The observational data was made available for this study by *Galeazzi Orthopedic Institute*, reporting the medical information and anagraphical personal data about 11814 hospitalizations (4593 patients) from January 2019 to May 2019, in particular: patient's age, sex, hemoglobin levels (g/dl), International Classification of Diseases (ICD9) code representing the patient's disorder, the urgency of the operation and the information about whether the transfusion occurred or not. An extended description of the features used for modelling is shown in *table 1*.

Table 1. Features of the dataset provided by IOG

Features	Description	Data Type
Age	Age at operation date	Numerical
Sex	Patient gender	Categorical
Hemoglobin (HGB)	HGB (g/dL)	Numerical
ICD9	Operation code	Categorical
Urgency	Urgent operation	Categorical
Transfused	Transfusion received	Categorical

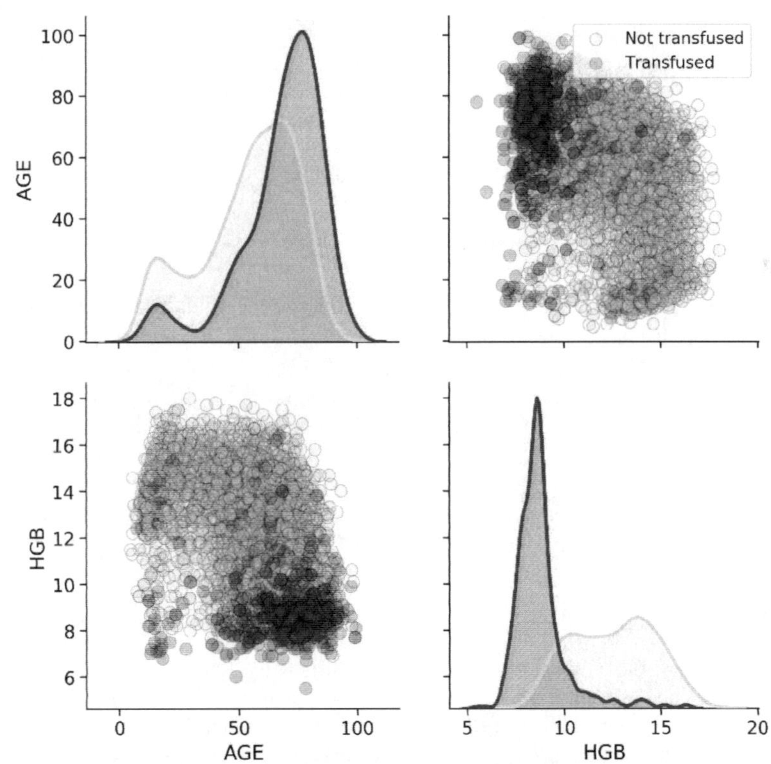

Figure 3. Visualization of the distribution of age and HGB levels in IOG's patients. Color of traces indicates whether the patient was transfused or not, as indicated in the figure's legend

Figure 3 shows the pairwise distribution plots and scatterplots of "*Age*" and "*HGB*" variables. Plots denote a negative correlation between "*Age*" and "*HGB*" (Pearson correlation: $r = -0.43$, p-value < 0.001). In addiction, transfused patients show an "*HGB*" distribution with lower mean and variance than non-transfused patients.

The outcome of our analysis is the prediction of whether a hospitalization will require a transfusion (or more) or not during the surgery.

2.2 Data Manipulation

The dataset features several samples with the hemoglobin values detected after the transfusion (n=7530, 64%). This information was made available by the metadata coming with the case-study dataset, which reports the date of surgery and the date of the hemoglobin value detection. To avoid a biased model training, those samples were removed since this work aims to predict possible transfusion, based only on preoperative data.

Missing data was imputed using K-Nearest Neighbor (KNN) imputation (Zhang, 2012), with K=1, meaning the imputation value is taken from the closest patient. The resulting dataset presents 4284 hospitalizations. The categorical variable ICD9, as it contains multiple categories, has been represented using *one-hot encoding* (Ramaravind et al. 2020).

The target feature, *Transfused*, presents a class imbalance, with 499 positive cases (11.6%) and 3785 negative cases (88.4%). To deal with this issue, the division of the dataset into training and validation sets was performed in a stratified manner and a *SMOTE* (Chawla et al., 2002) was applied in order to balance the classes. *Smote* is a type of data augmentation, which consists generating new minority class instances conditioned on the distribution of known instances.

2.3 Sensitivity Analysis

Firstly, we would like to perform a sensitivity analysis in order to optimize the existing method, and to find an optimal threshold that decreases the operation costs. After an analysis performed at IOG, we found that the cost of a False Negative, meaning a patient is classified as not requiring a transfusion when in reality he will, is about six times the cost of a False Positive (FP), meaning when a patient is given the drugs in order to decrease the transfusion risk but he really does not require it. We show the potential cost of misclassification of each possible HGB threshold, and then choose the best overall threshold in terms of cost savings.

2.4 Machine Learning Classifier

In order to create the classifier model, we tested a set of machine learning techniques:

- *heuristic* approaches, e.g. *Decision Tree* (Safavian, 1991)*, Random Forest* (Kam, 1995) and *K-nearest neighbors* (Altman, 1992)
- *Regression based* approach, e.g. *Logistic regression* (Hosmer et al 2013),
- *Separative* approach, e.g. *Support vector machine* (Schölkopf et al. 2002)

10-fold Nested cross-validation has been used to provide an unbiased estimation of model performances. In a nested cross-validation, an inner loop chooses the best set of hyperparameters (chosen from the most common ones used for each model) using a set grid search, and the outer loop evaluates the model performance.

2.5 Evaluation of Methods

Due to the imbalance of the *Transfused* target feature, the evaluation metrics were chosen as the precision, recall, *f1-score*, which is the harmonic mean of precision and recall, and Area Under the Curve (AUC). However, what is most important to us (after, of course, patient's health), is the overall decrease in costs in comparison to the existing baseline. This entails, since the cost of FN is six-fold of FP, that in terms of classical evaluation measures the *recall* is the most important amongst those.

3. RESULTS

3.1 Sensitivity Analysis Results

We have performed the sensitivity analysis as described in the "Methods" section, and the results are shown in *Figure 4*, which visualizes the behavior of the adopted metrics by varying the hemoglobin threshold. The normalized cost is also reported, representing the cost in the [0, 1] interval, in order to provide a better graphical comparison with the other performance measures considered. The costs normalization is given by:

$$cost_{std} = (cost - cost_{min}) / (cost_{max} - cost_{min}),$$
$$cost_{normalized} = cost_{std} * (cost_{max} - cost_{min}) + cost_{min}.$$

We can see that for each HGB threshold, the various metrics change correspondingly. In particular, the threshold that minimizes the potential costs is 9.4 g/Dl, and for this threshold the precision is 0.56 and the recall is 0.86. The overall cost for each patient is equal to 18.43 euros.

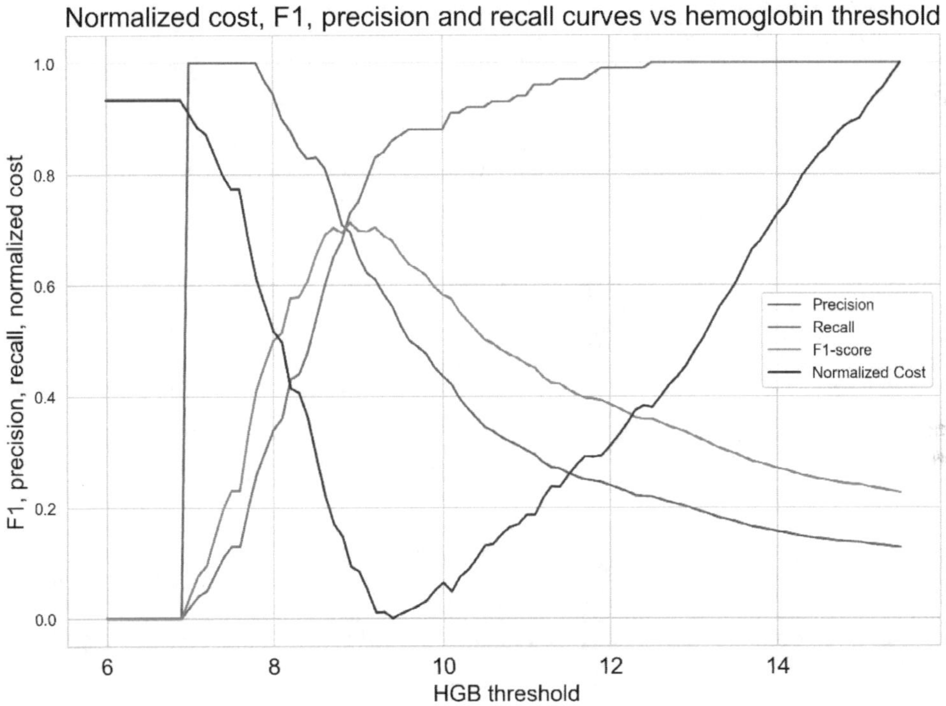

Figure 4. Metrics and costs variation in comparison to the changing HGB threshold. It is shown that the threshold with the lowest cost (red line) is 9.4 g/dL HGB

3.2 Evaluation of Costs from Machine Learning Models

Table 2 shows the machine learning classification models performances, in form of 95% confidence interval for each computed metric. The best overall performing model is the Random Forest: in terms of both F1 score and costs. The Support Vector machine also achieves high AUC score; however, our choice is still the Random Forest.

Table 2. Models performances. Lower and upper bounders of 95% confidence interval are shown. CIs are computed on 10-folds results

	Precision	Recall score	F1-score	AUC	Cost per patient (average)
Decision Tree	[0.68, 0.85]	[0.58, 0.72]	[0.57, 0.70]	[0.91, 0.93]	29.81 euros
K-nearest neighbor	[0.70, 0.78]	[0.59, 0.66]	[0.63, 0.7]	[0.88, 0.92]	25.61 euros
Logistic regression	**[0.87, 0.97]**	[0.57, 0.62]	[0.66, 0.71]	[0.92, 0.95]	21.76 euros
Random Forest	[0.75, 0.84]	**[0.64, 0.7]**	**[0.69, 0.76]**	**[0.93, 0.95]**	**18.26 euros**
SVM	[0.78, 0.86]	[0.60, 0.65]	[0.68, 0.72]	**[0.93, 0.95]**	20.77 euros
Baseline method	[0.17, 0.21]	[0.95, 0.99]	[0.29, 0.35]	[0.70, 0.72]	46.32 euros
Proposed method	[0.51, 0.61]	[0.81, 0.87]	[0.63, 0.71]	[0.86, 0.9]	**18.43 euros**

The best performing model, i.e. the Random Forest, has also been trained with the unbalanced dataset (without applying SMOTE) to test the effectiveness of the rebalancing method. The nested-CV showed the following 95% confidence intervals: [0.7, 0.82] for *precision*, [0.63, 0.68] for *recall*, [0.62, 0.75] for *f1-score* and [0.92, 0.94] for *AUC*. The rebalancing method led to slightly higher performances even if there are no statistically significant differences.

A graphic overview of the models results coming from the *nested-CV*, presented by *table 2*, is given in *figure 5*, that shows the distribution of each performance measure over ten folds. Since *nested-CV* also performs the hyperparameters search, the models are tested with their best hyperparameters.

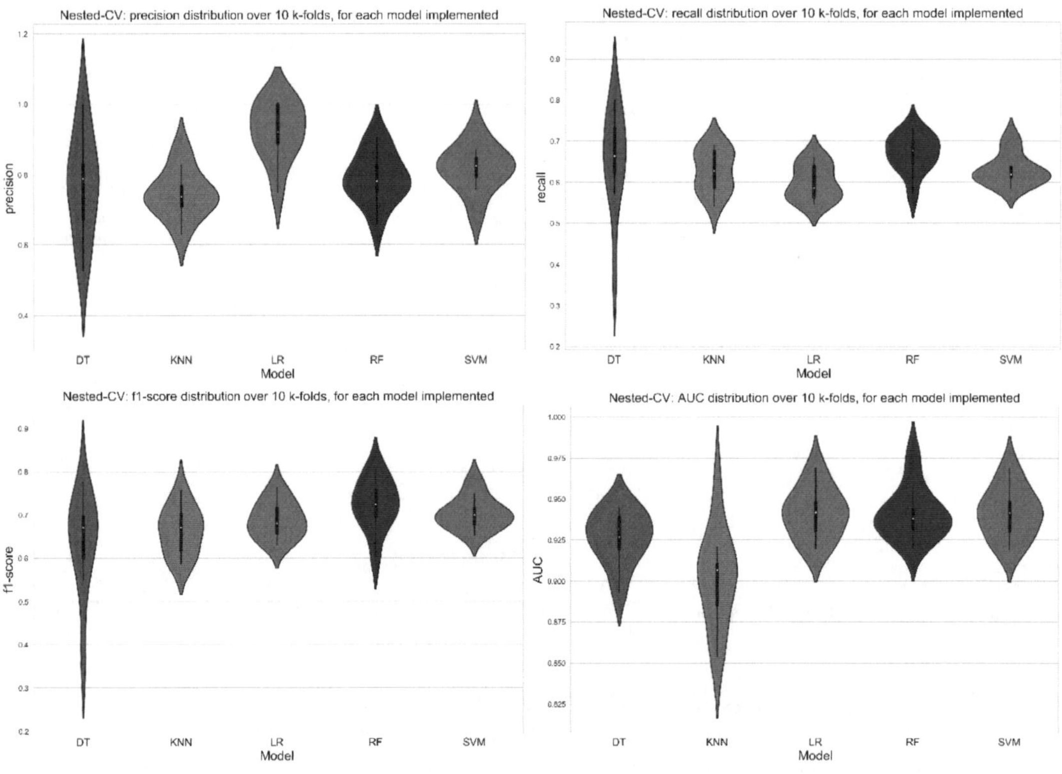

Figure 5. Violin plots representing the distribution of model performances over ten folds

4. DISCUSSION AND CONCLUSION

Our methods show improvement in terms of cost savings in our setting.

In particular, the sensitivity analysis method allowed us to choose an optimal threshold, that is, a data driven decision rather than one that comes from experience. Clearly, this threshold must be validated in terms of clinical expertise, and we would like to validate our results in the future considering the medical knowledge of IOG's experienced staff. The threshold method can be interpreted in an intuitive way and it can also be used in an "offline setting".

On the other hand, the machine learning techniques are more of a black box model approach, so they cannot be used by just remembering one simple threshold. Instead, they require a computer interface that considers a multitude of variables. This second approach is different from the former and requires the use of an electronical device (such as, a computer or a tablet) for entry point to the models. However, the cost savings of a more complex model are higher than the simpler methods, and as such, one should think about whether the explainability and simplicity are more important to them rather than the increased cost savings.

Overall, we proposed two different methodologies for decreasing the costs of patient blood management. We explained each method's advantages and weaknesses, and both could lead to approximately 25€ of savings per patient (roughly 54% of initial cost per patient).

We would like to discover in the future which one is more appealing in a real-world setting, and our future work will build upon this foundation to propose a method that is applicable to real medical cases and to assess its performance and whether there is a real demand for it from the clinical staff or not.

ACKNOWLEDGEMENT

The authors are grateful to Alberto Cappelli from Healthcare Control Systems, Inc. (HCS), and to Francesco Magro, Chief Operating Officer of the IOG, for providing the dataset and useful information to conduct this research.

REFERENCES

Altman NS, 1992 "An introduction to kernel and nearest-neighbor nonparametric regression." The American Statistician 46(3):175-185

Chawla, Nitesh V., et al., 2002. "SMOTE: synthetic minority over-sampling technique." *Journal of artificial intelligence research* 16: 321-357.

Hayn, Dieter, et al., 2017. "Development of multivariable models to predict and benchmark transfusion in elective surgery supporting patient blood management." *Applied clinical informatics* 8.02 (2017): 617-631.

Hosmer Jr DW, Lemeshow S, Sturdivant RX (2013) "Applied logistic regression", vol 398. John Wiley & Sons

Kam HT, 1995. "Random Decision Forests." In: Proceedings of the 3rd International Conference on Document Analysis and Recognition, Montreal, Canada, August, vol 1416, p 278282

Ramaravind K. Mothilal, Amit Sharma, and Chenhao Tan., 2020. "Explaining machine learning classifiers through diverse counterfactual explanations". In *Proceedings of the 2020 Conference on Fairness, Accountability, and Transparency*: 607-617

Sadana, Divyajot, et al., 2000. "Promoting high-value practice by reducing unnecessary transfusions with a patient blood management program." *JAMA internal medicine* 178.1: 116-122.

Safavian SR, Landgrebe D, 1991. "A survey of decision tree classifier methodology". IEE transactions on system, man, and cybernetics 21(3):660-674

Schölkopf B, Smola AJ, Bach F, et al., 2002. "Learning with kernels: support vector machines, regularization, optimization and beyond." MIT press

Zhang, S., 2012. Nearest neighbor selection for iteratively kNN imputation. *Journal of Systems and Software*, *85*(11), 2541-2552.

SHORT PAPERS

ONCONET - A DIGITAL SERVICE FOR REMOTE MONITORING AND SUPPORT OF CANCER PATIENTS ON CHEMOTHERAPY

Irina G. Kargalskaja[1], Sergei A. Shinkarev[2], Boris V. Zingerman[3], Inna A. Fistul[3], Arkadiy V. Nozik[3] and Sergei B. Arseniev[4]

[1]*Committee "Patient Oriented Telemedicine", 109390, Moscow, Yunih Lenintsev 12, Russian Federation*
[2]*Lipetsk Oncological Hospital,* 398005, Lipetsk, *Admiral Makarov, 1-e, Russian Federation*
[3]*TelePat LLC, 115533 Moscow, Andropov Avenue 22, Nagatinsky Business Center, Russian Federation*
[4]*Clinical and Research Institute for Emergency Pediatric Surgery and Trauma (CRIEPST)*
119180, Moscow, Bolshaya Polyanka 22, Russian Federation

ABSTRACT

Background: Cancer patients suffer from many complications caused by both the disease itself and the side effects of anticancer treatment. Many symptoms go unnoticed by doctors because of the patient's remote location, significant interruptions between face-to-face visits to the doctor and differences in the individual course of the disease.In a pandemic caused by Covid 19, cancer patients are at great risk. The practical impossibility of patient's visit to a doctor, conduct clinical trials determines the highest risk for this group of patients. Thus the relevance of remote support for these patients in isolation increases many times. Objective: The goal was to create and implement in a pilot project a digital service (ONCONET system) for remote monitoring of cancer patients outside the hospital as well as adaptation to COVID 19 pandemic. Methods: The developed platform is a cloud service hosted in a data processing center which provides healthcare organizations with a ready-made service. The system is based on specialized questionnaires developed by oncologists for remote assessment of the patient's condition. The important component of the system are content libraries for patients and doctors, information and reference materials as the collection of video-schools from leading experts of nosology, rules of life, nutrition, rehabilitation and care. Results: To evaluate the effectiveness of the ONCONET system, 2 groups of patients were formed: ONCONET group - 206 patients and retrospective control group - 209 patients. Statistical analysis revealed significant differences between groups as a whole, as well as between individual factors of complications. Conclusion: The study demonstrates the usefulness and high medical efficiency of the ONCONET system for remote monitoring of cancer patients, especially against the background of a pandemic and the necessary self-isolation of cancer patients.

KEYWORDS

Remote Monitoring, Cancer Patients, Telemedicine, Tele-Rehabilitation, COVID 19

1. INTRODUCTION

Cancer patients suffer from many complications caused by both the disease itself and the side effects of anticancer treatment. Many symptoms go unnoticed by doctors because of the patient's remote location, significant interruptions between face-to-face visits to the doctor and differences in the individual course of the disease. Obviously it is the result of inadequate communication between a doctor and a patient. (Reilly et al., 2013). In general, worsening conditions, missed increase in symptoms indicate cancer progression or serious adverse events and cause a decrease in quality of life and survival rates (Trajkovic-Vidakovic et al., 2012).

Recently, in the United States and Europe, digital tools have been increasingly used, involving patients in collecting additional information about their condition and transmitting it electronically to the attending physician, which allows for timely and continuous collection and analysis of automatic data on the well-being of a patient outside medical facility (Jensen et al., 2014).

If such a collection of typical symptoms is also accompanied by the possibility of automatic emergency notifications ("alarms") sent in dangerous situations to the patient and his doctor, this gives the patient a chance to respond to the problem in time, contact the doctor and reduce the possibility of developing urgent conditions or on time call an ambulance. Currently, there is convincing evidence of the effectiveness of such monitoring (Basch et al., 2017).

In a pandemic caused by Covid 19, cancer patients are at great risk. The practical impossibility of patient's visit to a doctor, conduct clinical determines the highest risk for this group of patients. Thus the relevance of remote support for these patients in isolation increases many times.

The aim of this work was to develop the ONCONET digital service for remote monitoring of cancer patients outside the hospital, as well as its pilot implementation and adaptation to the conditions of the COVID 19 pandemic.

2. MATERIALS AND METHODS

For remote monitoring of patients at various stages of cancer treatment, a specialized ONCONET platform developed by Telepat LLC was used (https://telepat.online/).

2.1 ONCONET Technology Platform

The ONCONET platform is a cloud service hosted in a data processing center and provides healthcare organizations with a ready-made service. A medical provider is registered on the platform, receives a management cabinet through which the organization registers its doctors, and each doctor then in his personal account connects his patients to monitoring process.

Technically, the platform consists of the content management module, an interactive services module, a questionnaire management and analytical modules and administrative control module.

The content management module of the platform consists of the doctor's and patient's libraries. The doctor's library includes reference, methodological and other materials useful for doctors. The patient's library contains 1,250 pages of patient-oriented content and more than 230 videos of patient schools from leading experts of nosology, rules of life, nutrition, rehabilitation and care.

The interactive service module consists of personal accounts of doctors and patients, through which they interact with the ONCONET system.

The questionnaire management module provides a centralized mechanism for electronic distribution of questionnaires as well as a special tool for designing questionnaires, with which you can quickly create new questionnaires and new monitoring schemes offered by oncologists.

One of the most important element of the system is the analytical module, which ensures the construction of so-called "heat maps" of the patient's condition and ranking patients monitored by a doctor according to the severity and urgency of the necessary attention."Heat map" is a clear color-graphic tool for a quick assessment by the doctor of the severity and dynamics of the patient's condition "at one glance". Answering the questionnaire, the patient chooses the color of the answer (from green to red), depending on the degree of a particularsymptom. All patient's responses are displayed in a table with the observation dates vertically and the controlled symptoms horizontally. Each cell indicates the "color" of the condition selected by the patient when answering the question or selected patient's score.

2.2 Questionnaires

The ONCONET system is based on specialized questionnaires developed by oncologists for remote assessment of the patient's condition. One of the main problems in compiling the questionnaires was the conversion of assessments into a language that was understandable to the patient. Any criteria that the patient cannot communicate were excluded from the questionnaires. When developing questionnaires, representatives of patient communities were involved to assess their comprehensibility. Clarity of questionnaires was tested on large groups of patients.

It is important to note that with the onset of the COVID 19 pandemic and the particular vulnerability of patients undergoing chemotherapy, a COVID 19 symptom questionnaire was added to the system. By a special mailing list, the doctors were notified of the possibility of connecting the COVID 19 questionnaire to their patients. Groups of questionnaires and number of questions are presented in Table 1.

Table 1. Groups of questionnairesand number of questions

Symptoms/Topic	Number of questions
Covid 19	11
Drug therapy	109
Colorectal cancer	51
Additional questionnaires	51
Mental health monitoring	56
Monitoring of surgical complication	198
Total	476

There are 2 possible variants of answers: descriptive response options, ranking from normal (marked in green) to critical (marked in red) and as a score from 0 to 10 (selected by the patient using a slider). By default, all response values are set to normal (green).However, if the patient felt problems due to any of the symptoms, then he could open this question and indicate the value that most closely matches his condition. This approach was chosen to exclude self "overdiagnosis" by patients.

2.3 ONCONET Functional Description

The attending physician registers the patient in the remote monitoring system, assigning him a set of specialized questionnaires and the frequency (calendar) of sending them to him. The patient receives an e-mail link to the questionnaire in accordance with the schedule set by the doctor, and answers questions in it. The questionnaires are adapted to mobile phones and are patient-oriented, accessible to the patient regardless to their level of education).

After completing the entire questionnaire, the patient receives a set of recommendations in the form of links to information materials or video schools posted on the same ONCONET portal. These materials contain tips on care and lifestyle management to reduce the impact of non-critical symptoms reported by the patient. If the patient indicated a dangerous or critical symptom, he will immediately receive a notification “Call an ambulance immediately!” or "Contact your doctor Immediately!".

The system also allows the patient to attach additional documents for example, analyses performed in another laboratory, and write a message to the doctor describing a symptom or problem not included in the questionnaire. In addition to the questionnaires, the doctor can also assign the patient one or more times a day to monitor physiological parameters (pressure, temperature, glucose, weight), as well as reminders about taking medications, with confirmation of reception.

The results of the comprehensive monitoring are sent to the remote cloud server for automatic processing and then sent to the doctor in the form of a so-called "heatmap" showing the dynamics of the patient's symptoms. Periodically (according to the established schedule), the doctor also receives a summary of all his patients who are on remote monitoring. The summary is ranked by the overall complex severity of the patient's symptoms, so that the doctor can more quickly assess the condition of the most severe patients and take appropriate measures.

If necessary, the doctor can write an additional recommendation to the patient directly in the ONCONET system, invite them to a face-to-face appointment, or send a link to additional material with recommendations.

2.4 Clinical Study

A pilot project for the implementation of the ONCONET system was conducted in 2018-2019 in the Lipetsk region, in the Regional Oncology Hospital. The project involved 61 doctors and 206 patients on remote monitoring. Patients with breast cancer, lung cancer, cervical and ovarian cancer, prostate, thyroid, kidney, stomach and colorectal cancer and Hodgkin's lymphoma were included in the study.

To evaluate the effectiveness of the ONCONET system, 2 groups of patients were formed: the ONCONET group - 206 patients and theretrospective control group - 209 patients. The groups did not differ statistically in terms of diagnoses and stages of diseases, treatment protocols used, gender, and age.

The groups were compared by the frequency of complications and adverse reactions during treatment, as well as the most important indicator - the transfer or postponement of the next course of chemotherapy.

Chi-square values, two-sided, statistical difference between frequencies were calculated. A statistically significant difference was considered at $p < 0.05$. Statistical analysis was performed using the STATISTICA 8 package.

3. RESULTS AND DISCUSSION

The results of comparing the groups are presented in Table 2.

Table 2

Complications/ Chemotherapy date change	ONCONET group %	Reference group %	Significance
Hematological	32.5	41.6	p=0.0227 (<0.05) YES
Hepatotoxic	29.1	23.0	p=0.1022 (>0.05) NO
Nephrotoxic	3.4	13.4	p=0.001 (<0.05) YES
Total	65	78	
Chemotherapy date change	7.8	12.9	p=0.0177 (,0.05) YES

The total number of the three most significant complications decreased by 13% (65 vs 78), hematological complications significantly decreased by 9% (32.5 vs 41.6), and the number of nephrotoxic complications decreased by 11% (3.4 vs 13.4).

There were no significant differences between the groups in hepatotoxic complications. Moreover, the percentage of these complications was higher in the ONCONET group (29% vs 23%). This phenomenon is probably associated with a more sensitive diagnosis of hepatotoxic abnormalities in the ONCONET system, due to adequate questionnaires and the methodology of their application for monitoring.

Among other less quantitative complications, it should be noted a 3-times reduction in gastrointestinal disorders (0.7% vs. 2.6%), neurotoxic and skin manifestations are practically minimized.

One of the most important results is a statistically significant decrease, almost one and a half times, of the postponement of the date of the next course of chemotherapy (7.9% vs 11.9%) in the ONCONET group. As a result, this made it possible to maintain the necessary intervals between courses of chemotherapy, and as a result, improve survival rates.

In total, the ONCONET system was tested during 2018-2020 in 22 medical organizations in 10 regions of Russia. The system was used by 174 doctors and 382 patients. Initially, 460 patients were connected to the system but only 382 (92%) out of them actually participated in the monitoring. This indicates a high level of demand for the system by patients. A total of 21936 questionnaires were completed by patients at the time of publication (an average of 57 per patient) and 193248 symptoms were reported (more than 500 on average per patient).

The use of heatmaps and algorithms for ranking patients according to the severity of the condition enable the doctor to quickly respond to negative dynamics and thereby increase patient safety. In the current version of the ONCONET system, estimation algorithms based on weight coefficients proposed by oncologists are used, however, further development of these algorithms based on machine learning is also planned in the future.

Doctor's and a patient's libraries are of independent value. These informational and reference materials and video schools are constantly being developed and can be used not only within the ONCONET system, but also in any other means of content delivery.

Patient questionnaires are also an essential part of the system. Not only oncologists, but also psychologists and linguists participated in the development of the questionnaires, whose task was to make the questionnaires simple and understandable for a wide range of patients, while preserving their diagnostic value.

The addition of the COVID 19 block was due to the corona virus pandemic. Undoubtedly, this block remains relevant and in demand even after the end of the pandemic. The COVID 19 questionnaire, when further developed, can serve as an independent tool for remote rehabilitation of patients who have patients who have undergone the infection.

4. CONCLUSION

This study demonstrates the unconditional usefulness and high medical efficiency of remote monitoring of cancer patients, especially against the background of a pandemic and the necessary self-isolation of cancer patients. Remote monitoring and doctor support is in demand by cancer patients. This significantly improves the quality of life, satisfaction with the treatment, loyalty and adherence to the treatment, maintaining the right lifestyle recommended by the doctor, which also contributes to the increase in the effectiveness of rehabilitation.

It is extremely important to make the provision of remote monitoring services to cancer patients free of charge. This requires the full support of the state and local authorities, as well as public and non-profit organizations. It is also necessary to encourage oncologists and medical organizations to widely implement this service.

The developed ONCONET system is a universal platform, which, through appropriate content filling, can be adapted for other areas of distance rehabilitation. Currently, work is underway to create a system for remote monitoring of children with the consequences of a craniocerebral and spinal injury "Kids-rehab".

ACKNOWLEDGEMENT

The authors are deeply grateful to the oncologists of the Lipetsk region, as well as to all cancer patients who participated in the ONCONET pilot project.

REFERENCES

Basch E. et al, 2017, Overall survival results of a trial assessing patient-reported outcomes for symptom monitoring during routine cancer treatment, *JAMA,* Vol. 318, No. 2, pp 197–198.

Jensen R.E. et al, 2014, Review of electronic patient-reported outcomes systems used in cancer clinical care, *J Oncol Pract* Vol. 10, No 4, pp e215–e222.

Reilly C. M. et al, 2013 A literature synthesis of symptom prevalence and severity in persons receiving active cancer treatment, *Support Care Cancer,* Vol. 21, No. 6, pp1525–1550.

Trajkovic-Vidakovic M. et al, 2012, Symptoms tell it all: a systematic review of the value of symptom assessment to predict survival in advanced cancer patients, *Crit Rev Oncol Hematol*, Vol. 84, No 1, pp130–148.

https://telepat.online/

DIGITAL SUPPORT FOR A HUB-AND-SPOKE DISEASE NETWORK

Paolo Locatelli[1], Federica Cirilli[1], Eliana Panno[1], Salvatore Provenzano[2], Roberta Sanfilippo[2] and Paolo G. Casali[2-3]
[1]*Fondazione Politecnico di Milano, Italy*
[2]*Medical Oncology Unit 2, Fondazione IRCCS Istituto Nazionale dei Tumori di Milano, Italy*
[3]*Department of Oncology and Haemato-Oncology, University of Milan, Italy*

ABSTRACT

A disease network is a collaborative healthcare network focusing on a single disease, or a group of diseases, whether professional or formalized. A disease network may follow a "hub and spoke" model. Providing these networks with an efficient digital support poses some challenges. This paper describes the methodological approach to design the information system supporting such a disease network. After identifying a general model with network features, a use case related to the Italian Rare Cancer Network is reported.

KEYWORDS

Rare Cancer, Collaborative Networks, Digital Healthcare, Teleconsultation

1. INTRODUCTION

A collaborative healthcare network is an alliance of professionals and/or institutions to achieve common or compatible goals by sharing their expertise and/or resources. A network may behave as a professional network, being a flexible alliance of health professionals, or may behave as a formal network of healthcare institutions. In general, the goal is to provide patients with fair and equal opportunities to access healthcare. Health networks often focus on a single disease, or a group of diseases (Morando, 2014; Camarinha-Matos et al., 2015). Networks must be served by an information system and thus require a strong digital support (Camarinha-Matos et al., 2008)0.

An area in which the effectiveness and the efficiency of health interventions can be substantially improved by networking has to do with rare diseases, including rare cancers. Rare cancers are defined by an incidence of $\leq$6/100,000/year. They include 10 families of rare adult solid cancers, all childhood cancers and a group of haematological neoplasms. Overall, they represent more than 20% of all new cancer cases. Networking is largely seen as a key factor to improve effectiveness, limit health migration and reduce costs (both health and social costs). In the European Union (EU), the *European Reference Networks* (ERNs) have the task of tackling complex and rare diseases. They are mainly peer-to-peer networks. However, they should cooperate with national or regional networks, generally conceived as hub-and-spoke networks. One such network is the Italian Network on Rare Cancers, "Rete Nazionale Tumori Rari" (RNTR) (Casali, 2019).

The project of RNTR was approved in 2017 by the Italian government and regions, building on professional networks developed in the previous decades in Italy in the areas of rare adult solid cancers, childhood cancers and haematological neoplasms. Thus, the network is conceived as a formalized networks of healthcare institutions, whether hubs or spokes, supported by three professional networks, each covering one of those main groups of rare cancers. The network should help rationalize patient referral from throughout Italy towards its centers. Then, each professional network should coordinate a system of teleconsultations by which hubs share their expertise and resources with spokes (being thus named, respectively, "Providers" and "Users") (Locatelli et al., 2019; Elrod & Fortenberry, 2017).

This paper is meant to describe first the methodological approach to design the information system supporting such a disease network and secondly the general requirements of the information system identified following this approach. Then, starting from this general model, functional features which are more specific of this Italian disease network in the area of rare cancers are defined.

In literature, other similar works on digital support for collaborative networks have been reported (Costa et al., 2009; Noran, 2009); however, they focused on network decision support systems, not on systems supporting the real network collaboration. Other works focused on describing in detail methodologies and architectural models for planning, designing and implementing generic health information systems (Buranarach et al., 2011; Botheju, 2018). Moreover, in other papers (Thimm & Rasmussen, 2010; Lenz & Reichert, 2005), a tool for a collaborative network has been designed; however, a clear and general methodological approach on how the technical solution was achieved is not defined.

2. THE METHODOLOGICAL APPROACH AND THE APPLICATION MODEL

According to the Business Process Reengineering (BPR) methodology (Hammer & Champy, 1993)0, a structured approach is described in Figure 1.

AS-IS process and digital support (if present) analysis. The first step consists in the analysis of the processes and, if present, the digital tools through which hubs and spokes collaborate. The team dealing with the digital support design needs to learn in detail the collaboration process.

Processes mapping, with related digital support (if present). The second step is mapping the collaboration process between hubs and spokes with the related digital tools, if present. This step allows the team to catch a rough summary of its actual functioning.

Identification of critical points and possible improvements. The third step consists in identifying weak points and critical issues within the current process, in order to define organizational, as well as technological, improvements.

Definition of the network needs. This step is aimed to define the network needs, including digital support ones, according to the process mapping made in the second step and to the weak points within the process identified in the previous step. The needs have to be defined in terms of organizational improvements and functionalities of the digital solution, in order to effectively support the collaboration within the disease network.

Design of the digital support. The final step is aimed to design the system specific functionalities with the technological provider in charge of the new digital support development. It is important to gain an alignment with the digital support needs and the developed functionalities.

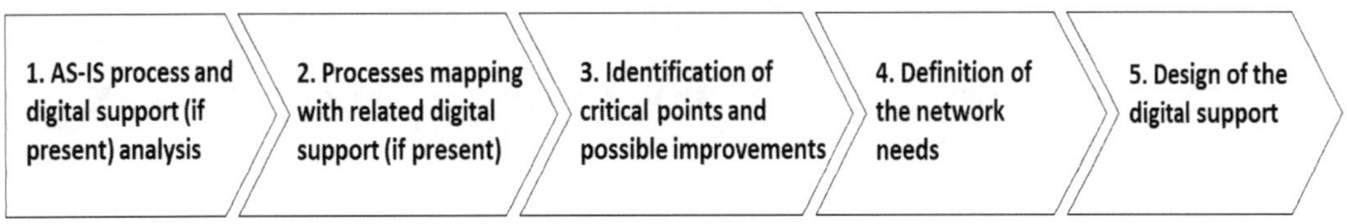

Figure 1. Methodological approach to design the digital support for a disease network

Following this approach, it is possible to generalize some aspects to any healthcare disease network.

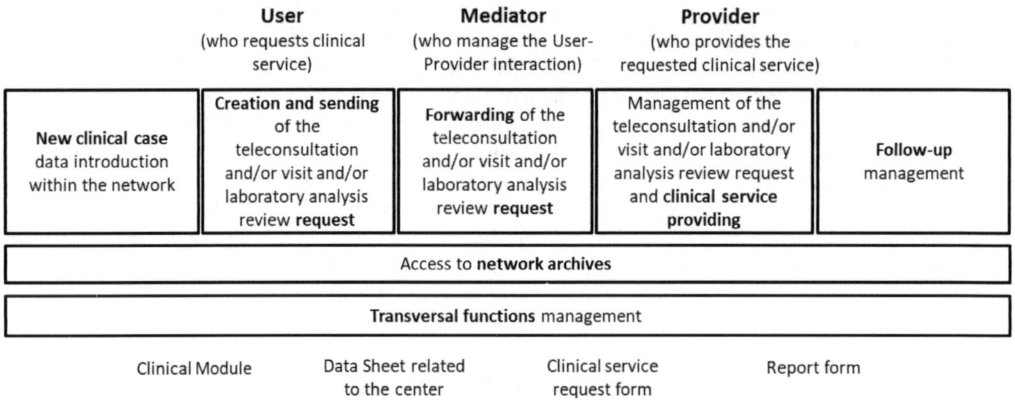

Figure 2. Schematic representation of the disease network features

Figure 2 shows a schematic representation of the disease network features, in terms of network users, system macro-functionalities and digital available forms.

Firstly, an information system supporting a disease network should ensure a User to submit a new clinical case, through the Clinical Module filling. Then, the User should be able to create and send a teleconsultation and/or visit request, using the appropriate clinical service request form. Moreover, he should be responsible for the follow-up by updating patient data within the Clinical Module. The User should also be authorized to access to the network archives, to search clinical cases for care purposes and to generate the teleconsultation report in a pdf format, possibly to be uploaded onto an online patient record. The system should allow the Mediator to manage the User-Provider interactions by automatically forwarding the clinical service request form, with the teleconsultation and/or visit request, to an adequate Provider within the network. On the other hand, the digital solution should support the Provider in managing the Data Sheet related to the centre availability and specialization, in visualizing the teleconsultation and/or visit and/or laboratory analysis request - with the possibility to request more details/further clarifications - and in providing the final report. Finally, more general and transversal system functionalities are essential to check the compliance with clinical practice guidelines, to assess the network efficiency through quality questionnaires for clinicians/patients and periodic monitoring of indicators.

Within an ongoing project, funded by the Italian Ministry of Health, "Italian Rare Cancer Network – Process Monitoring and System Impact Assessment", the methodological approach above-described was applied to design the new digital support for RNTR. The specific results are summarized in the next paragraph.

3. DESIGNING A DIGITAL SUPPORT FOR THE ITALIAN RARE CANCER NETWORK

Following the methodological approach, an analysis of the current management processes of patients in the disease network, building on the ongoing experience of the professional network in place as from 1997, was carried out. When the analysis was performed, the professional network already used a digital support. According to the methodological approach described above, we analysed this system, obsolete though it was, and we mapped the analysed processes. Then, according to the third and the fourth steps of the approach, we identified weak points and possible organizational/digital improvements and defined the network needs.

The Italian Rare Cancer Network includes different clinical centres with two main roles: Providers, who are the experts in a given rare cancer; Users, who request clinical services (e.g. teleconsultations) to the Providers; Mediators, who manage Provider-User interactions.

First, the patient is seen by the nearest local cancer centre. If this is a Provider, the patient will be managed at the centre, his/her case will be recorded in the information system, while only exceptionally will his/her case be shared with another Provider (e.g. the Provider does not own all clinical facilities needed to treat the patient). If the centre is a **User**, the oncologist can decide to **request a teleconsultation, including a**

pathologic review, and/or a visit. In this phase, the digital support should allow the User to introduce and manage new clinical case data, through: automatic creation of a unique identification code, associated with the patient's name and surname; patient's signed consent archiving; patient's personal data entry; easy data entry of clinical information; reports and images loading, with data for the biological sample traceability. Moreover, the digital system should allow the User to create and send a teleconsultation request, filling the: teleconsultation request form, with the diagnostic question object, or the histological review request form.

Then, the **Mediator** makes sure that an appropriate Provider is involved and **periodically checks the timeliness of interactions**. A big challenge is to make such a function as much automatic as possible. For this reason, the system should allow the Mediator to manage and forward the teleconsultation and/or visit request, through automatic assignment of the case to the Provider and automatic monitoring of the User-Provider interaction time. Moreover, the system should send periodic alerts to the Provider.

Once the **Provider** is in charge of the patient, it can ask for further clinical information (e.g. clarifications on patient history, exams, etc.). It is very important to make sure that this interaction process between Users and Providers is allowed by the information system to proceed as swiftly as possible. In this context, the digital support should ensure the Provider to **manage the received teleconsultation request and provide the service**, through: visualization of the teleconsultation request and sending the reply ('Request accepted') to the Provider; request form filling asking for more details/further clarifications; visualization of the clinical data of the patient, with the possibility to integrate it (the system should guarantee the changes traceability); reports and images uploading (in the case of visit providing); teleconsultation report filling and sending, with notification of the adherence between the therapeutic path and the updated clinical guidelines.

Once received the Provider consultation, the system should guarantee the **User** to **manage the follow-up** through: visualization of the Provider's teleconsultation report; patient's clinical outcome data updating; generation of the teleconsultation report in the pdf format.

In general, both User and Provider should be able to access to the network archives searching for anonymized cases (in case of external patients) or searching for specific by using identification code or patient's name-surname (in case of internal patients). Finally, the system should allow the management of other functionalities, such as: visualization of updated clinical guidelines; questionnaires filling for clinicians or patients and some indicators monitoring, in order to verify the network efficiency and impacts.

In defining the solution development, the compliance with non-functional requirements, such as data accessibility, accuracy, availability and privacy, must be guaranteed.

We shared these requirements with the technological provider, and we are currently dealing with the final step of the methodological approach. Once completed this step and when the new digital support is developed, the project will test the system and a first set of data on the application will be available at the end of the project (September 2021).

4. CONCLUSIONS

In this paper we described first the methodological approach to designing the information system supporting such a hub-and-spoke disease network and secondly the general requirements of the information system identified following this approach. Then, starting from this general model, functional features of the Italian RNTR digital support are defined.

In order to assess the impacts of applying the network model and its new digital system, an appropriateness indicators panel has to be defined (Locatelli et al., 2014). Starting from the process analysis, supported by a literature search (Italian Ministry of Health, 2014; Frey, 2014), we identified a panel of indicators designed to monitor the appropriateness of a generic hub-and-spoke disease network:

Effectiveness indicators: managed network volumes (such as total number of cases introduced in the network by all Users or total number of consultations provided by a specific Provider); performance of the network (such as the average consultation provision time); clinical appropriateness (such as survival rate); healthcare migration; patients' and professionals' satisfaction.

Efficiency indicators:_social costs for healthcare migration; social costs for psychological assistance to the patient; healthcare costs for clinical inappropriateness (such as cost for inappropriate surgery).

These indicators will be monitored specifically for the RNTR as the project progresses.

ACKNOWLEDGEMENT

The authors would like to thank the project partners: Fondazione IRCCS Istituto Nazionale Tumori di Milano, Istituto Europeo di Oncologia and Fondazione Politecnico di Milano. This work was supported by the Italian Ministry of Health (Grant No. RF-2016-02363386 to P.G.C.).

REFERENCES

Botheju D., "Smart eHealth" – A Holistic Approach to Implementing an Integrated Healthcare Information System (HIS) in Sri Lanka, 2018.

Buranarach M. et al., Design and Implementation of an Ontology-Based Clinical Reminder System to Support Chronic Disease Healthcare, IEICE Transactions on Information and Systems, (2011):432-439.

Camarinha-Matos L. M. et al., Classes of collaborative networks, Enciclopedia of networked and virtual organizations, (2008):193-198.

Camarinha-Matos L. M. et al., Risks and Resilience of Collaborative Networks, 16th IFIP WG 5.5 Working Conference on Virtual Enterprises 2015.

Casali P.G., Trama A (eds). Rare Cancers Agenda 2030. EU Joint Action on Rare Cancers, 2019.

Costa R. et al., VirtualECare: Group Support in Collaborative Networks Organizations for Digital Homecare, in Handbook of Digital Homecare, Springer, ISBN: 978-3-642-01386-7, (2009):151-178.

Elrod J. K. & Fortenberry J. L., The hub-and-spoke organization design revisited: a lifeline for rural hospitals, BMC Health Services Research (2017): 795.

Frey S., The economic burden of schizophrenia in Germany: A population-based retrospective cohort study using genetic matching, European Psychiatry, (2014):479-489.

Hammer M. & Champy J., Reengineering the corporation: a manifesto for business revolution, Harper Business, New York, (1993).

Italian Ministry of Health, Telemedicine – national guidelines, 2014.

Lenz R. & Reichert M., IT Support for Healthcare Processes, Business Process Management: Proceedings of the 3rd International Conference, Nancy, France (2005):354-363.

Locatelli P. et al. The role of the ICT in the process monitoring and system impact assessment of rare disease collaborative networks, Multi Conference on Computer Science and Information Systems, MCCSIS 2019 - Proceedings of the International Conference on e-Health.

Locatelli P. et al., Effectively introducing mobile solutions in hospitals: The importance of process perspective and KPIs, Proceedings - IEEE Symposium on Computer-Based Medical Systems, (2014): 417-420.

Morando V., Cancer network configurations and regional health systems governance: a European comparison, Transatlantic Dialogue 2014 "From Public Administration to XXI Century Collaborative Administration: The role of public networks".

Noran O., A decision support framework for collaborative networks, International Journal of Production Research, vol. 47, no. 17, (2009):4813-4832.

Thimm H. & Rasmussen K. B., Information Support Services for Intermediation Tasks of Collaborative Networks, Journal of Universal Computer Science, vol. 16, no. 13, (2010):1776-1800.

HOW CAN QUALITY REQUIREMENTS FOR MEDICAL SOFTWARE BE SYSTEMATICALLY FULFILLED BETTER?

Karin Messer-Misak[1], Jeroen de Bruin[2] and Sten Hanke[3]
eHealth Institute, FH Joanneum, University of Applied Science, Graz, Austria
[1]+43 316 5453 6521
[2]+43 316 5453 6515
[3]+43 316 5453 6526

ABSTRACT

Due to a variety of factors, among others medical and legislative requirements, the quality requirements on medical software become increasingly demanding. Several quality standards and guidelines are already in place, e.g., ISO/IEC s9126, 14598 and 25000 (SQaRE). However, there is a debate on whether these standards are not specific enough for medical software. Furthermore, their effective application for requirement gathering, analysis and monitoring can be difficult. In this paper, we propose a workflow for the application of the ISO/IEC 25010:2011 standard for the systematic collection, analysis and evaluation of requirements for various stakeholders in order to systematically better fulfill quality requirements on medical software.

KEYWORDS

Software Quality Standards, Requirements Engineering, Medical Software Engineering

1. INTRODUCTION

Current developments in healthcare Information and Communication Technology (ICT) include the design and development of increasingly complex integrated systems comprising both software and hardware. Technological advances in healthcare have encouraged the development of new technologies that drive connectivity across the healthcare sector. As a result, systems are more difficult to certify, more difficult to abstract from other software and hardware components, and are more difficult to understand with respect to descision making. With increasing healthcare demands, these systems will inevitably outgrow medical professionals' capabilities to deliver safe, quality care in a timely manner.

Technological innovations like mobile health solutions are met with great interest by healthcare professionals, but at present evidence on their medical efficacy, as well as quality criteria frameworks for their development and use in medical settings is still lacking. An underlying factor that complicates medical application of such technologies resides in the legislative landscape, which is currently undergoing a process of continuous definition and re-definition of guidelines for the development and application of medical devices integrating complex software and hardware components.

A central focus that connects the legislative and medical perspectives on medical software is the requirement of delivering high quality systems, software and services. To fulfill this prerequisite, a framework is needed that comprises the following: i) quality criteria for medical software and hardware development, ii) a process to validate products against these criteria and iii) a process to monitor the development process and to track the quality criteria during the project.

In the end, the assessment of the effect of software quality on patient treatment quality, healthcare institution efficiency and certain applications like clinical trials is essential. The aim of the paper is to introduce a method how to apply well-established methods from quality management as a process, which might also be used in the medical domain to control, manage and evaluate systems against certain requirements.

2. CURRENT GUIDELINES AND STANDARDS

Although standards that define quality criteria for software and hardware have already been constructed, their applicability to medical software and systems is still a matter of debate; there is a discussion ongoing about the criteria in general, because application and software quality is a multifaceted concept determined by a number of properties (Aghazadeh, S. et al., 2015).

Over time, different qualitative models have been proposed to describe and delineate the different system and software quality attributes. The ISO/IEC standards 9126, 14598 and 25000 (SQaRE) are widely referred to as standard frameworks for developing complementary or alternative models for evaluating specific software or developing custom models.

The ISO/IEC 9126 standard has been introduced in 1991 by International Standard Organization (ISO) and International Electro-technical Commission (IEC). This model is used for the definition and integration of various characteristics of software quality, and has been used to asses impact of software quality characteristics on healthcare outcome (Ordozgoiti, A. V. et al., 2016). More recently, ISO/IEC 25000 standard is considered more suitable for evaluating the quality of ICTs for healthcare use from the perspective of institutional acquisition (Ordozgoiti, A. V. et al., 2016). ISO/IEC 25000 attempts to harmonize, unify and update the international standards for evaluating software quality.

Additionally, there are some non-ISO models that mostly standardize reference works and most of them are impractical and inappropriate for healthcare institutions, since they base their models on life cycle of software under development.

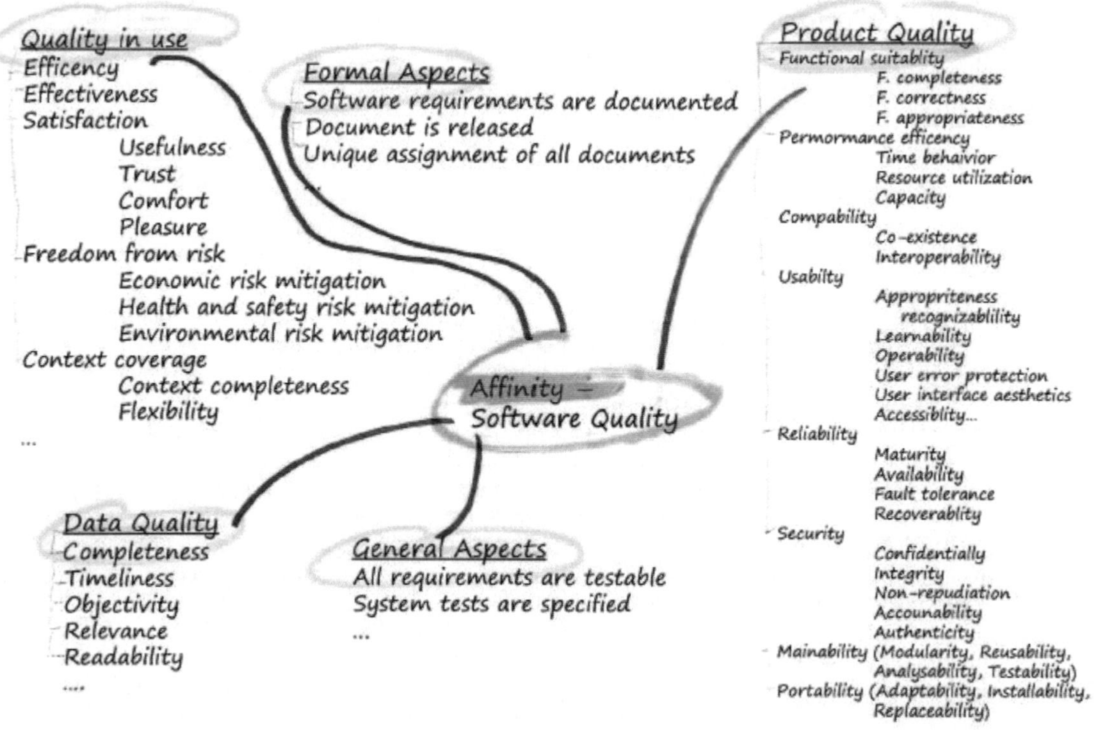

Figure 1. Affinity Diagram

Software quality requirements can be complex and their application and priority depend on the application under consideration, as well as the target application environment and stakeholders. These criteria are normally assorted in functional and non-functional requirements. For example Aghazadeh et al. (2015) indicated sixhealthcare indicators in relation to software quality characteristics based on a questionnaire with health experts. The characteristics are 1) User satisfaction, 2) Quality of patient care, 3) Clinical Workflow and Efficiency, 4) Care providers communication and information exchange, 5) Patient satisfaction and

6) Care costs (Aghazadeh, S. et.al., 2015). Nonetheless, although requirements can vary depending on the application and application field, a process is needed to evaluate the system against these requirements.

Identifying software quality requirements early in the software life cycle as a part of the requirements specification is considered crucial. At present, quality requirements are commonly elicited from literature and experts. The standard ISO/IEC 25030 (as part of the ISO/IEC 25000), which is part of the SQuaRE series of International Standards, aims to improve the value of software quality requirements by providing guidelines and recommendations for quality requirements, and guidance for the processes used to define and analyze these requirements (Idri, A., Bachiri M., and Fernández-Alemán J.L., 2016). In particular, ISO/IEC 25030 provides a guide with which to identify software quality requirements, to validate the completeness of requirements specification and to identify quality assurance and acceptance criteria for a software product. Examples of the application of this guide include *Alves et al.* (2016, pp. 30-35) who validated a poison central system as well as *Kadi et al.* (2016) who evaluated the quality requirements of a pregnancy monitoring system, both using the ISO/IEC 25010 standard.

3. USE OF QM METHODS IN PRACTICE

The following activities can benefit from the use of tools known from quality management (BSI Standards Publication, 2011, p.1):

- Identification of software and system requirements;
- Validation of the completeness of a requirement definition;
- Identification of software and system design objectives;
- Identification of software and system test objectives;
- Definition of quality characteristics and control criteria in the context of quality assurance;
- Identification of acceptance criteria for a software product and/or software-intensive computer system;

These activities, methods and tools must be carefully coordinated and comnistently used throughout the entire product development process in order to ensure that the required quality requirements can be verified in the final product.

In order to identify the optimal set of software and system requirements and verify them for completeness, a comprehensive analysis of stakeholder requirements must be carried out. These requirements are to be specified as criteria, i.e. described in concrete terms, and measurement methodologies or metrics need to be defined for evaluation purposes. In order to achieve accurate and comprehensive results, it is recommended to work on this process in a stepwise fashion in interdisciplinary teams, and to employ quality management tools and methods .

4. STEP-BY-STEP IMPLEMENTATION USING A PRACTICAL EXAMPLE

In this section, the process will be presented step by step and proven tools that support the process will be described.

First, requirements need to be collected from all stakeholders, as they are likely to vary between stakeholders. We can distinguish between primary users (person who interacts directly with the system), secondary users (i.E. support or content provider, system manager, security manager, analyzer, installer) and sometimes indirect users (person who receives output without interacting with the system) (BSI Standards Publication, 2011, pp. 5-6).

Usually s survey of the requirements of the various stakeholders is employed, for example, with the help of an affinity diagram. Complete sets of requirements are most likely to be achieved with a collaborative working method that is carried out step-by-step. This enables user and system requirements, design requirements, etc. to be comprehensively identified, named, and categorized. The relationships between the individual information groups become transparent. In software development, it makes sense to use the models

"Quality in Use" and "Product Quality" and their attributes mentioned in ISO/IEC 25010:2011 and then to extend and supplement them accordingly (see Figure 1).

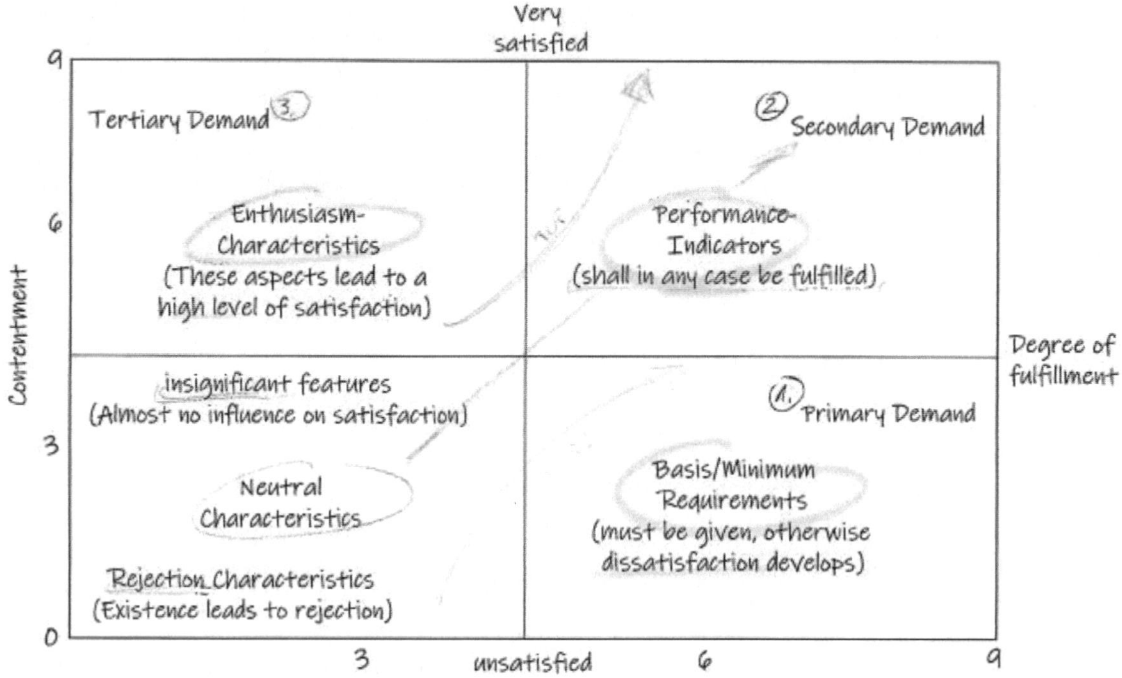

Figure 2. Kano Analysis

After requirements are collected, the information gained from this is supplemented by an affinity diagram, in which the requirements are further specified from the stakeholder/user perspective using a Kano analysis. In this process, the relationship between the achievement of certain characteristics of the requirements and the satisfaction expected by the stakeholder/user is established.

After this information supplementation phase, the requirements can be divided into 5 categories. As implicit must criteria, which are not articulated concretely by the stakeholders, but are tacitly presupposed by this, are so-called basic characteristics. Performance characteristics are requirements explicitly demanded by the stakeholders, which have a high influence on satisfaction. Enthusiasm characteristics donate felt or actual auxiliary use, which is not expected by the stakeholders explicitly, but whose satisfaction increases to a large extent. Further one distinguishes between insignificant characteristics (thus characteristics, whose existence leads neither to satisfaction nor to the dissatisfaction) and rejection characteristics (these are characteristics, which, if they exist with the user/client dissatisfaction cause, however with not being present also no satisfaction create). Each requirement is assigned to a category before it is further specified (see Figure 2).

Next, stakeholders requirements are translated into specifiable and measurable quality characteristics. In this process, the target criteria to be achieved are defined, control criteria for quality assurance are defined and the acceptance of the software product is presented in a matrix.

The tool "Quality function deployment" or "House of Quality" is suitable for planning. Especially when interfaces and contradictions occur, these can be easily identified by correlations. It shows "what" the stakeholders want and "how" the requirements are technically implemented. The method is supplemented by target values to be defined, which are defined by fixed values in a corresponding unit.

Here, too, a systematic step-by-step approach is recommended (see Figure 3):

- After the structuring and weighting of the requirement analysis results, these are displayed on row level. By weighting these entries (e.g. between 1-9) from the Kano analysis, the relevance from the point of view of the stakeholders is represented.

- The column level lists how these requirements are met. In the next step, the links between rows and columns are displayed. For simplicity's sake, the rating is bucketed in 4 relationship levels (no relationship, weak relationship, medium relationship or strong relationship) between the requirements and the functions or technical specifications. The values are determined using the correlation matrix (weighting * relevance) both horizontally and vertically. As a result, the evaluation of the stakeholder requirements with regard to the functions on the horizontal side and the evaluation of the functions on the vertical side can be seen at a glance.
- In order to check the functions for mutual influences and dependencies, both the stakeholder requirements and the functions can be correlated with themselves.

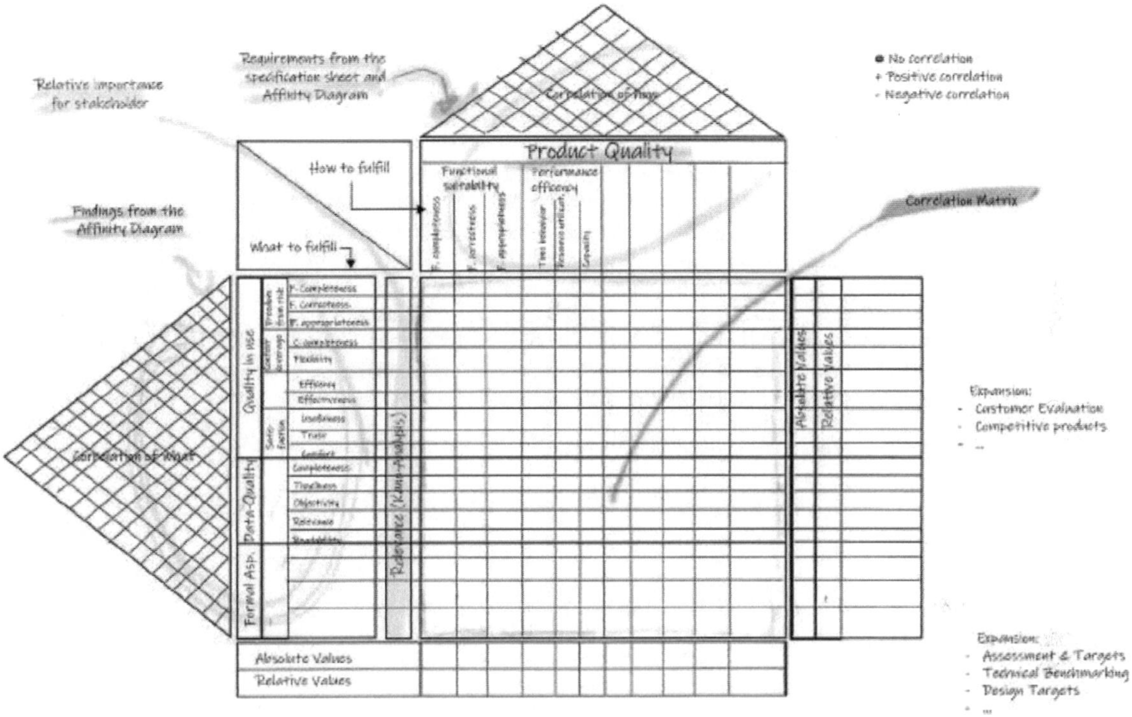

Figure 3. Quality Function Deployment

The method can be extended as required and can be flexibly adapted to suit specific requirements. Yoji Akao - one of the developers of the method - coined the sentence "Copy the spirit, not the form" (Möhrle, M.G. et al., 2007, p. 146). Herzwurm, et al. (2003) show that it makes sense to adapt existing methods and tools accordingly and to adapt them to the requirements.

5. SUMMARY

The demand that high-quality software systems to be realized within the planned time and at the estimated costs led to the adaptation of the methods in software development (Hohler, B., 2014, p. 831) The quality aspect is an essential factor in ensuring security and stakeholder satisfaction, i.e. success in the use of software products, especially in health care. Potential negative effects should be excluded from the outset and positive aspects should be identified and promoted. Quality characteristics must be known, specified, measured and evaluated.

REFERENCES

Aghazadeh, S., Pirnejad, H., Aliev, A. and Moradkhani A. (2015). "Evaluating the effect of software quality characteristics on health care quality indicators," *Indicators*, vol. 2, no. 3, pp. 67–73.

Alves J.M. et al. (2015). "Quality evaluation of poison control information systems: A case study of the DATATOX system," in *Proceedings - IEEE Symposium on Computer-Based Medical Systems*, vol. 2016-August, pp. 30–35.

BSI Standards Publication (2011). Systems and Software engineering - *Systems and Software Quality Requirements and Evaluation (SQuaRE)* - System and software quality models, BSI, 31.3.2011, p. 1.

BSI Standards Publication (2011). *Systems and Software engineering - Systems and Software Quality Requirements and Evaluation (SQuaRE)* - System and software quality models, BSI, 31.3.2011, pp. 5-6.

Herzwurm, Schockert, Dowie u. Breidung (2003). *Requirements Engineering for Application Development in Volatile Project Environments Using Continuous Quality Function Deployment (CQFD)*, in Proceedings of the International Conference on Software Engineering Research and Practise, Vol. 1, pp. 440-447.

Hohler, Bernd (2014). Qualitätsmanagement bei der Softwareentwicklung, pp. 827-866 in: *Masing Handbuch Qualitätsmanagement*, Carl Hanser Verlag, München, Wien, p. 831.

Idri, A., Bachiri, M. and Fernández-Alemán, J.L. (2016). "A Framework for Evaluating the Software Product Quality of Pregnancy Monitoring Mobile Personal Health Records," *J. Med. Syst.*, vol. 40, no. 3, pp. 1–17.

Kadi, I., Idri, A. and Ouhbi, S. (2016). "Quality evaluation of cardiac decision support systems using ISO 25010 standard," in *Proceedings of IEEE/ACS International Conference on Computer Systems and Applications, AICCSA*, 2016, vol. 0.

Möhrle, Martin. G., Hartung, Knud and Spilgies Wulf-Dieter (2007). Quality Function Deployment for bundled services., in *Innovation Management in the Service Industry. Basics, practical examples and perspectives*. Edited by Klaus Schmidt/Roland Gleich/Ansgar Richter. pp.139-160, Haufe Publishing House, Freiburg. 2007, p. 146 - quotes Akao 1992, p. 19.

Ordozgoiti, A.V., Hito, P.D., Guix Comellas, E.M., Sanchez, C.M.F., Hernandez M.G., and Canut, T.L. (2016). "Software quality evaluation models applicable in health information and communications technologies. A review of the literature," in *Studies in Health Technology and Informatics*, vol. 226, pp. 169–172.

AUDAPRET: TOWARDS A CLOUD ARCHITECTURE SUPPORTING MULTI-DEVICE RESEARCH STUDIES

Nadine von Frankenberg, Felix Matschilles and Stephan Jonas
Department of Informatics, Technical University of Munich, Germany

ABSTRACT

Recent improvements in the functional range and measurement accuracy of commercially available biosignal measurement devices enable cost-effective human trials with large numbers of participants. Biosignals can be tracked outside the research institution, with only minor restrictions in the daily activities of the study participants, compared to stationary measuring devices. However, there is a lack of data acquisition systems supporting the use of several different devices needed, so that the collected data from measurement devices have to be processed manually. Researchers must also abide by strict legal regulations when handling health-related data of study participants, which is often cumbersome and requires additional time-consuming efforts. To address these challenges, this paper proposes the AUtomatic DAta Processing of REsearch Trials (AUDAPRET) architecture, which enables researchers to conduct trials requiring health-related data of participants in a less time-consuming and legally compliant manner. AUDAPRET is designed in such a way that study participants retain full control over their data and the access rights of anyone requesting access.

KEYWORDS

Human Study Management, Data Collection, eHealth Architectures, Health Information Systems

1. INTRODUCTION

With the rise of mobile technologies in the last decade, mobile applications and wearable devices were developed to continuously assess biosignals with increasing accuracy. These advancements have turned health tracking devices into assets for scientific research (Cox et al., 2018), in particular in the health sector. Such research studies often involve a variety of different data sources, especially when the study's focus is on human health: many health-related research studies have successfully been using wearables, the majority of which are activity tracking wristbands (Hartwell et al., 2019). Prominent examples are devices of the brand FitBit, which have been successfully tested in research studies to assess and promote physical activity in cancer and diabetes patients (Coughlin et al., 2020; Zheng et al., 2020). However, studies that require data from multiple different wearables and other sources, such as questionnaires or smart home devices, usually require these data to be collected in a structured and organized manner. Data sources do not necessarily offer uniform interfaces, which makes data collection a challenge. Existing standards, e.g. FHIR or Continua, do not support communication with the participants. Many research institutions follow manual data collection processes, resulting in an additional source of error and workload that increases linearly with the number of participants. Others have used customized systems that were designed specifically for a single study (Aledavood et al., 2017). Developing customized systems for individual studies involves a considerable expenditure of time and budget. While wearable devices appeal to participants due to their minimal restriction on freedom of movement, they also raise privacy concerns (Angeletti et al., 2018). Such devices constantly collect data and may create data profiles about the respective participant's health (Paul and Irvine, 2014). This requires that the data is stored securely and in accordance with the prevailing data protection laws. In reality, this is not always adhered to sufficiently. In Europe, the processing and storage of personal data is regulated by the General Data Protection Regulation (GDPR) (European Commission, 2018): Study participants must be able to give their consent to the use of their data voluntarily, revoke it at any time, and have the right to demand the immediate deletion of their data [GDPR, Art. 7], unless long-term archiving is mandatory, as in clinical trials. The study conductor is obliged to follow such requests immediately [GDPR,

Article 16]. Health-related data have other specific requirements that must be considered when conducting a research trial, e.g., getting approval of the institutional review board, or follow data processing guidelines.

In this paper, we present the cloud-based AUDAPRET architecture which supports the automation of a data collection process from various data sources for research studies, in a privacy-compliant manner. It is applicable to research studies in any field. The proposed architecture focuses on modularity; additional data sources can be integrated flexibly. To validate the system design and to show its feasibility, a prototype for an exemplary study that integrates data from FitBit, LimeSurvey, and Netatmo has been developed.

2. RELATED WORK

In order to conduct research studies, researchers often develop customized systems or workflows that fit to their studies in order to organize and collect multi-source data, which is often time-consuming and cumbersome (Aledavood et al., 2017; Rutkove et al., 2019). Few approaches aim at supporting a multitude of studies and the devices they use to collect and store study-relevant data. Wong et al. conducted a pilot study to test the non-profit software platform Tidepool (https://www.tidepool.org) to automate the data collection in a research study in which diabetic patients were monitored with blood glucose meters. Tidepool focuses on diabetes data and currently only supports data collection wide range of blood glucose meters, making it unsuitable for studies that require data from other wearables, such as EEG headbands or ECG measuring activity trackers. EEG headbands are promising in, but not limited to, studies that aim to record brain activity during epileptic seizures or to investigate cognitive load. Because EEG headbands are unobtrusive, they can be worn outside the clinical environment, which is beneficial because the incidence of seizures is unpredictable and requires continuous monitoring (Byrom et al., 2018).

Niima, the Non-Intrusive Individual Monitoring Architecture platform by Aledavood et al., targets the automation of data collection in research studies to detect and diagnose mental disorders. Niima can be applied to several research areas because of its modular structure. The platform stores data in a raw format and anonymizes these data, before they are accessible to researchers. Study participants do not have access to the data. This restrictive access control leads to a lack of transparency for the participants as to what data are collected in the study as all study participants' data are stored in a central storage (Aledavood et al., 2017).

Angeletti et al. present a software to automate the data collection process in clinical trials. They argue for the need of digital measurements in clinical trials in general, as they noted a decline in people's willingness to participate in studies. This decline indicates that potential study participants often find the time required to participate in studies too long. Their approach specifies that only anonymized or pseudonomized data are stored in their central data storage. Study participants can access the raw sensor data, whereas researchers access the central data storage. However, Angeletti et al.'s solution does not specify the data collection process itself. They argue that the devices used in clinical studies must be certified as clinically useful, but do not propose a solution on how to integrate the devices into their architecture (Angeletti et al., 2018).

The aforementioned approaches focus on different aspects, such as GDPR compliance or the flexibility of integrated data sources. To the best of our knowledge, no existing solution allows researchers to manage data from multiple devices while complying with data protection laws and ensuring transparency for participants.

3. AUDAPRET

Our goal was to develop an architecture to assist researchers during data collection and management of research studies, which has to (1) be applicable to different studies with multiple different data sources, (2) comply with data privacy regulations, (3) provide transparency about who can access a participant's data, and (4) allow both researchers and participants to interact. This requirements analysis results from literature research, own experiences, and in consultation with a medical researcher from (REMOVED DUE TO BLIND REVIEW). To address these requirements, AUDAPRET is a decentralized, extensible, cloud-based architecture. It encompasses multiple users of types *Researcher* and *StudyParticipant*, each contributing study-relevant data: StudyParticipants provide data from sources such as *Wearables* or *Questionnaires*, Researchers contribute data from shared data sources, such as environmental data. AUDAPRET is decomposed into two levels of abstraction: First, the single-user level defines the interaction between each user and their dedicated instance of the *SingleUserSubsystem*, which collects the user's study relevant data and allows researchers to access this data. In addition to their SingleUserSubsystem instance, Researchers use the *StudyAdministrator* to manage a study and to request data from StudyParticipants. Second, the multi-user level covers the interaction between multiple SingleUserSubsystem instances and the StudyAdministrator.

The SingleUserSubsystem consists of three subsystems (see Figure 1): The *StudyDataManager* stores and administrates study-relevant data of a user. The user interacts with the *StudyDataManagerClient* to view, delete or anonymize StudyData of the respective StudyDataManager. An *Adapter* binds a DataSource to the StudyDataManager; it independently collects data from its DataSources, processes it, and sends it to the StudyDataManager. The StudyDataManager passively receives and stores data from all its Adapters and returns all data on request. Adapters are bound to the study's domain, whereas the StudyDataManager and the StudyAdministrator operate independently and can therefore be reused in studies of any domain.

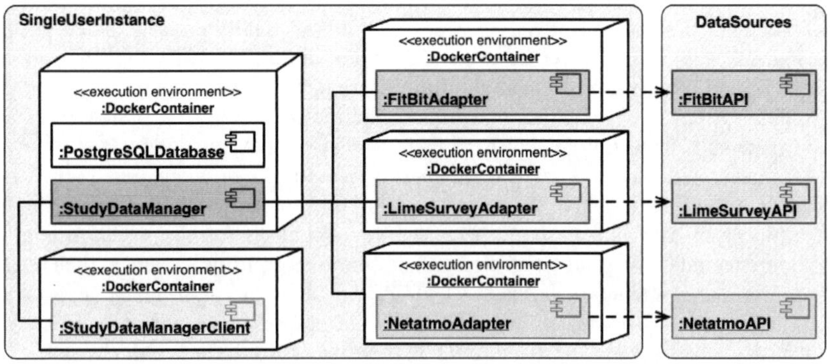

Figure 1. Subsystem decomposition and prototype adapters of the SingleUserInstance

We have developed a prototype of AUDAPRET, where all individual subsystems are implemented in TypeScript. The prototype integrates Adapters for activity trackers (FitBit), questionnaires and surveys (LimeSurvey), and indoor weather stations (Netatmo). It addresses a concrete instance of the SingleUserSubsystem as *SingleUserInstance*, consisting of a StudyDataManager, StudyDataManagerClient, and the domain-dependent Adapters *FitBitAdapter*, *LimeSurveyAdapter*, and *NetatmoAdapter*. Each Adapter is deployed in a separate docker container so that the Adapters will continue to collect data even in the event of an API malfunction. The containers of the StudyDataManager, StudyDataManagerClient, and Adapters can be composed to an isolated network, in which each Adapter's container is linked to the StudyDataManager. Each subsystem server of a SingleUserInstance can be deployed in a web cloud in its own container, and composing it into a container network, the SingleUserInstance can be duplicated to any number with container orchestration tools, such as Kubernetes.

At the multi-user level, AUDAPRET consists of multiple independently deployed SingleUserInstances, one per user. Additionally, Researchers use the StudyAdministrator which allows them to administrate a study and to request data from all StudyParticipants of this study. The StudyAdministrator communicates with the SingleUserSubsystem instances of the respective StudyParticipants (see Figure).

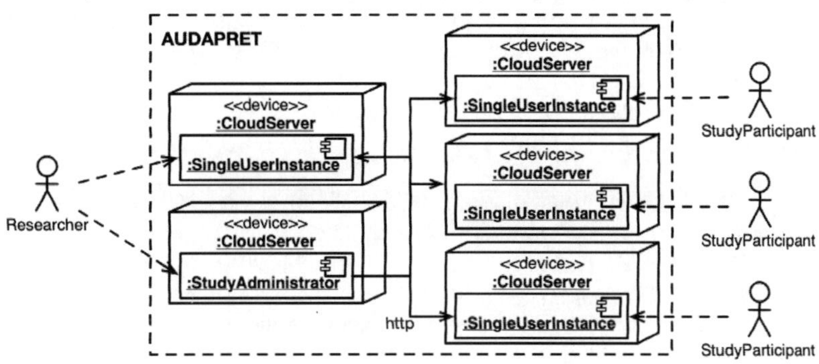

Figure 2. Subsystem decomposition of an exemplary study on the multi-user level as an UML diagram

If personal data is collected and stored as part of a research trial, StudyParticipants have the right to access and delete their own data at any time (European Commission, 2018). Furthermore, Researchers who store data are liable for violations against the GDPR. By using localized or personalized instances, data ownership is delegated to the StudyParticipant. Thus, each StudyParticipant has their own database which stores their respective study data, and which is initially only accessible to the respective StudyParticipant. If a Researcher wishes to access a StudyParticipant's data for evaluation purposes, they can request access to read the StudyParticipant's data. This enables the StudyParticipant to view, delete, and anonymize their data at any time and therefore does not require the Researchers to offer this service to the StudyParticipants.

4. CONCLUSION

AUDAPRET supports the conduct of human studies: It automatically collects data from multiple sources, e.g., questionnaires, wearables, and environmental sensors. It can be easily extended to collect data from additional sources and can be used in studies from different fields that require data from different sources. The AUDAPRET system is designed so that data collection, processing, and storage is GDPR-compliant, while Study-Participants have full control over their personal StudyData. StudyParticipants can share their StudyData voluntarily with Researchers and are able to view, delete or anonymize their data at any time. As proof of concept, we have implemented a prototype of the system which supports AUDAPRET's main functionality: collecting data and sharing it with the Researcher on the participant side and requesting this data on the researcher side. The prototype collects data from FitBit, LimeSurvey, and Netatmo.

For future work, we intend to validate AUDAPRET in a multi-participant experiment, where each participant contributes data from multiple sources. The current design of a participant-centered data storage allows StudyParticipants to retain their StudyDataManager and StudyData after the end of the study. Since they own their personal StudyData, they could share them with other Researchers for other research projects. We envision a platform that enables StudyParticipants to share their StudyData with other Researchers using the same access grant logic as designed for a single study. Other existing mHealth applications, whose core functionality is the collection of health data, could also be integrated into AUDAPRET as data sources.

REFERENCES

Aledavood, T., Hoyos, A.M.T., Alakörkkö, T., Kaski, K., Saramäki, J., Isometsä, E., Darst, R.K., 2017. *Data collection for mental health studies through digital platforms: requirements and design of a prototype*. JMIR Res. Protoc. 6, e110.

Angeletti, F., Chatzigiannakis, I., Vitaletti, A., 2018. *Towards an architecture to guarantee both data privacy and utility in the first phases of digital clinical trials*. Sensors 18, 4175.

Byrom, B., McCarthy, M., Schueler, P., Muehlhausen, W., 2018. *Brain monitoring devices in neuroscience clinical research: the potential of remote monitoring using sensors, wearables, and mobile devices*. Clin. Pharmacol. Ther. 104, 59–71.

Coughlin, S.S., Caplan, L.S., Stone, R., 2020. *Use of consumer wearable devices to promote physical activity among breast, prostate, and colorectal cancer survivors: a review of health intervention studies*. J. Cancer Surviv. 1–7.

Cox, S.M., Lane, A., Volchenboum, S.L., 2018. *Use of wearable, mobile, and sensor technology in cancer clinical trials*. JCO Clin. Cancer Inform. 2, 1–11.

European Commission, 2018. *2018 reform of EU data protection rules*.

Hartwell, L., Ross, H.M., La Belle, J.T., 2019. *Project honeybee: Clinical applications for wearable biosensors*. Biomed. Microdevices 21, 37.

Paul, G., Irvine, J., 2014. Privacy implications of wearable health devices. In: *Proceedings of the 7th International Conference on Security of Information and Networks*. pp. 117–121.

Rutkove, S.B., Qi, K., Shelton, K., Liss, J., Berisha, V., Shefner, J.M., 2019. *ALS longitudinal studies with frequent data collection at home: study design and baseline data*. Amyotroph. Lateral Scler. Front. Degener. 20, 61–67.

Zheng, Y., Weinger, K., Greenberg, J., Burke, L.E., Sereika, S.M., Patience, N., Gregas, M.C., Li, Z., Qi, C., Yamasaki, J., 2020. *Actual Use of Multiple Health Monitors Among Older Adults With Diabetes: Pilot Study*. JMIR Aging 3, e15995.

IS PRIVACY INSPIRED BY WELLBEING? A PRELIMINARY STUDY

Ana Ferreira[1,2], Cristina Jácome[1,2], Rute Almeida[1,2], Filipa Lopes[3], João A. Fonseca[1,2,3] and Pedro Vieira-Marques[1,2]
[1]*CINTESIS - Center for Health Technology and Services Research*
[2]*Departamento Medicina da Comunidade, Informação e Decisão em Saúde (MEDCIDS)*
Faculty of Medicine, University of Porto, Portugal
[3]*Medicina, Educação, I&D e Avaliação, Lda (MEDIDA)*
Porto, Portugal

ABSTRACT

Most patients with chronic diseases aim to successfully manage their disease and treatment in everyday life. Although many mobile applications are available to help them achieving this, there is still evidence of the low adherence from chronic patients to those applications. Obstacles include poor usability, lack of resources or low IT literacy, as well as the lack of trust in the privacy and security of mobile applications. Beyond the technical measures, the perception of privacy may not be a fixed concept but influenced by factors such as context, emotions and wellbeing. The aim of this paper is to investigate how health outcomes can alter the perception of privacy and trust in the use of mhealth applications, by patients with chronic respiratory diseases. Findings show that there may be significant links between patients' characteristics, their privacy perception and the type of personal data involved. The preliminary results motivate us to further research this topic.

KEYWORDS

Privacy, Trust, Patients' Perception, Wellbeing, Mhealth Apps, Chronic Patients

1. INTRODUCTION

Many mobile applications are available to help patients with chronic diseases to continuously and successfully manage their condition, however, barriers such as poor usability, lack of resources or low IT literacy, as well as the lack of certification and trust in mobile applications, still exist (Kruse, 2019; Zhou, 2019; Larson, 2018; Woldeyohannes, 2017). Ultimately, these obstacles hinder adherence to health enhancing behaviours, such as regular medication adherence, which can negatively affect health outcomes. In relation to users' privacy perception, most research works on mhealth applications focus on the factors that may promote/hinder the use of mhealth and, amongst these, are privacy and trust (Saheb, 2020; Wei, 2020; Vo, 2019; Majid, 2019). Privacy is closely related with users' trust in the technology, which can improve users' adherence to treatment as well as health outcomes (Hong, 2018; Serenko, 2013), as already evidenced in the trust relationship between patients and healthcare providers (Birkhäuer, 2017; Polinski, 2014). But what about the reverse? Can health outcomes influence the perception of privacy/trust in technology? Beyond technical measures of mobile applications, the perception of privacy may not be a fixed concept, but influenced by many factors (Rooy, 2010). Is patients' wellbeing one of them? The authors could only find a work studying how the severity of a disease can influence their assessment of health information's trust (Vervier, 2018). The aim of this paper is to investigate if patient health outcomes can alter patients' perception of privacy in mhealth applications.

2. THE MOBILE APPLICATION: *INSPIRERMUNDI*

The *InspirerMundi* application (Figure 1) is a health serious game with various monitoring functionalities to improve inhaled medication adherence in adolescents and adults with asthma (Jácome, 2017). For ubiquitous verification of adherence, the application incorporates an image-based tool for inhalers and, via image processing techniques, confirms the inhaler presented to the camera and locates the dose counter. This application was developed grounded in research and in close cooperation with patients and physicians. The application has been tested for usability (Ferreira, 2019) and feasibility in 26 hospital outpatient asthma clinics (Jácome, 2018). The survey presented here was applied to patients using this mobile application (Section 3).

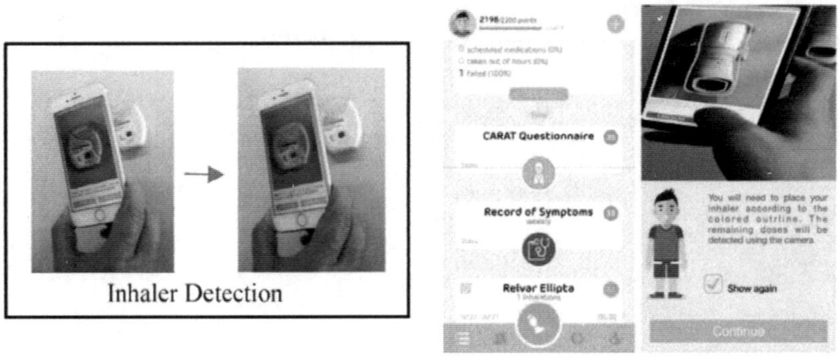

Figure 1. Example of screenshots of the *InspirerMundi* mobile application

3. METHODS

A literature review was performed with the queries presented in Table 1. Inclusion criteria comprised works from the past 5 years, written in English, which focused on the perception and attitudes towards privacy in relation to mhealth. A review of their titles and abstracts was performed and selected full papers were read and analysed. In order to search for more sources of related works, a Google query, with a combination of the terms "health, condition, influence, patient, privacy, trust", was applied, but with no further results on the subject.

For the survey, during an initial visit, patients with persistent asthma (active prescription for an inhaled controller medication) were invited by their physician to use the *InspirerMundi* application daily. One month later, through a telephone interview with the aim of assessing patients' feedback about the application, they were also questioned about their perception of privacy, when using mobile applications. This survey, conducted in health institutions in Portugal, included 77 participants and was approved by the ethics committees of all participating institutions and by the National Data Protection Committee (no 8825/ 2017). More details are described here (Jácome, 2018). The survey included parts such as: demographics, asthma and allergic rhinitis symptoms, medication adherence, anxiety and depression symptoms, use of social networks, mobile devices and applications, as well as four questions, evaluated with a 3 point Likert scale (never, sometimes or always), which asked if participants **authorized** the applications they installed, to access their: 1) Location, 2) Contacts, 3) Photographs, and allowed those applications to 4) send them Notifications. The interviews were performed between March and June 2018, and analysed in May 2020 with IBM SPSS Statistics version 26. All variables collected in the interview were tested for association/correlation with the four security questions. The a) Spearman rank test was used to calculate the correlation between categorical ordinal variables; and the b) Eta^2 Coefficient test (which is the percentage of variance in the dependent variable accounted for by the independent variable) was used to determine the strength of non-linear associations, between categorical and scale variables. The level of significance was set at 0.05.

4. RESULTS

The review results show that the analysed works try to understand the reasons for low patients' adherence to eHealth technologies (n=10; 77%), resulting that privacy and trust can influence the continued use of applications (n=10; 77%). The sample includes articles from: 2020=2; 2019=8; 2018=1; 2017=1; and 2016=1.

Table 1. Results from the literature review of published works for the past 5 years

Database Engines	Queries	Before titles/ abstracts	After titles/ abstracts
SCOPUS	TITLE-ABS-KEY ((health OR disease or chronic) AND (privacy OR trust) AND (perception OR attitude) AND (mhealth OR mobile OR app))	186	4
PUBMED	((health OR disease OR chronic) AND ("privacy" OR "trust") AND ("perception" OR "attitude") AND (mhealth OR mobile OR app))	160	1
ISI WOK	TS=((health OR disease OR Chronic) AND ("privacy" OR "TRUST") AND ("perception" OR "attitude") AND (mhealth OR mobile OR app))	49	8
TOTAL		**395**	**13**

A total of 66 participants answered the complete survey with the security questions. The mean age of the sample who responded the security questions is 27.7 years (N=65), with a standard deviation of 12.9 and the youngest participant being 13 years old and the oldest 70. The majority of respondents were women (59%), with secondary education (70%), 48% were students, 39% employed and 9% unemployed. Also, 98% of respondents use social networks, with 66% all the time or several times a day (89% Facebook; 38% Instagram; 51% another network). Moreover, 75% of participants use smartphones and 23% use tablets or iPad, to: Access GPS (32% never or once a month); Browse the Internet (36% all the time and 42% several times a day); Take pictures (16% several times a day and 27% sometimes per week); Use mobile applications (63% all the time or several times per day); and Play games (30% several times per day or per week and 27% never).

In addition, participants have already downloaded and used health and fitness applications (50%) but only a minority did the same for asthma applications (1%). However, 89% of participants would like to use an asthma application. Table 2 describes the answers to the four security questions and the median of the "health status score" (0-100) they gave themselves on the day they were enquired, in relation to the security answers. Tables 3 and 4 present the results from the application of Spearman and Eta statistical tests, respectively, the authors believe can be analysed in future research.

Table 2. Answers to the four security questions by the participants (N=66) with health status' score median (Hsm)

	NEVER		SOMETIMES		ALWAYS		Don't know
Security questions	**n (%)**	**Hsm**	**n (%)**	**Hsm**	**n (%)**	**Hsm**	**n**
Lets app access to the location	24 (36)	84	37 (56)	85	5 (8)	85	0
Lets app access photographs	34 (51)	85	28 (42)	84	3 (5)	60	1
Lets app send notifications	13 (20)	90	31 (47)	83	20 (30)	81	2
Lets app access the contacts	46 (70)	83	19 (29)	87	0 (0)	0	1

Table 3. Relations and statistical significance found in the sample using Spearman rank correlation test

Demographics and health	Security variables	Value	Sig. (2-tailed)
Last week: felt very anxious with fear and a tight stomach	Contacts	.245	**.047**
Education level	Photographs	-.358	**.003**
Last week: felt fear as if something terrible is going to happen	Contacts	.228	.066 ns
Last week: felt restless and cannot stop	Notifications	-.221	.074 ns
Last week: felt tense or nervous	Notifications	-.220	.076 ns
Pain/Feeling bad	Photographs	-.215	.083 ns

In bold, the statistically significant associations (p <.05); ns=not-significant (p >.05)

Moreover, each one of the continuous variables "health status score" and "adherence to inhaler" was tested for association with the security variables (Table 4).

Table 4. Eta coefficient test for security (dependent), "health status" and "adherence to inhaler" (independent) variables

Independent/Dependent	Location	Photographs	Notifications	Contacts
Health status	.477	.531	.349	.571
Adherence to inhaler	.619	**.652**	**.692**	.613

In bold, Eta^2 value with moderate to strong association (Eta^2 value > .7 is a strong association)

5. DISCUSSION

In general, participants answer the options "Never" or "Sometimes" to the four security alternatives, so they appear to be a moderate/high IT literate and privacy-aware sample, because the most permissive answers are related to the application sending "Notifications", with the highest number of "Always", and the least permissive is accessing the "Contacts". However, although the participants chose to block most accesses to applications when generically asked, they allowed access to those features to the specific *InspirerMundi* application, to participate in the study. This underlines the idea that the goal, context and needs relating to their well-being can influence their privacy perception. Still, a connection between privacy awareness and good, to very good, health status, may be worth further testing, with a more balanced and larger sample. Also, most "Sometimes" selections are for letting access to "Location" and send "Notifications" and the most "Never" selections are for letting access to "Photographs" and "Contacts", maybe because the later are considered to be closer to private matters, and location is not yet a concept for people to closely link with their privacy. With the current pandemic and the increase use of contact tracing applications, this perception may change overtime.

In terms of statistically significant findings (Table 3), the only correlation between patients' well-being and their security options is between "Feeling very anxious and with fear", which positively correlates with "Contacts". This needs further clarification as "Contacts" have more "Never" selections, from the four options.

Although not directly related to healthcare, the other statistically significant finding relates to the "Education level", which is negatively correlated with "Photographs" (p-value<.001). If this can be confirmed, then different types of personal data, context and experience may affect different types of privacy perception.

While all the other associations are not statistically significant, it may be worth noting relations that have a p-value close to .05, such as: i) the "Feeling something terrible is going to happen" is positively correlated with "Contacts" (which corroborates the statistically significant relation already described in Table 3, 1st row); ii) the "Feeling of restlessness and unable to stop" is negatively correlated with "Notifications"; and iii) the "Feeling pain/bad" is negatively correlated with "Photographs". These need to be verified with a larger sample.

In terms of relations between the "health status score" and "adherence to inhaler" variables, with the security variables (Table 4), stronger associations were found between the later and the security variables, than with the "health status score" variable. And from these stronger associations, the strongest relate to "Notifications" and "Photographs", which may be explained by the specificities of the tested application, which includes notifications to remind patients to take medication and images to register inhaler medication intake. For the "health status score", there are two moderate associations with the two variables "Photographs" (corroborating with results from last row, Table 3, where people with "Pain/feeling bad" associate negatively with "Photographs") and "Contacts", with much less association with "Notifications".

Other particular relations that may require some attention are with participants: 1) that "Never sit or relax, or feel some pain" choose the highest number of "Never"; 2) who are "Not in pain" give many times access to "Photographs" (already discussed); 3) with regular "Good mobility", give "Sometimes" accesses to all data, except for the "Contacts"; and 4) participants with "More interest in their appearance" contradict most of the analysed sample and are the ones that give more "Sometimes" accesses to "Contacts".

Limitations: the small sample size; the self-reported health status score of the analysed sample which is, in general, good to very good, and does not provide a balanced sample in terms of well-being to offer a wider range of comparisons and associations; the literature review was performed by only one reviewer; and, due to space constraints, a more detailed presentation of results and discussion was not possible.

6. CONCLUSION

This preliminary study focuses on understanding if there is a link between patients' health outcomes and their perceptions of privacy and trust in mhealth applications. Many questions were raised that need further investigation; however, findings show that privacy perception may change according to demographic factors or type of personal data being used/accessed together with context and patients' well-being. Results motivate the authors to proceed research in this area because, if strong links are found between well-being or health related outcomes and privacy related data or actions, providing mobile applications with features that can nudge chronic disease patients to improve adherence to medication and therapeutics, and be more confident to trust those applications, can become a possibility.

ACKNOWLEDGEMENT

This work has been funded by TagUBig - Taming Your Big Data (IF/00693/2015) from Researcher FCT Program funded by National Funds through FCT; and has been supported by ERDF (European Regional Development Fund) through the operations: POCI-01-0145-FEDER-029130 (mINSPIRERS—mHealth to measure and improve adherence to medication in chronic obstructive respiratory diseases—generalisation and evaluation of gamification, peer support and advanced image processing technologies) cofunded by the COMPETE2020 (Programa Operacional Competitividade e Internacionalização), Portugal 2020 and by Portuguese Funds through FCT (Fundação para a Ciência e a Tecnologia).

REFERENCES

Birkhäuer, J., Gaab, J., Kossowsky, J., Hasler, S., Krummenacher, P., Werner C, et al, 2017. Trust in the health care professional and health outcome: A meta-analysis. *In PLoS ONE*, 12(2): e0170988.

Ferreira, A., Almeida, R., Jácome, C., Fernandes, J., Fonseca, J.A. and Vieira-Marques, P., 2019. How inspiring is your app? A usability take on an app for asthma medication adherence. *Proceedings of the International Conference on e-Health, EH 2019*, pp. 225-229.

Hong, Z., Deng, Z. and Zhang, W., 2018. Examining factors affecting patients trust in online healthcare services in China: The moderating role of the purpose of use. *In Health Informatics Journal*, 25.

Jácome, C., Almeida, R., Teixeira, J.P., Vieira-Marques, P., Vilaça, R., Fernandes, J.M., Ferreira, A. and Fonseca, J.A., 2017. Inspirers: an app to measure and improve adherence to inhaled treatment. *Proceedings of the International Conference on E-Health, EH 2017*, 135-139.

Jácome, C., et al., 2018. mINSPIRERS – Feasibility of a mobile application to measure and improve adherence to inhaled controller medications among adolescents and adults with persistent asthma: Protocol for a multicentre observational study. *In Rev Port Imunoalergologia*, 26:1 47-61.

Kruse, C., Betancourt, J., Ortiz, S., Valdes Luna, SM., Bamrah, IK and Segovia, N., 2019. Barriers to the Use of Mobile Health in Improving Health Outcomes in Developing Countries: Systematic Review. *In J Med Internet Res*, 21(10):e13263.

Larson, RS., 2018. A Path to Better-Quality mHealth Apps. *In JMIR Mhealth Uhealth,* 6(7):e10414.

Majid, H., Jetzabel, S. and Kai, R., 2019. Revealing the unrevealed: Mining smartphone users privacy perception on app markets. *In Computers & Security*, Volume 83, Pages 332-353.

Polinski JM., Kesselheim, AS., Frolkis, JP., Wescott, P., Allen-Coleman, C. and Fischer, MA., 2014. A matter of trust: patient barriers to primary medication adherence. *In Health Education Research*, Volume 29, Issue 5, Pages 755–763.

Saheb, T., 2020. An empirical investigation of the adoption of mobile health applications: integrating big data and social media services. *In Health Technol.* (2020).

Serenko, N. and Fan, L., 2013. Patients' perceptions of privacy and their outcomes in healthcare. *In Int. J. Behavioural and Healthcare Research*, Vol. 4, No. 2.

van Rooy, D. and Bus, J, 2010. Trust and privacy in the future internet—a research perspective. *In IDIS 3*, 397–404.

Vervier, L., Valdez, A. and Ziefle, M., 2018. "Should I Trust or Should I Go?" or What Makes Health-Related Websites Appear Trustworthy? - An Empirical Approach of Perceived Credibility of Digital Health Information and the Impact of User Diversity. *Proceedings of the 4th International Conference on Information and Communication Technologies for Ageing Well and e-Health (ICT4AWE 2018)*, pages 169-177.

Vo, V., Auroy, L. and Sarradon-Eck, A., 2019. Patients' Perceptions of mHealth Apps: Meta-Ethnographic Review of Qualitative Studies. *JMIR Mhealth Uhealth*, 7(7):e13817.

Wei, J., Vinnikova, A., Lu, L. and Xu, J., 2020. Understanding and Predicting the Adoption of Fitness Mobile Apps: Evidence from China [published online ahead of print, 2020 Feb 10]. *In Health Commun*, 1-12.

Woldeyohannes, H.O. and Ngwenyama, O.K., 2017. Factors Influencing Acceptance and Continued Use of mHealth Apps. *In: Nah FH., Tan CH. (eds) HCI in Business, Government and Organizations. Interacting with Information Systems. HCIBGO.* Lecture Notes in Computer Science, vol 10293. Springer, Cham.

Zhou, L., Bao, J., Watzlaf, V. and Parmanto, B., 2019. Barriers to and Facilitators of the Use of Mobile Health Apps from a Security Perspective: Mixed-Methods Study. *In JMIR Mhealth Uhealth*, 7(4):e11223.

REFLECTION PAPER

APPLICATION OF DATA LAKE APPROACH FOR DATA ANALYTICS IN E-HEALTH ISSUES OF ARCHITECTURES AND ORGANIZATION

Bálint Molnár[1] and Ádám Tarcsi[2]
[1]*Eötvös Loránd University, Budapest, Hungary*
Department of Information Systems
[2]*Eötvös Loránd University Faculty of Informatics,*
Data Science and Engineering Department
IoT&Data Innovation Lab

ABSTRACT

By a case study approach, the paper analyzes a project and information ecosystem of the Clinical Research environment within a Medical School at a University. There are two types of data sources, a medical information system based on a transactional database for recording patients and serving as the basis of nursing and treatments of patients and medical and other medical IoT systems (like ECGs, etc). There exists neither Data Warehouse solutions nor any other data collection solutions dedicated to utilizing Data Science methods and tools. The emerging technologies provide opportunities for synergy between the traditional Data Warehouse and the most recent Data Lake approaches. Therefore, it seems feasible and reasonable to integrate these two architecture approaches to support data analytics on the research of medicine and biomedical areas.

KEYWORDS

Data Warehouse, Data Lake, Enterprise Architecture, Data Science

1. INTRODUCTION

The healthcare and the healthcare industry is one of the major sources for generating huge data collections. In healthcare, the exploitation of data efficiently and effectively is a critical issue. Data analytics looks back a long history as the analysis of data is a long-practiced activity that started with descriptive statistics then, later on, continued with second-order statistics. The huge volume of data that are generated in Health-Care Information Systems comes from sources as Electronic Health Record, IoT (Internet of Things) devices monitoring physiologic states, medical pieces of equipment, genomic data, radiology, etc.

2. DATA LAKE AND DATA WAREHOUSE

Generally, organizations are full of data that are stored in existing databases, produced by various information systems, data streams are coming from mobile applications, social media, web information systems, and other devices that are linked to the internet (Internet of Things, IoT). This statement is true in an e-Health environment too as healthcare organizations cope with data that are gathered by information systems dedicated to patients' medical records, transcripts of audio records of Medical Doctors. The literature makes difference among Electronic Medical Record (EMR), Electronic Health Record (EHR), Personal Health Record (PHR), Computerized Provider Order Entry (CPOE), Picture Archiving and Communication System (PACS) (Shachak & Reis, 2011) . The collected data can be considered as heterogeneous in both structure and content, i.e. there are structured, semi-structured, and unstructured data items, some of them are accompanied by meta-data.

There is an essential difference regarding data capture in Business Information Systems and Information Systems of Healthcare. In a healthcare environment, data gathering is typically manual, and even the structured data contains plenty of textual information. The input data are created typically by human agents who can make errors in data either intentionally or unintentionally. Examples of error-prone data entry are as follows: name, address, identifiers (social security numbers). Thereby, the method of data entry leads to imprecise data that makes difficult any data mining exercises, e.g. association relationships that may have been found prove to be erroneous. While inputting data into the Information System of Electronic Health Record, the Medical Doctors frequently use either templates or copy-paste commands to fill in various forms as e.g. medical flow forms to comply with policies, standards, and other regulations. The problem with that practice is that it conceals the subtle differences among the electronic patients' records that would be precious for data analytics and knowledge discovery. Information that is produced by automated systems along with auto-fill and edit check options as e.g. speech to text, OCR (optical character recognition) that computerize patients' data may introduce systematic and random inaccuracies that differ from physicians to physicians, software tool to software tool. These errors are difficult to quantify and forestall. Moreover, there can be uncertainty in explanations of diagnosis and procedures among Medical Doctors, and then the encoding of semantic contents may differ as well.

To ensure, that there will be a centralized location that would serve as the single source of truth, the data with different types and structures from various sources should be collected and loaded into a Data Lake. The most recent technologies can yield opportunities for the application of data analytics and models of Data Science. The results of running data analytic algorithms could be actionable knowledge in the clinical research environment.

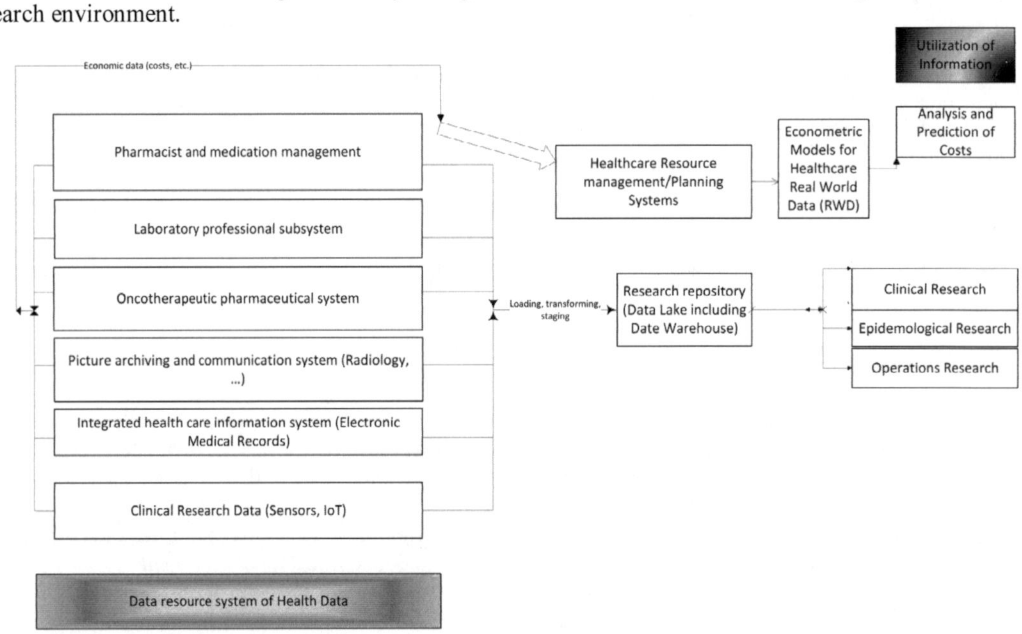

Figure 1. Sources of data and their usage

Generally, it assumed that the data coming from source systems are in good quality however, the market and administrative forces have not enforced a high level of standards of data quality in health care data. The typical life history of data can be seen in Figure 1. On the left part of the diagram, the various major source systems of healthcare data can be found. The data are represented as patients' electronic personal/health/medical records, pharmaceutical events at hospitals/clinics, clinical/diagnostic laboratory data, geo-codes for geographical information, and other loosely coupled data related to management, and business administration.

Our case study deals with a moderate size healthcare environment within a Medical School so that the Vs (volume, velocity, variety, veracity, variability, value) of the Big Data are as follows. A population of a million patients may generate electronic health records in the order of terabytes yearly the clinical images (PACS) could be a magnitude higher. Primarily, the variability of structured, semi-structured, and unstructured data increases the complexity thereby the difficulty of ensuring the single point of truth within the data collection.

Within a hybrid Data Lake that contains a robust Data Warehouse as well, the life cycle of data commences with transformation, cleansing, and integration. The objective to build up a Data Lake is to separate the daily operation, transactional data from the non-production data collection. Historically, the Data Warehouse technology has been employed for that purpose. During the data staging phase, the data are cleansed and filtered for the target data structure within the Data Warehouse, i.e. the fact table and dimensions. The phase of data staging includes data migration, data integration, translation of codes used for data representation, the transformation between database management systems. The Data Warehouse served as a basis for data analysis traditionally. The ETL (Extract, Transform, Load) procedure is applied for feeding data into the Data Warehouse. During that step the general data cleansing and transformation happens, e.g. amputation of trailing and leading whitespaces, superfluous zeros, standardization of identifiers/identifying numbers, inflicting constraints on data fields, converting English units into metric units.

While the before mentioned data-transformation is carried out relationships among entities may be dropped or harmed. Similar way, the data integration from multiple source systems, can lead to errors that are transmitted into the Data Warehouse.

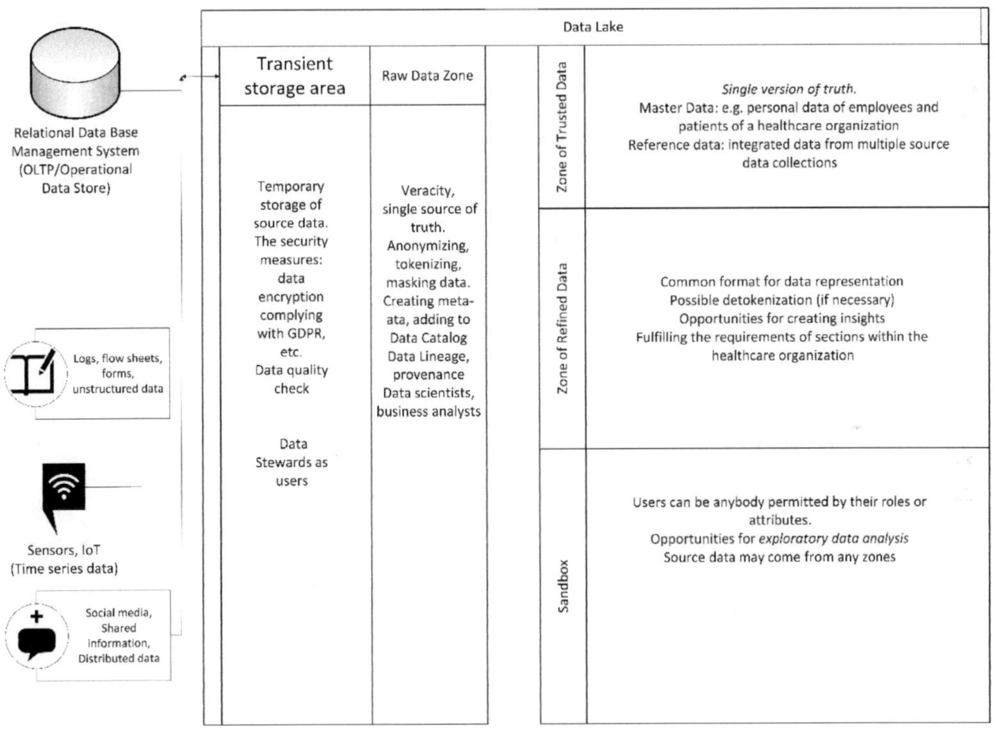

Figure 2. Sources of data and their usage

To overcome the data quality limitations of Data Warehouse, the idea of the Data Lake is conceptualized. The touted idea of Data Lake is that it deposits data in their *original* form, i.e. the *Transient* and/or *Raw Data* zone (see Figure 2; (LaPlante & Sharma, 2014)) contains the data after an ingestion phase so that the Data lake collects structured data from RDBMS (Relational DataBase Management Systems), semi-structured data (XML, binary XML, JSON, BSON, etc.), and unstructured data along with meta-data that are represented typically in a semi-structured format. The *ingestion phase* may mean loading, batch processing, data stream processing of source data while the necessary quality checks are carried out utilizing the MapReduce capability (Duggal, et al., 2015). The essential property of the raw data zone is that it is considered as "a single source of truth" as it keeps the data in their original form, however, the masquerading, tokenizing of the data may happen in that zone. Data scientists, business/data analysts can return to that zone when they look for associations and relationships that may have lost during data conversion, transformation, encoding, and encrypting.

The *Trusted Zone* executes the procedures for data alteration as quality assurance, compliance with standards, data cleansing, and data validation. In that zone, several data transformations happen to correspond to prescribed local and global policies whereby the data can be considered as a "single version of the truth". This zone may contain master data and the fact data that are governed through the data catalog that is filled in by meta-data automatically or semi-automatically. The data in the *Refined Zone* undergo several further changes that aim at the usability of data in algorithms of Data Science. These transformations include formatting, potential detokenization, data quality control to fulfill of requirements of algorithms so that models of the subject area (e.g. healthcare) and data analytics can be created. Thereby, knowledge discovery exercises can be carried out and understanding of data collections may be achieved. The access rights of users within each zone should be strictly maintained in the form of Role-Based Access Control, and for the temporary deviations from the baseline can be solved by Attribute-Based Access Rights.

For researchers, managers, and other subject area experts who would like to pursue exploratory data analytics (Myatt, 2007), the *Sandbox* provides opportunities to set up models, discover associations and relationships among attributes without involving members of IT department and other extra costs. The researcher can bring data into the Sandbox from any other zones within a controlled environment. It is even allowed to export interesting results back to the raw data zone for re-use.

The purpose of the Data Lake and Data Warehouse dedicated to research within e-Health Informatics is to lay the foundation of scientific workflows. The architecture should support several requirements as follows: (1) achieving processing speed through adjusting configuration parameters, (2) exploitation the distribution of data among cluster nodes, (3) usage of provenance data, (4) data placement and scheduling algorithm for input data to prepare the efficient data processing.

3. CONCLUSION

As a result of our researches our the proposed Data Lake architecture consists of the above mentioned data zones and a Data Warehouse as well, to support or the information requirements and the workflows of the organization.

ACKNOWLEDGMENT

The research was supported by the project No. 2019-1.3.1-KK-2019-00011 financed by the National Research, Development and Innovation Fund of Hungary under the Establishment of Competence Centers, Development of Research Infrastructure Programme funding scheme.

REFERENCES

Duggal, R., Khatri, S. K., & Shukla, B. (2015). Improving patient matching: single patient view for Clinical Decision Support using Big Data analytics. In B. S. Kapur (Ed.), 4th International Conference on Reliability, Infocom Technologies and Optimization (ICRITO)(Trends and Future Directions) (pp. 1–6). Amity University Uttar Pradesh, Noida, India: IEEE.

LaPlante, A., & Sharma, B. (2014). Architecting Data Lakes. O'Reilly Media Sebastopol.

Molnár, B., & Tarcsi, Á. (2013). Design and architectural issues of contemporary web-based information systems. Mediterr. J. Comput. Netw, 9, 20–28.

Myatt, G. J. (2007). Making sense of data: a practical guide to exploratory data analysis and data mining. John Wiley & Sons.

Shachak, A., & Reis, S. (2011). The Computer-Assisted Patient Consultation. In Clinical Technologies: Concepts, Methodologies, Tools and Applications (old.: 160–171). IGI Global.

POSTER

ACQUIRING BREATHING MOVEMENT AND HEART RATE FROM INERTIAL MEASUREMENT UNITS

Martin Csongor László[1], Ádám Tarcsi[2] and Zoltán Istenes[3]
[1]*Eötvös Loránd University, Budapest, Hungary*
IoT&Data Innovation Lab
[2]*Eötvös Loránd University Faculty of Informatics,*
Data Science and Engineering Department
IoT&Data Innovation Lab
[3]*Eötvös Loránd University, Budapest, Hungary*
3in Research Group, Martonvásár, Hungary

ABSTRACT

Wearable devices can raise suspicion about disorders, so action can taken in a timely fashion. However, wearable devices are still not widespread and inaccessible for many. Using smartphone's and smartwatch's inertial measurement units people's breathing movement and heart rate can be extracted. Breathing movement can be measured by device rotation heart rate can be measured by abdominal vibration. The devices are capable of acquiring signals very similar to Siesmocardiography, Cardiography and slightly similar to Ballistocardiography.

KEYWORDS

Smartphones, Seismocardiography, Gyrocardiography, Screening, Sleep Disorders, Wearables

1. INTRODUCTION

Siesmocardiography and Gyrocardiography are very sensitive instruments. They are based on accelerometers and gyroscopes therefore any type of movements influences the signal. However, during sleep there is minimal motion to be found. Also, any displacement or sleeping position changes can influence them.

2. MATERIALS AND METHOD

During signal acquisition the temperature of the smartphone can increase to a level which is inconvenient and can disturb sleep. To minimize it and to reduce electromagnetic radiation during recording the airplane mode can be activated. We tested the devices on several different positions: forehead, sternum, under the sternum, epigastric, left chest, right chest, epigastric region, back, limbs. Probably one of the best placement is the epigastric region where abdominal breathing can be measured by device rotation and also the abdominal aorta's vibrations are measurable, from which pulse rate (heart rate) is computable. This placement is the least obtrusive and could provide sufficient information. However, it needs to be assessed how the abdominal vibrations change with age, cardiac function, and body morphology.

The problem with Seismocardiography that it measures chest vibrations through subtle rotations and linear acceleration as seen on figure 2. The output is the sum of two phenomenon (D'Mello et al., 2019). The accelerometers output also are dependant on the orientation they are in. In comparison Gyrocardiography which measures just the rotational component of the vibration as seen on figure 1. (D'Mello et al., 2019).

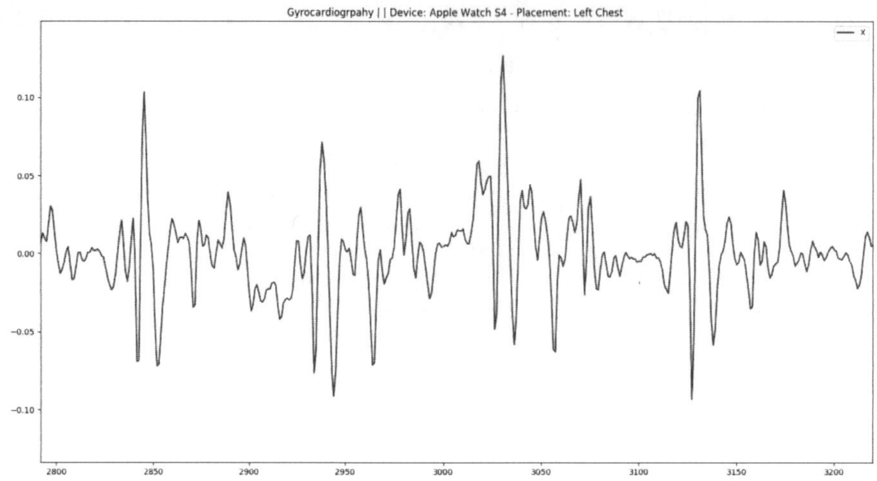

Figure 1. Approximated Gyrocardiography signal measured on the left chest with Apple Watch S4

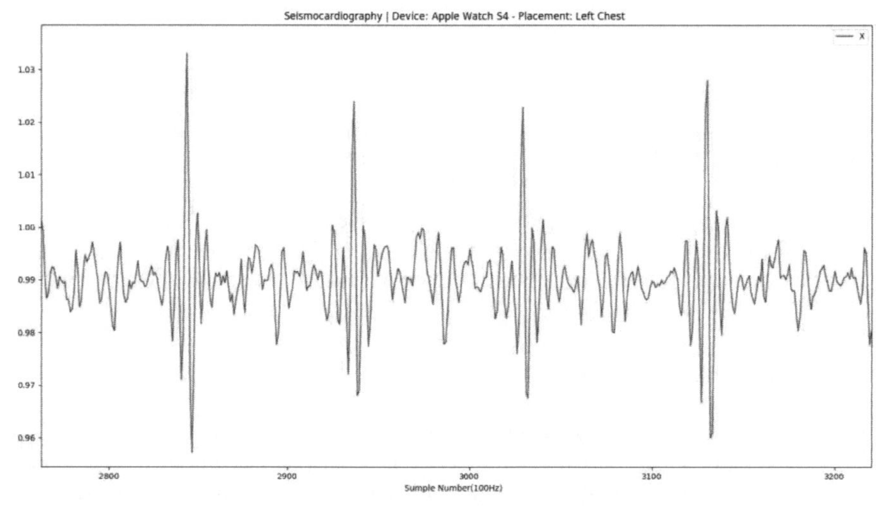

Figure 2. Approximated Seismocardiography signal measured on the left chest with Apple Watch S4

2.1 Objectives

Our research goals are the followings for the next few months:

- Assess the variability in abdominal vibration and exact nature. One option is that Respiratory effort belts may contain pulse vibrations, usually recordings also contain an ECG therefore reference beats are available.
- Investigate normalization algorithms for handling overnight body morphology changes.
- Augment our breathing detector application with an SCG/GCG processor.

3. CONCLUSION

The data collected by the iPhone6/Apple Watch Series 4 sensors are sufficient to approximate respiratory effort(similar to respiratory effortbelts), detect and identify paradoxical breathing, periodic breathing, central apnea episodes and obstructive apnea episodes, and in addition arrythmias can also be detected from beat-to-beat intervals either.

In iPhone6 gyroscope is better for breathing movement and heart rate extraction however in certain situations the rotation speed can be very slow, so the breathing movement will be closer to the noise level, in this situation algorithms can fall back to accelerometers. The gyroscope depends firstly on the person's body morphology then the momentary sleeping position which may changes body morphology.

ACKNOWLEDGEMENT

The research was supported by the project No. 2019-1.3.1-KK-2019-00011 financed by the National Research, Development and Innovation Fund of Hungary under the Establishment of Competence Centers, Development of Research Infrastructure Programme funding scheme.

REFERENCES

D'Mello, Y.; Skoric, J.; Xu, S.; Roche, P.J.R.; Lortie, M.; Gagnon, S.; Plant, D.V. Real-Time Cardiac Beat Detection and Heart Rate Monitoring from Combined Seismocardiography and Gyrocardiography. *Sensors* 2019, *19*, 3472.

László M. Cs. Istenes Z and Tarcsi A. Extracting Physiological Signals from Smartphone Sensors. *11th International Conference on Applied Informatics*. Eger, Hungary.